COMPUTATIONAL AND EXPERIMENTAL CHEMISTRY

Developments and Applications

COMPUTATIONAL AND EXPERIMENTAL CHEMISTRY

Developments and Applications

Edited by

Tanmoy Chakraborty, PhD, Michael J. Bucknum, PhD, and Eduardo A. Castro, PhD

Apple Academic Press

TORONTO NEW JERSEY

Apple Academic Press Inc.
3333 Mistwell Crescent
Oakville, ON L6L 0A2
Canada

Apple Academic Press Inc.
9 Spinnaker Way
Waretown, NJ 08758
USA

International Standard Book Number-13: 978-1-926895-29-1 (hbk)

Library of Congress Control Number: 2013943606

Library and Archives Canada Cataloguing in Publication

Computational and experimental chemistry: developments and applications/edited by Tanmoy Chakraborty, PhD, Michael J. Bucknum, PhD, and Eduardo A. Castro, PhD.

Includes bibliographical references and index.
ISBN 978-1-926895-29-1
1. Chemistry, Physical and theoretical. I. Castro, E. A. (Eduardo Alberto), 1944-, editor of compilation II. Chakraborty, Tanmoy, editor of compilation III. Bucknum, Michael J., editor of compilation

QD453.3.C64 2013 541 C2013-904417-5

Apple Academic Press also publishes its books in a variety of electronic formats. Some content that appears in print may not be available in electronic format. For information about Apple Academic Press products, visit our website at **www.appleacademicpress.com** and the CRC Press website at **www.crcpress.com**

ABOUT THE EDITORS

Tanmoy Chakraborty, PhD

Dr. Tanmoy Chakraborty, PhD, received his degree from the University of Kalyani, West Bengal, India, in QSAR/QSPR methodology in the bioactive molecules. He is working as Assistant Professor in the Department of Chemistry at Manipal University Jaipur, India. He is an International Editorial Board Member of the International Journal of Chemoinformatics and Chemical Engineering. He is also the reviewer of the World Journal of Condensed Matter Physics (WJCMP). He received the prestigious Paromeswar Mallik Smawarak Padak, from Hooghly Mohsin College, Chinsurah (University of Burdwan), in 2002.

Michael J. Bucknum, PhD

Michael J. Bucknum, PhD, received his degree in crystallography from Cornell University in Ithaca, New York. He currently resides in San Diego, California, and holds an academic appointment as corresponding investigator with Consejo Nacional de Investigaciones Cientificas y Tecnicas (CONICET) based in Argentina. Michael has also been appointed to the Editorial Advisory Board of Chemistry Central, the open-access chemistry publisher based in the United Kingdom.

Eduardo A. Castro, PhD

Eduardo A. Castro, PhD, is full professor in theoretical chemistry at the Universidad Nacional de La Plata and a career investigator with the Consejo Nacional de Investigaciones Cientificas y Tecnicas, both based in Buenos Aires, Argentina. He is the author of nearly 1000 academic papers in theoretical chemistry and other topics, and he has published several books. He serves on the editorial advisory boards of several chemistry journals and is often an invited speaker at international conferences in South America and elsewhere.

CONTENTS

LIST OF CONTRIBUTORS

I. Alcalde-Segundo
Universidad Autónoma del Estado de México. Centro Universitario Texcoco, Unidad Tianguistenco. Paraje El Tejocote s/n, San Pedro Tlaltizapan. Tianguistenco, México. CP. 52640

A. K. Bandyopadhyay
Murshidabad College of Engineering and Technology, W. B. University of Technology, Berhampur 742102, Murshidabad, India

S. Basu
Chemical Sciences Division, Saha Institute of Nuclear Physics, 1/AF, Bidhannagar, Kolkata 700 064, India. E-mail: samita.basu@saha.ac.in; Telephone: +91-33-2337-5345, Fax: +91-33-2337-4637

A. Biswas
Department of Electronics and Communication Engineering, National Institute of Technology, Durgapur 713209, India

J. Carmona-Espíndola
Universidad Autónoma Metropolitana-Iztapalapa. Departamento de Química. San Rafael Atlixco 186, Col. Vicentina, México D. F. C.P. 09340

A. Chakraborty
Department of Chemistry and Center for Theoretical Studies, Indian Institute of Technology, Kharagpur – 721 302, India

T. Chakraborty
Department of Chemistry, Manipal University Jaipur, Jaipur, 302026, India. Email: tanmoychem@gmail.com; tanmoy.chakraborty@jaipur.manipal.edu

P. K. Chattaraj
Department of Chemistry and Center for Theoretical Studies, Indian Institute of Technology, Kharagpur – 721 302, India. E-mail: pkc@chem.iitkgp.ernet.in

S. Das
Department of Materials Science and Engineering, University of North Texas, Denton, TX 76207, USA

R. Das
Department of Chemistry and Center for Theoretical Studies, Indian Institute of Technology, Kharagpur – 721 302, India

H. Dave
FCIPT, Institute for Plasma Research, Gandhinagar, India

S. Dhail
Department of Chemistry, Manipal University Jaipur, Jaipur – 302026, India

J. Garza
Universidad Autónoma Metropolitana-Iztapalapa. Departamento de Química. San Rafael Atlixco 186, Col. Vicentina, México D. F. C.P. 09340. E-Mail: jgo@xanum.uam.mx

L. Ledwani
Department of Science, Manipal University Jaipur, Jaipur, India

S. R. Mane
Polymer Research Centre, Department of Chemical Sciences, Indian Institute of Science Education and Research Kolkata, India

S. K. Nema
FCIPT, Institute for Plasma Research, Gandhinagar, India

S. Naskar
Department of Chemistry, Indian Institute of Chemical Biology, Council of Scientific and Industrial Research, 4 Raja S.C. Mullick Road, Jadavpur, Kolkata, 700032, India

S. Roy
Department of Chemistry, Malda Govt. Polytechnic College, West Bengal State Council of Technical Education, Malda–732102, India

M. K. Sarangi
Chemical Sciences Division, Saha Institute of Nuclear Physics, 1/AF, Bidhannagar, Kolkata 700 064, India

R. Shunmugam
Polymer Research Centre, Department of Chemical Sciences, Indian Institute of Science Education and Research Kolkata, India. E-mail: sraja@iiserkol.ac.in

R. Vargas
Universidad Autónoma Metropolitana-Iztapalapa. Departamento de Química. San Rafael Atlixco 186, Col. Vicentina, México D. F. C.P. 09340

N. Vijaykameswara Rao
Polymer Research Centre, Department of Chemical Sciences, Indian Institute of Science Education and Research Kolkata, India

LIST OF ABBREVIATIONS

AFM	atomic force microscopy
AIM	atoms-in-a-molecule
APGD	atmospheric pressure glow discharge
APPJ	atmospheric pressure plasma jet
CDC	carbide derived carbon
CE	correlation energy
DAC	diamond anvil synthesis cell
DB	discrete breathers
DBD	dielectric barrier discharge
DC	direct current
DCC	dicyclohexylcarbodiimide
DFT	density functional theory
DMF	dimethylformamide
DOS	density of states
EEP	electronegativity equalization principle
FWHM	full width at half maximum
HF	Hartree-Fock
HIPIMS	high power impulse magnetron sputtering
HMDS	Hexamethyldisilane
HMDSO	hexamethyl disiloxane
ILM	intrinsic localized modes
LDA	local density approximation
MF	magnetic field
MFE	magnetic field effect
MO	molecular orbital
MSTO	modified Slater type orbitals
MW	microwave
NPs	nanoparticles
P	pressure
PAA	polyacrylic acid

PDR	percept desizing ratio
PET	photo-induced electron transfer
PVA	polyvinyl alcohol
PVD	physical vapor deposition
QBs	quantum breathers
RF	radio frequency
RIPs	radical ion pairs
ROMP	ring opening metathesis polymerization
S	singlet
SAR	structure activity relationship
SEM	scanning electron microscopy
SPM	scanning probe microscopy
SPR	structure property relationship
SRR	split-ring-resonator
T	temperature
TE	total energy
TEM	transmission electron microscopy
TFA	trifluoroacetic acid
THF	tetrahydrofuran
TPBS	two-phonon bound states
TPc	critical treatment power
TTMSVS	tris(trimethylsilyloxy)vinylsilane
VB	valence bond
VQAS	vinyl quaternary ammonium salt
XPS	X-ray photoelectron spectroscopy

LIST OF SYMBOLS

B	bulk modulus
R_c	radius of a sphere
Z	atomic number
$\hat{J}_i$	Coulomb operators
$\hat{K}_i$	exchange operators
F	Fock matrix
S	overlap matrix
$Y_{l,m}$	spherical harmonics
w_0	ratio of molar concentration of water to AOT
V	voltage
P_s	saturation polarization in C/m^2
E_c	coercive field
t_c	critical time scale
W_L	domain wall width
E_0, E_1 and E_2	eigen values
k	wave vector
a_n^+ and a_n	creation and annihilation operators
hc	Hermitian conjugate
c	velocity of light
R	lattice parameter
1	measure of the form of the structure
r′	elastic chemical bond deformation
a^2	area of a crystalline plane normal to that elastic chemical bond deformation

Greek Symbols

μ	chemical potential
η	hardness of the system
ε_{HOMO}	orbital energies of the highest occupied orbitals
ε_{LUMO}	orbital energies of the lowest unoccupied orbitals
ξ	coefficient of hydraulic resistance

υ	dimensionalities of energy
ν	electromagnetic wave frequency
l	interaction term
L	diagonal matrix

PREFACE

Computational and Experimental Chemistry: Developments and Applications provides an eclectic survey of contemporary problems in theoretical chemistry and applied chemistry. The problems addressed in its pages vary from the prediction of a novel spiro quantum chemistry edifice of carbon-based structures to applications of many body perturbation theory in helium-like ions, and also from the elucidation of a novel dynamic elasticity theory applied to carbon, to the description of equalization principles in chemistry. The book is divided in to two main parts. Part I entitled "Exotic Carbon Allotropes" is in four chapters and describes the theoretical work of Bucknum et al. applied to the enumeration of novel carbon patterns and their properties. Part II entitled, "New Developments in Computational and Experimental Chemistry" comprises the last eight chapters of the book and provides an interesting survey of contemporary problems in theoretical chemistry and applied chemistry.

In the first part of *Computational and Experimental Chemistry: Developments and Applications.* the book provides a popular as well as technical presentation of the ideas surrounding the emergence of a synthetic, analytical, and theoretical spiro quantum chemistry edifice of carbon-based structures, as well as a chemical topology scheme that successfully describes molecules and patterns, including the hydrocarbons and allotropes of carbon. Thus the purpose of Part I of this book is to describe the generalization and realization of the organic chemistry concept of spiro-conjugation into 1-, 2- and 3-dimensions. Chapter 1 covers the theory of spiroconjugation and a lattice is described, the so-called "glitter" lattice, that exhibits the spiroconjugation phenomena in fully 3-dimensions. Also described is the corresponding 1-dimensional substructures of this lattice that exhibit spiroconjugation delocalized in 1-D.

In Chapter 2, experimental evidence is provided and assessed for the prediction of the synthetic realization of the "glitter" allotrope of carbon by shock-based synthetic methods. This is the allotrope of carbon that is spiroconjugated in 3-D. Part of this analysis includes the prediction of the so-called 3-dimensional resonance structures of glitter, these are resonance forms that can be drawn over the tetragonal glitter unit cell. Chapter 3 includes a description of the chemical topology of the hydrocarbons and crystalline allotropes of carbon. The chemical topology principles are enunciated, based upon the Euler equations for graphs and patterns, and the results are applied to the glitter allotrope of carbon and the other crystalline allotropes of carbon including fullerenes, grapheme and diamond. This chemical topology scheme enables one to map the C allotropes in a topology space allowing them to be classified.

Chapter 4 is the final chapter of Part I and it develops a novel dynamic elasticity theory based upon the work of Feynman et al. The dynamic elasticity theory addresses questions like: What is the ultimate strength of materials? And What is the ultimate strong material? The theory is applied to cubic diamond and tetragonal glitter and the results of an approximate strength calculation show that glitter should actually rival cubic diamond in strength along its longest axis. Other allotropes of carbon, as models, including the hexagonite pattern are analyzed approximately as well.

Part II of *Computational and Experimental Chemistry: Developments and Applications* is entitled "New Developments in Computational and Experimental Chemistry", and this part of the book narrates contemporary problems in theoretical chemistry and applied chemistry. Thus, Part II of this book contains several chapters that cover a wide range of topics from theoretical and computational chemistry to experimental applied chemistry. Part II, therefore, comprises the last eight chapters of the book and provides an interesting survey of contemporary problems in theoretical chemistry and experimental chemistry that include contemporary perturbation theory results for helium and helium-like ions, the investigation of so-called quantum breather ferroelectric materials by second quantization techniques, the study of magnetic field effects in small molecule organic systems, plasma surface modification in textile processing, the equalization principle in theoretical chemistry, and so on.

In Chapter 5, a successful application of second order many-body perturbation theory to estimate the CE for confined many-electron atoms has been depicted. The second order many-body perturbation theory proposed by Møller and Plesset and detailed implementation of code in GPU proposed by Gazra et al. has also been discussed vividly. Finally the authors have made a comparative study between their results for helium-like atoms with more sophisticated techniques to find out the percent of correlation energy recovered by the MP2 method.

In Chapter 6, distance dependence of magnetic field effect (MFE) on donor-acceptor (D-A) pair inside a confined heterogeneous environment has been emphasized. In this study N, N-dimethyl aniline (DMA) and the protonated form of Acridine (Acr) have been treated as electron donor and electron acceptor respectively. In this chapter, Basu et al. have elaborately reported the mechanistic features of excited state proton transfer with the photoinduced electron transfer between Acr and DMA.

Nowadays, plasma surface modification is an active field of research. In Chapter 7, the various experimental approaches of plasma surface modification of polyester textile have been described. The improved quality plasma treatment mentioned in this chapter has potential to solve problems with synthetic polyester textile to expand their usefulness and this method is also very ecofriendly.

From the early days of theoretical chemistry, the equalization principle and its application in real field are very popular among researchers. In the Chapter 8, Chattaraj and his colleagues have analyzed the three important equalization principles in chemistry *viz.*, electronegativity equalization principle (EEP), hardness equalization principle (HEP), and principle of electrophilicity equalization (PEE). The qualitative nature and validation of three principles have been concluded nicely at the end of this chapter.

In Chapter 9, the authors have described the ring opening metathesis polymerization procedure that has potential application in cancer therapy. The synthesis and complete characterization of both norbornene derived doxorubicin (mono 1) and polyethylene glycol (mono 2) monomers have been clearly described. Secondly their copolymerization by ring-opening metathesis polymerization (ROMP) to get the block copolymer has been vividly elaborated. Similarly the NDB monomer as well as its homopolymer (NDBH) synthesis and complete characterization have been thoroughly discussed.

In Chapter 10, Chakraborty et al. have studied theoretically biological activity of newly synthesized mesoionic heterocycles. Invoking global and local reactivity based descriptors, the experimental mechanistic features of mesoionic molecules have been nicely correlated with their theoretical counterparts.

In Chapter 11, Bandyopadhyay et al. have reported on the non-linear Klein-Gordon equation that is based on their discrete Hamiltonian in a typical array of ferroelectric domains. The effect of second quantization, in a particular environment, toward the nonlinearity has been described. This is considered useful for a future study in this new field of investigation of quantum breathers in ferroelectrics.

Finally, in the last chapter (Chapter 12) of Part II of this book, Ramon has studied the molecular quantum similarity (QS) measures involving three density functions. The necessary algorithms have been described here. General theory and computational feasibility of a hypermatricial or tensorial representation of molecular structures associated to any molecular quantum object set (MQOS) have been nicely explained in this chapter. Secondly, generalized Carbó similarity indices (CSI) have also been studied. The theoretical and computational approaches have been supported by various suitable applicative examples.

Computational and Experimental Chemistry: Developments and Applications is a collection of chapters that has a wide and eclectic range of subject matter regarding the applications of theoretical and experimental chemistry in realistic, everyday settings. The research articles comprising this book are based upon established methods of theoretical and computational chemistry, and applied chemistry, and this subject matter is thus very important in the context of contemporary problem solving in chemistry. Finally, two of the chapters of this book, Chapters 7 and 9, have been included so as to cover important research in applied chemistry of textiles and polymers, and these topics were thus chosen for their importance in the everyday applications of chemistry.

— Tanmoy Chakraborty, PhD, Michael J. Bucknum, PhD, and Eduardo A. Castro, PhD

PART I
Exotic Carbon Allotropes

CHAPTER 1

SPIRO QUANTUM CHEMISTRY

MICHAEL J. BUCKNUM and EDUARDO A. CASTRO

CONTENTS

SUMMARY

A recent paper has described the structure of a hypothetical 3-,4-connected net termed glitter. This is a model of an allotrope of carbon in the form of a synthetic metal. That paper pointed to the importance of through-space p_σ interactions of adjacent olefin units in the net in understanding the electronic structure at the Fermi level. The present communication elucidates the role of spiroconjugation in understanding features of the electronic band structure and density of states of glitter. With this analysis of spiroconjugation in the 1-dimensional polyspiroquinoid polymer and the 3-dimensional glitter lattice, the foundations have been laid for a new type of quantum chemistry herein called spiro quantum chemistry. Spiro quantum chemistry complements traditional quantum chemistry which is focused on linear polyenes, circular annulenes, polyhexes, 2-dimensional graphene sheets and related structures including fullerenes, by focusing on spiroconjugated hydrocarbon structures in 1-,2-, and 3-D, including linear spiro[n]quinoids and polyspiroquinoid in 1D, circular cyclospiro[n]quinoids, spiro[m,n]graphene fragments and spirographene in 2D and [m,n,o] glitter fragments and glitter in 3D.

1.1 INTRODUCTION

In 1994, the structure of the all-carbon "glitter" lattice [1] was described along with B_2C and CN_2 phases adopting this structure-type. Glitter is a hypothetical tetragonal allotrope of carbon. The geometrical structure of the lattice (space group $P4_2/mmc$, # 131) is shown in Fig. 1.

The dimensions of the lattice are a = 2.53 Å and c = 5.98 Å, these are constrained by the geometry of the 1,4-cyclohexadiene molecule upon which the structure is based [2]. It is a 3-,4-connected net [3] containing trigonal and tetrahedral atoms in a ratio of 2:1, with a calculated density of 3.12 g/cm^3; intermediate between graphite and diamond. Apparent in the structure are the ethylenic columns, which run perpendicular to each other to generate the 3-dimensional lattice.

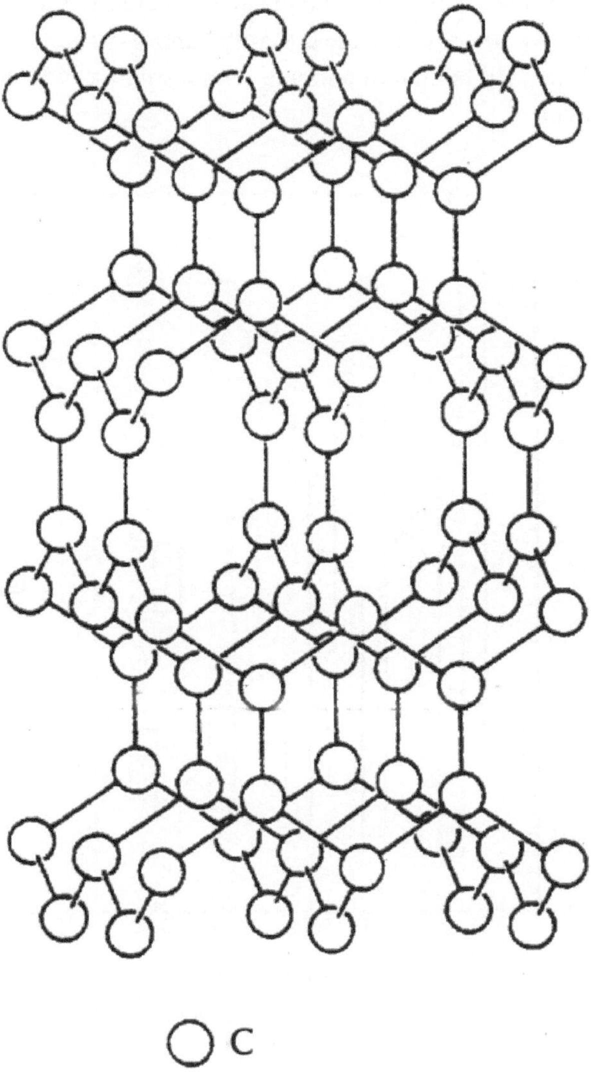

FIGURE 1 Structure of the glitter model in space group $P4_2/mmc$.

Merz et al. [4] have described the electronic band structures of several related 3-,4-connected nets each possessing stacked olefin units. Evolution of π–π* band overlap with the separation between adjacent units in an olefin stack is presented in their paper, this diagram is shown in Fig. 2. Note the touching of the π and the π* bands at an interaction distance of about 2.5 Å. From this diagram it is clear that the approximate density of states profile of glitter will be that of a metal, with a small π–π* band overlap at the Fermi level.

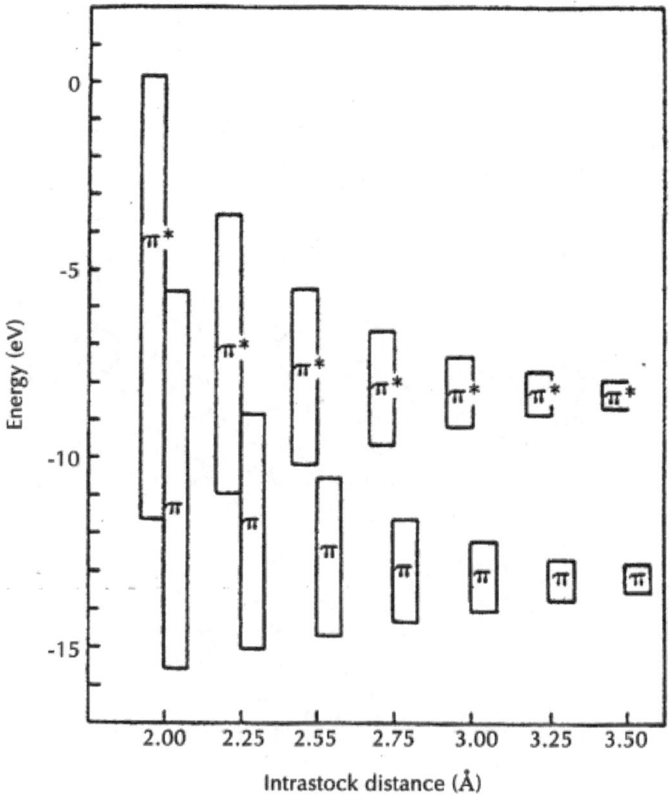

FIGURE 2 Evolution of π–π* overlap with the separation between adjacent units in an olefin stack.

The structure shown in Fig. 1 also contains the 1,4-cyclohexadiene-oid molecular units linked through their tetrahedral vertices to adjacent rings. These linkages form chains. Each such chain is termed a "polyspiroquinoid" substructure of the "glitter" lattice. The electronic structure of a polyspiroquinoid model is described in connection with the concept of spiroconjugation.

Band structure calculations for this net were carried out using the extended Hueckel method adapted for application to extended structures. [5] Electronic band structures of various sublattice models, including the polyspiroquinoid one, were presented in the original report of the glitter structure [1] in order to clarify the importance of through-space interactions in these lattices.

1.2 SPIROCONJUGATION IN 1-DIMENSION AND POLYSPIROQUINOID

Spiroconjugation [6] is claimed to occur in the electronic structure of the polyspiroquinoid chain. [1] This type of through-space interaction can be considered in terms of 4 p_π atomic orbitals held "spiro" to each other about a tetrahedral C atom (see Fig. 3). These four orbitals interact to form 4 combinations of differeing symmetry with respect to the two perpendicular mirror planes dividing the tetrahedral C atom. One combination is symmetric under reflection in each mirror plane, labeled (SS), and there are also (SA), (AS) and (AA) combinations. The (AA) combination is shown, it is of the proper symmetry for a bonding spiro-type interaction.

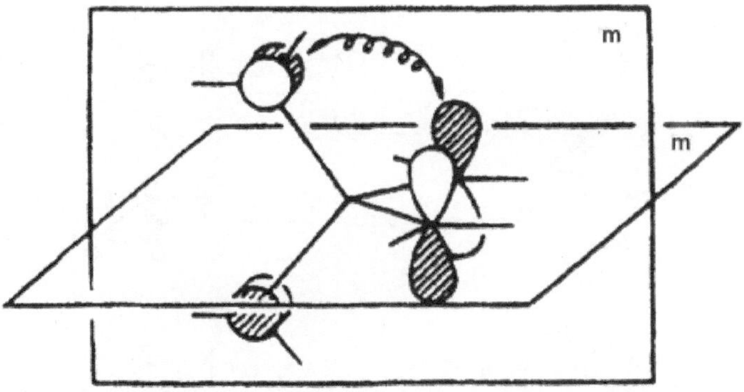

FIGURE 3 "Spiro" molecular orbitals derived from 4 p_{spiro} combinations.

From the original analyses of the effect of spiroconjugation in discrete molecular systems, including spiro [4.4] nonatetraene [7] the importance of it in stabilizing (and destabilizing) a frontier molecular orbital was of primary interest. This work was based upon electronic spectra, chemical reactivity and electronic structure calculations [6]. From this background, in the original paper on the glitter structure [1], only the lowest-lying π* band, the LUCO (lowest unoccupied crystal orbital) was considered in assessing the importance of through-space p_{spiro} interactions in the 1-dimensional polyspiroquinoid substructure of glitter.

The interaction diagram for spiro[4.4] nonatetraene, reported by Duerr and Kober [7h], is shown in Fig. 4. The diagram reveals that in a situation involving 8 p_{π} orbitals (four interacting π systems as opposed to four interacting p orbitals, as in Fig. 3), there are generated two sets of four butadieneoid π systems. These four butadieneoid π systems are of (AS), (AA), (AS) and (AA) symmetry, unlike the archetypal system diagrammed in Fig. 3. Interaction of these four π systems through the spiro C atom, yields eight molecular orbitals over the skeleton of the spiro [4.4] nonatetraene molecule.

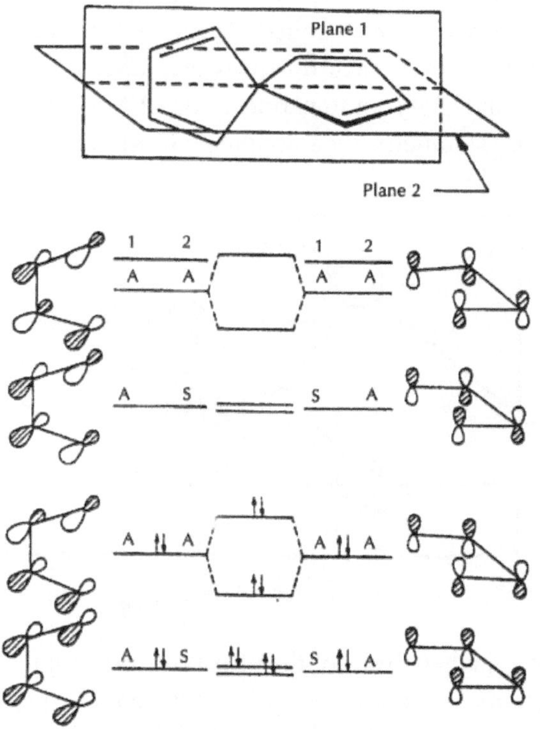

FIGURE 4 Molecular orbitals of spiro[4.4]nonatetraene.

Note the presence of the two (AA) combinations, one derived from the interaction of the HOMO (highest occupied molecular orbital) π levels of butadiene and one derived from the interaction of the LUMO+1 (next highest to the lowest unoccupied molecular orbital), π* levels of butadiene. There are thus fourlevels in which the effects of spiroconjugation could be important, twodown in the bonding manifold and the other pair in the anti-bonding manifold. The splitting of these pairs of (AA) combinations results from one spiro [4.4] nonatetraene orbital carrying a bonding p_{spiro} interaction and its sibling carrying an anti-bonding p_{spiro} interaction.

An analogy exists between the interaction diagram of the spiro [4.4] nonatetraene molecule and the unit cell of the polyspiroquinoid chain. Certain of the energy levels in both of these systems are comprised of 8 p_{π} atomic orbitals present in each structure. This analogy will be followed,which describes the stabilizing effects of spiroconjugation in the electronic band structure of polyspiroquinoid.

Figure 5 shows the structure of the model polyspiroquinoid chain (H atoms added to the glitter fragment to reach a realistic structure) and Fig. 6 is its electronic band structure. To the right in this diagram are appended the levels of the 3,3, 6,6-tetramethyl-1,4-cyclohexadiene molecule. This molecule is a reasonable model for the polymer, but one lacking the spiroconjugation interaction. The 2 π and 2 π* levels in the molecule have been indicated alongside the 4 π and 4 π* bands of polyspiroquinoid.

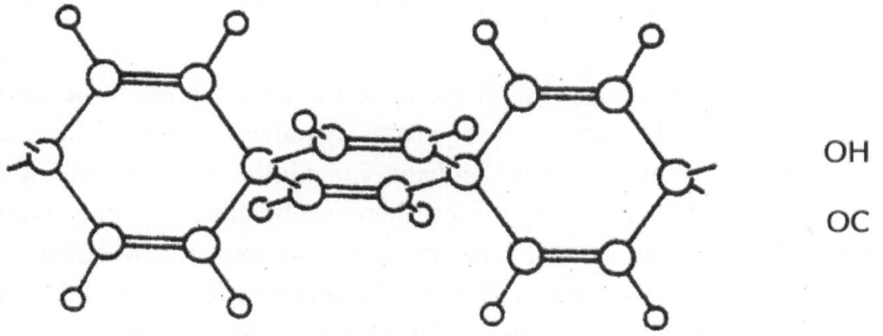

FIGURE 5 Structure of the polyspiroquinoid model.

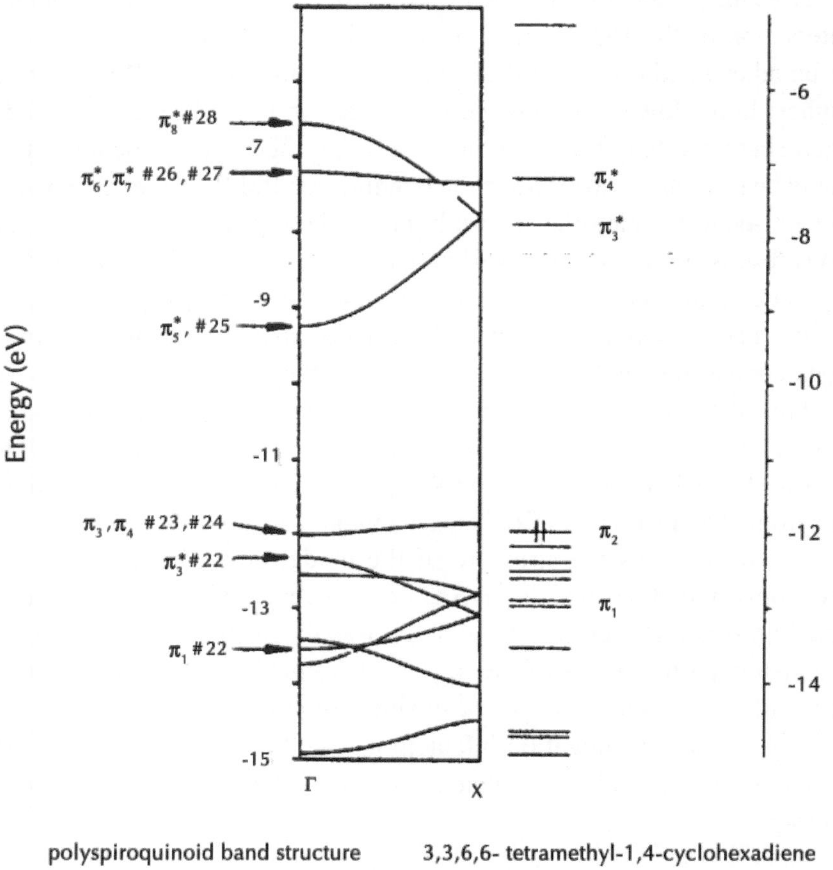

FIGURE 6 Electronic band structure of polyspiroquinoid and molecular orbitals of 3,3,6,6-tetramethyl-1,4-cyclohexadiene with π and π* levels indicated.

As pointed out previously, [1] the lowest-lying π* band in the electronic structure of the chain is of the proper symmetry, (AA), for there to be a bonding p_{spiro} interaction in that band. In addition, the lowest-lying π band is also of (AA) symmetry, so that there is a bonding p_{spiro} interaction in that occupied band as well. The relative importance of the effects of spiroconjugation in the chain will be posed with respect to the stabilization of the occupied π band compared to the energy level of the discrete π molecular orbital of 3,3,6,6-tetramethyl-1,4-cyclohexadiene from which it is derived.

In analogy with spiro [4.4] nonatetraene, the 8 p_π atomic orbitals in the unit cell of polyspiroquinoid generate 4 π crystal orbitals and 4 π^* crystal orbitals in its electronic band structure. At the zone center, half of these eight crystal orbitals are constructed from real coefficients of the same magnitude at each p_π atomic site in the unit cell. Because of their importance in the analysis of spiroconjugation in polyspiroquinoid, these four (AA) crystal orbitals are sketched exactly as they occur at the zone center, in Fig. 7. These are analogous to the four (AA) combinations from the spiro [4.4] nonatetraene analysis. They occur as band #18 (p_1), band #22 (p_2), band #25 (π_5*) and band #28 (π_8*).

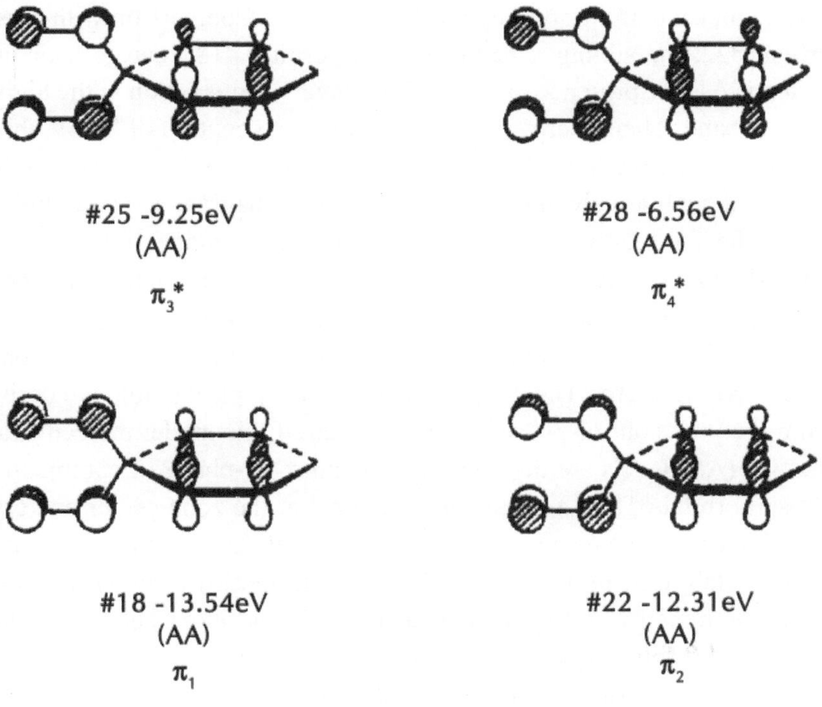

FIGURE 7 The p_π crystal orbitals of polyspiroquinoid sketched at the center of the Brillouin zone.

The four (AS) combinations occur in degenerate band pairs (compare to the orbital interaction diagram of spiro [4.4] nonatetraene) across the Brillouin zone of the chain. These are constructed from real coefficients of the same magnitude and sign at each p_π atomic site in one or the other ring of the polyspiroquinoid unit cell. Every other ring in the polymer possesses these π and π* type interactions. The π and π* interactions in half the unit cell are involved in σ or σ* interactions with the adjacent ring, which maintain the (AS) symmetry of the crystal orbital. At the zone center they occur as bands #23 (p_3) and #24 (π_4), (the HOCO, highest occupied crystal orbital, levels) and as bands #26 (π_6*) and #27 (π_7*), (the LUCO+1 levels).

One can analyze the attendant orbital interactions of the (AA) combinations graphically from the sketches. In crystal orbital #18, the 4 bonding p_π combinations in the unit cell also possess a bonding p_{spiro} interaction about each of the spiro carbons in the chain. Comparison with crystal orbital #22, the sibling π band to crystal orbital #18, reveals their only difference is the presence of an anti-bonding p_{spiro} interaction in the higher-lying π band. Their energy difference at the zone center is 1.23 eV, this is the actual magnitude of the splitting of the (AA) combinations; this was predicted qualitatively in the orbital interaction diagram shown in Fig. 4.

In the 3,3,6,6-tetramethyl-1,4-cyclohexadiene model molecule, the lower-lying π (p_1) level is the "out-of-phase" π combination, it is of (AA) symmetry with respect to the two mirror planes dividing the spiro carbon atoms in the ring. Its sibling level (p_2) is the "in-phase" pcombination, it is of (AS) symmetry. The 2 π (AA) levels in polyspiroquinoid are derived from the pure "out-of-phase" π combination of this model molecule, and the two (AS) levels are derived from the pure "in-phase" π combination. Crystal orbital #18 (p_1) occurs at –13.54 eV at the zone center and crystal orbital #22 (p_2) occurs at –12.31 eV. Therefore, the spiroconjugation band is stabilized by 0.55 eV with respect to the lower-lying π level in 3,3,6,6-tetramethyl-1,4-cyclohexadiene and its sibling π band, with anti-bonding p_{spiro} interactions, is destabilized by 0.68 eV.

Tracing the evolution of these bands along the symmetry line of polyspiroquinoid, one sees the twosibling bands meet at the zone edge where they are degenerate (*see* Fig. 6). This degeneracy occurs because of the presence of a 4_2 screw axis in the polyspiroquinoid unit cell. The band structure is "folded" in half due to the 4_2 screw axis. Across the Brillouin zone crystal orbital p_1, the spiroconjugation band, is stabilized with respect to the lower-lying π level in 3,3,6,6-tetramethyl-1,4-cyclohexadiene (p_1). At the zone edge, the degeneracy at −13.07 eV is just 0.08 eV below the corresponding molecular orbital (p_1) in the 3,3,6,6-tetramethyl-1,4-cyclohexadiene model molecule.

Figure 8 shows the density of states (DOS) diagram along with the band structure diagram (Fig. 6) of polyspiroquinoid. The shaded area is the contribution of p_π orbitals to the total DOS. It has a "spike" at an energy, which corresponds to the spiroconjugation band at the zone center (−13.5 eV). This spike indicates there is a large density of states in the electronic structure of the polymer at that energy. Evidently from the DOS diagram, there is not a corresponding spike coincident with the energy of the sibling pband, which possesses anti-bonding p_{spiro} interactions at the zone center. This is consistent with the small degree of dispersion (flatness) of band #18 in the neighborhood of the zone edge.

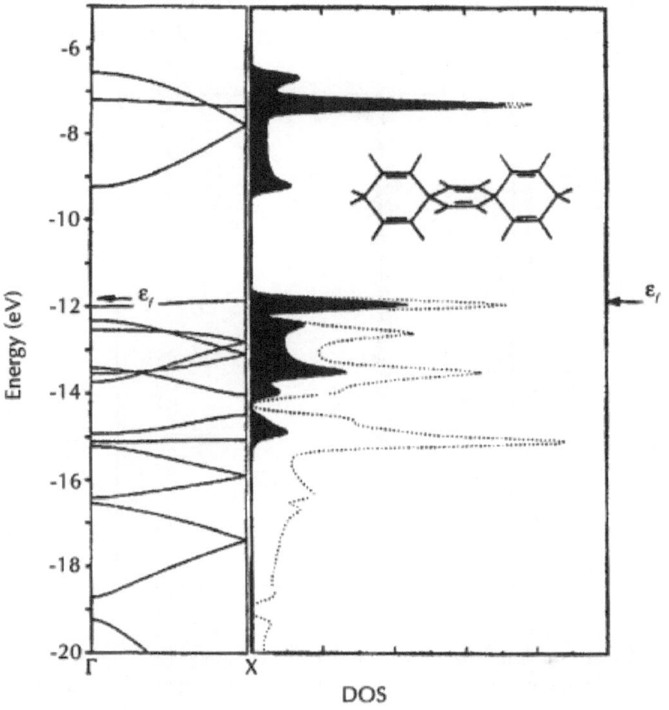

FIGURE 8 Electronic band structure and density of states (DOS) of the polyspiroquinoid model; shaded area indicates contribution of p_π orbitals to the total DOS.

1.3 SPIROCONJUGATION IN 3-DIMENSIONS AND GLITTER

In the full 3-dimensional glitter lattice shown in Fig. 1, it is obvious the polyspiroquinoid chains are linked together along the *[100]* and*[010]* directions of the lattice. Although through-space bonding p_s interactions are primary in the analysis of what occurs at the Fermi level of glitter, it turns out that the spiroconjugation effects seen in the 1-dimensional polyspiroquinoid substructure will carry over into the 3-dimensional structure.

As is implied in the model structural drawing on the inset of Fig. 9, through-space bonding p_σ interactions are responsible for reducing the band gap (see Fig. 2) in the 1-dimensional polycyclophane substructure, which possesses the p_σ interactions uniquely among the 1-dimensional substructures. Clearly in this substructure, which from its geometry is absent orbital interactions connected with spiroconjugation, the π and π^* crystal orbitals come in pairs, the HOCO-1 and the HOCO as the π bands, and the LUCO and the LUCO+1 as the π^* bands. These bands run down from the zone center, where they occur in nearly degenerate pairs, out to the zone edge, as shown in Fig. 9.

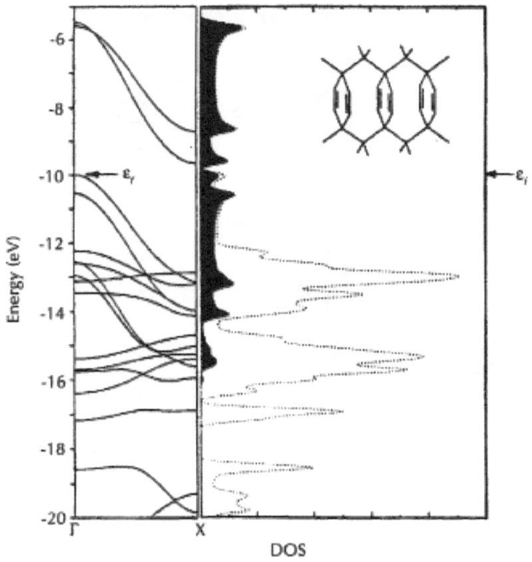

FIGURE 9 Electronic band structure and density of states (DOS) of the polycyclophane structure. Shaded area indicates the contribution of p_π orbitals to the total DOS. Structural model shown on onset.

Figure 10 shows the band structure of glitter, as there are 6 C atoms in the unit cell it consists of 24 bands, 12 of these bands are fully occupied. In order to assess the importance of through-space interactions in the lattice, including the importance of spiroconjugation, it is only necessary to consider the crystal orbitals derived from combinations of the $p_p$atomic orbitals present in the unit cell of glitter. These 4 p_π atomic orbitals, one such atomic orbital for each of the 4 trigonal planar C atoms in the unit cell, combine together to form 2 π crystal orbitals and 2 π* crystal orbitals in the electronic band structure of the glitter lattice.

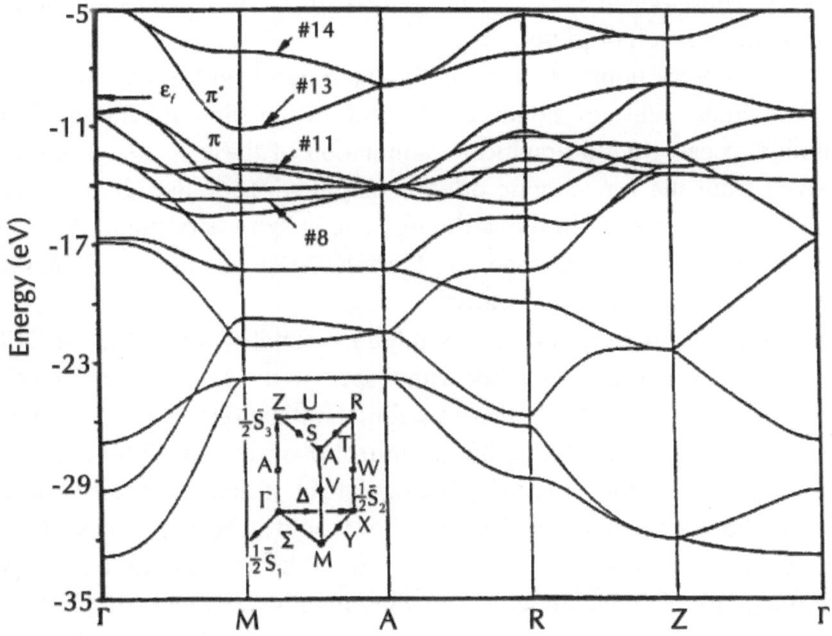

FIGURE 10 Electronic band structure of glitter. The π and π* bands are indicated at symmetry point M of the Brillouin zone.

In analogy with the polycyclophane substructure, it would be expected that these π and π* crystal orbitals would bracket the Fermi energy; that is they would occur as the HOCO-1, HOCO, LUCO and LUCO+1 (see Fig. 10). In the original report of the electronic structure of glitter, the authors pointed to the importance of through-space p_σ interactions, occurring together with through-bond p_π interactions, at the short interaction distance of 2.53 Å present in the unit cell.

Comparison of the Brillouin zone of glitter to the unit cell of glitter in direct space, shown on the inset of Fig. 10, reveals that symmetry point M is analogous to the zone center in the Brillouin zone of the polyspiroquinoid model. The symmetry line from M to A corresponds to the symmetry line followed from the zone center to the zone edge in the electronic structure of the polyspiroquinoid model. The π and π* bands in glitter, the π bands labeled as #8 and #11 and the π* bands labeled as #13 and #14, at symmetry point M in the diagram, become degenerate pairs at symmetry point A. These degeneracies occur because of the presence of a 4_2 screw axis in the unit cell of glitter.

At symmetry point M in the Brillouin zone of glitter these 4 crystal orbitals enter with real valued coefficients of the same magnitude. It is possible to carry out a graphical examination of these 4 crystal orbitals derived from the 4 p_π atomic orbitals of glitter. They may be sketched exactly as they appear in the Brillouin zone at symmetry point M. From these sketches, which are shown in Fig. 11, the attendant orbital interactions may be analyzed graphically.

Labeling the 4 π crystal orbitals at high symmetry point M in the Brillouin zone, using their symmetry with respect to the 4_2 screw axis and the **c** glide plane of the unit cell, there are fourcombinations: p_1 (SS), p_2 (AA), p_3* (AA) and π_4* (SS). Using the twomirror planes along *[001]* to assign the symmetries of the π crystal orbitals, instead of their behavior with respect to the space symmetry elements, they would all be of symmetry (SS), this represents an inversion of symmetry from the polyspiroquinoid (AA) combinations. Had the unit cell been chosen with *[100]* and *[010]* rotated through 45° from the present unit cell, all fourcombinations would be of (AA) symmetry with respect to the twomirror planes along [001]. This is entirely analogous to the four (AA) combinations of the polyspiroquinoid model, although not strictly homologous to the electronic structure of the polymer, as is discussed later. At any event, the present choice of unit cell makes visualization of the orbital interactions more clear to the reader.

From these sketches, one can see the presence of p_π interactions within the unit cell and also the presence of inter-cell and intra-cell p_σ and p_{spiro} interactions. Band #8 is the lowest-lying π crystal orbital (–14.87 eV), labeled p_1 (SS). It possesses through-space bonding p_{spiro} interactions and lies 1.67 eV below the next highest π crystal orbital, band #11 labeled p_2 (AA) (–13.25 eV) at symmetry point M. This splitting is analogous to the splitting of the bonding (AA) combinations in polyspiroquinoid. Bands #13 and #14 are the 2 π* crystal orbitals at M, they differ only in that the lower-lying of the pair has bonding p_{spiro} interactions, they have an energy separation of over 4 eV.

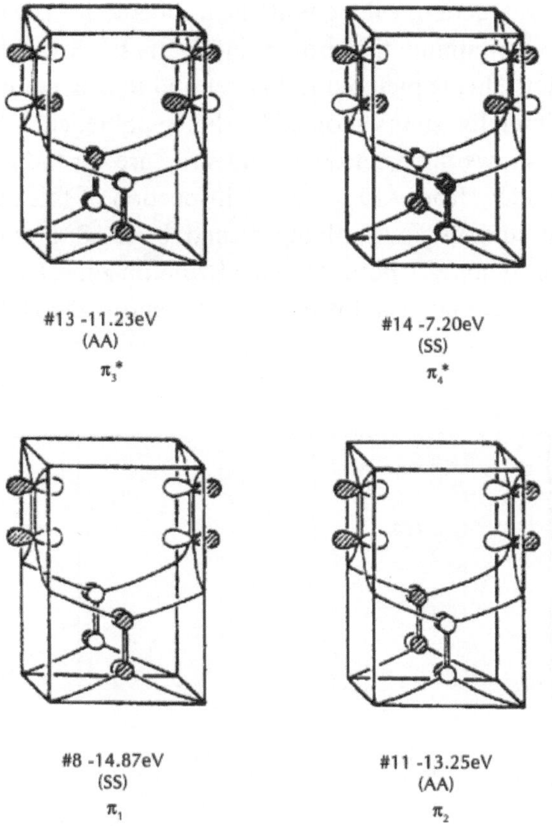

FIGURE 11 4 p_π crystal orbitals of glitter at symmetry point M in the Brillouin zone.

At this symmetry point in the Brillouin zone, the spiroconjugation effects present in the unit cell of glitter provide a stabilization energy of 1.93 eV compared to the energy of the "out-of-phase" π combination (p_1) of 3,3,6,6-tetramethyl-1,4-cyclohexadiene (see Fig. 6). The destabilized sibling π band (AA), at –13.25 eV, is itself 0.26 eV stabilized with respect to the "out-of-phase" π combination of the 3,3,6,6-tetramethyl-1,4-cyclohexadiene model molecule. Evidently, the p_σ interactions are quite important in stabilizing this p_2 (AA) crystal orbital as well.

An indirect band crossing of the lower-lying π* band, the LUCO, with the HOCO and twoother bands, is present in the band structure of glitter (see Fig. 10). Of the 2 π* crystal orbitals present in the electronic structure of glitter, the π* crystal orbital which is the LUCO has through-space bonding p_{spiro} interactions (along with the p_σ interactions) which cause this band to dip to a minimum that crosses indirectly the lower-lying occupied bands in glitter. This is pictured in Fig. 11 and thus explains the origin of the synthetic metallic status adopted by this hypothetical allotrope of C.

Figure 12 shows the electronic band structure for the glitter lattice in the energy window from –15 to –5 eV. Inspection of this calculated band diagram shows that the lowest-lying π band, band #8, occurs at its lowest energy point at symmetry point M in the Brillouin zone, as does the lower-lying π* crystal orbital. The lowest-lying π band is the HOCO-4, not the expected HOCO-1.

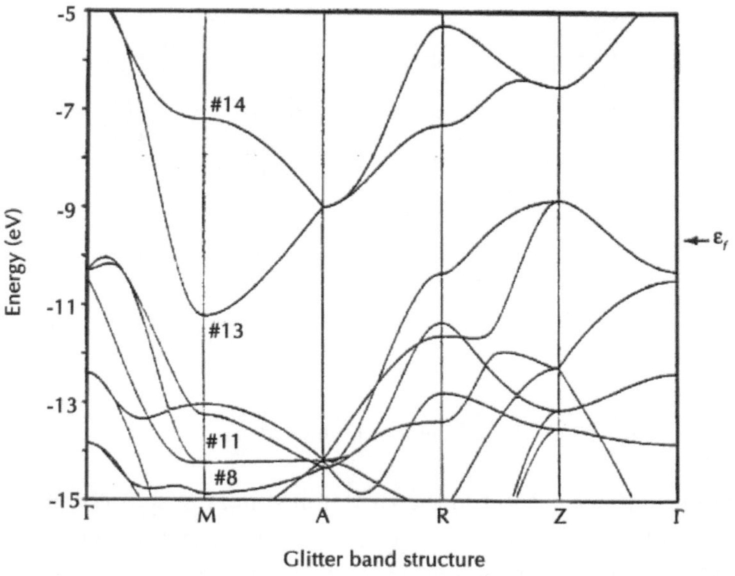

FIGURE 12 Electronic band structure of glitter in the energy window from –15 to –5 eV.

It occurs as such through the entire Brillouin zone of glitter. Band #11, refer to the corresponding sketch in Fig. 10, is the next highest π band, it forms a degenerate pair as the highest occupied crystal orbital along the symmetry line from Z to Γ in the Brillouin zone of glitter. Inspection of the Brillouin zone (Fig. 12) reveals that the 2 π crystal orbitals maintain an energy separation of about 1 eV throughout most of the reciprocal space of glitter

Note also in Fig. 12, the small dispersion (flatness) of band #8 about symmetry point M in the Brillouin zone, this gives rise to the secondary peak in the DOS of glitter (see Fig. 13), at an energy of about –15.0 eV. The shaded area indicates the contribution of p_π orbitals to the total DOS. The primary peak occurring at about –13.0 eV nearly coincides with the flatness of the sibling π crystal orbital, band #11, about this symmetry point in the Brillouin zone of glitter.

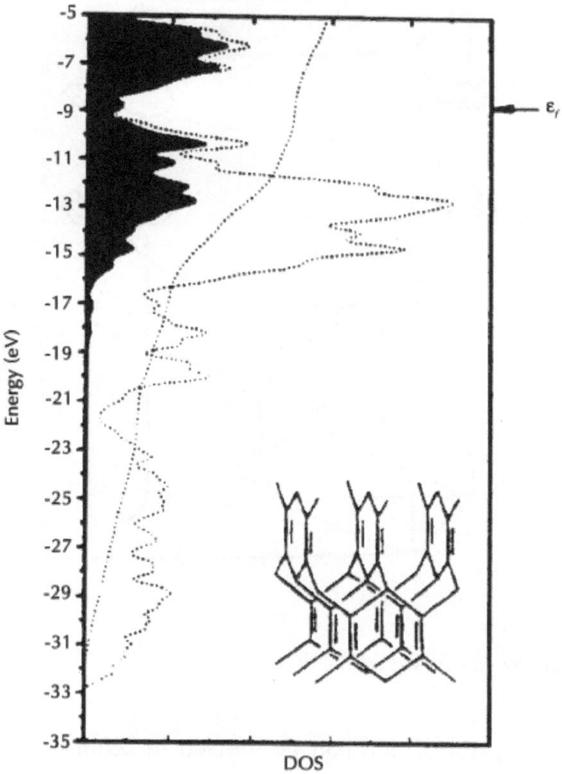

FIGURE 13 Density of states (DOS) of glitter. Shaded area indicates the contribution of p_π orbitals to the total density of states.

1.4 THE SPECIAL CRYSTAL ORBITAL OF GLITTER

Inspection of the atomic orbital composition of band #8 of glitter, the lowest-lying π crystal orbital, at symmetry point M in the Brillouin zone, reveals it has all fourp_π coefficients entering with the *same magnitude and sign*. Because of the presence of a bonding p_π interaction, twobonding p_σ interactions and fourbonding p_{spiro} interactions about *every* p_π atomic orbital in the unit cell of glitter, this is termed its special crystal orbital. This is diagrammed in Fig. 14. The effects of spiroconjugation along with the other interactions present in this 3-,4-connected network may confer it with an unusual electronic stability akin to the aromaticity observed in the planar polyhexes and 2D graphene sheet of ordinary quantum chemistry [8].

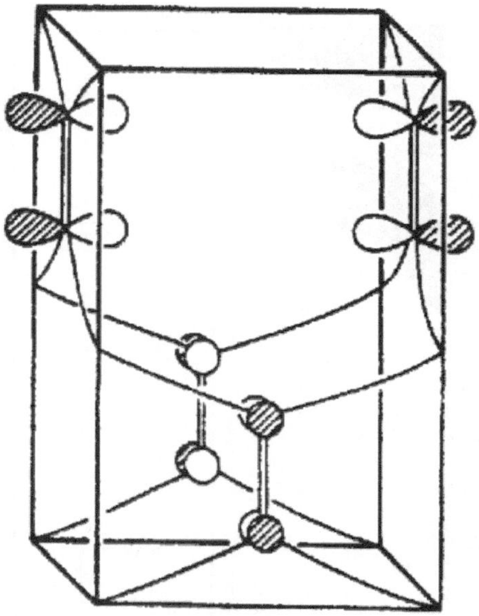

(SS)

FIGURE 14 The special crystal orbital of glitter.

The special crystal orbital can be viewed in sections through several unit cells perpendicular to the (001) plane of the tetragonal structure, and with the spiro atoms flattened out for clarity, as is shown in Fig. 15 (this

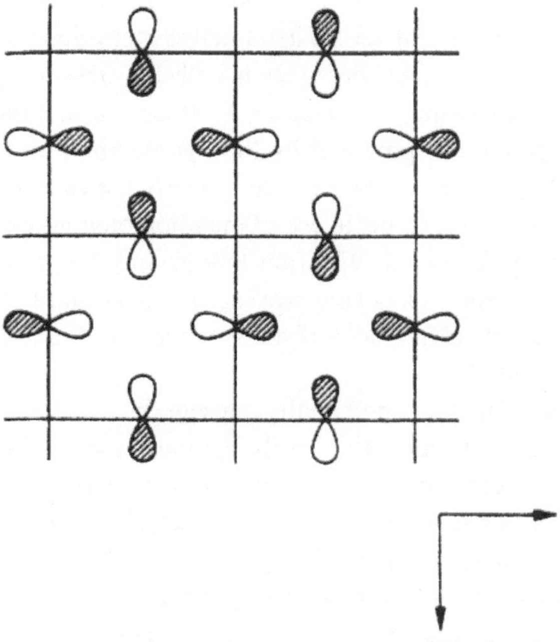

FIGURE 15 "Sections" of the special crystal orbital along *[100]* and *[010]* lattice directions. View is perpendicular to (001).

diagram is applicable to the view of the lower-lying π* crystal orbital, the LUCO, perpendicular to (001) as well, however alternating sheets would be out of phase along the 3rd dimension in the LUCO). Note the resemblance of this pattern to that of a standard checkerboard square, it arises as a consequence of the tetragonal symmetry of the glitter structure and from the nature of the p orbital interactions admitted by the glitter lattice. The constructive interactions present in the unit cell involve intra-cell interactions: p_π, p_σ and p_{spiro}; and inter-cell interactions of the same nature.

1.5 EXO- AND ENDO-SPIROCONJUGATION AND DIFFRACTION

Self assembly involving fourallene units per reaction, in the presence of a catalytic amount of the organometallic reagent bis-(triphenylphosphite)-

nickeldicarbonyl, has been reported to produce 1,3,5,7-tetramethylenecyclooctane [10]. Synthesis of 1,3,5,7-tetramethylenecyclooctane represents the first example of a molecular system with the possibility of exhibiting the effect of endo-spiroconjugation. Well known in structural chemistry are the homologies provided by the hydrocarbon molecules naphthalene and adamantane, which have very nearly the carbon–carbonbond lengths and bond angles, respectively, of their homologous extended structures: graphite and diamond [9]. From this perspective, the synthesis of 1,3,5,7-tetramethylenecyclooctane provides the opportunity to have a homologous hydrocarbon fragment of the structure of glitter, which can be studied experimentally.

Endo-spiroconjugation, potentially present in the 1,3,5,7-tetramethylenecyclooctane molecule, is precisely the mechanism of spiroconjugation that would occur intracellularly and extracellularly in the glitter lattice, please refer to the previous sections for details. Literature previously cited on aspects of spiroconjugation [7] concerned the related effect of exo-spiroconjugation, the interaction diagram of which is shown in Fig. 3 (interaction of 4 p orbitals) and Fig. 4 (interaction of 4 π orbitals). As discussed in Chapter 1, in the hypothetical polyspiroquinoid polymer, exo-spiroconjugation is operative.

Ideally, endo-spiroconjugation involves 4 p orbitals that point along lines, through space, which represent a pair of *edges* of a perfect tetrahedron, held at right angles to each other. Oppositely, exo-spiroconjugation, as pictured in Fig. 3, involves 4 p orbitals which are centered at each *vertex* of a tetrahedron and point at right angles, in two different pairs, respectively, each to one of a corresponding pair of orthogonal tetrahedral edges. Thus, the tetrahedron in exo-spiroconjugation is formed from 4 sp^3 hybrid orbitals extending from the spiro C atom to each of the atoms possessing the interacting p orbital. Quite in contrast, the p orbitals at a site of endo-spiroconjugation are cenetered on a point in space. From this subtle difference it is apparent that, site-by-site, the p orbital endo-spiroconjugated sees each of the 2 p_{spiro} interactions of its counterpart p orbital that is exo-spiroconjugated; and simultaneously it sees a p_{σ} interaction.

In the structure of 1,3,5,7-tetramethylenecyclooctane, ideally the effect of endo-spiroconjugation should manifest itself in the geometry of the skeletal cyclooctane ring system; the fourtrigonal atoms in the ring should inscribe a perfect tetrahedron inside the octagonal cradle. At the center of the inscribed tetrahedron, a significant π electron density should be present from the four-center interaction of endo-spiroconjugation. As the scattering power for x-rays depends upon the electron density, diffraction patterns obtained from molecular crystals of the 1,3,5,7-tetramethylenecyclooctane molecules may indicate weak reflections corresponding to the electron density $r_{spiro}(x,y,z)$ at the fractional coordinates of the unit cell specifying the center of the inscribed tetrahedron in the cyclooctane cradle (the center of symmetry of the molecule). Alternatively, electron density maps of the structure of 1,3,5,7-tetramethylenecyclooctane calculated by direct methods from the original diffraction data would reveal, quantitatively, the magnitude of π electron density at the fractional corrdinates corresponding to the center of symmetry of the 1,3,5,7-tetramethylenecyclooctane molecule.

Reflections due to π electron density concentrated at the center of symmetry of the 1,3,5,7-tetramethylenecyclooctane molecule would be weak for two principal reasons; the scattering power of the skeletal C atoms would be approximately six at low values of Θ; six being the number of electrons in C, whereas $r_{spiro}(x,y,z)$ could be at most approximately two at low values of Θ. The value 2 at low Θ is thus identified with the electronically unstable case that the π electron pair is cradled at the center of the inscribed tetrahedron. Intensity from $r_{spiro}(x,y,z)$ would be diminished substantially as the inscribed tetrahedron distorted from local T_d symmetry. As the non-ring C's in 1,3,5,7-tetramethylenecyclooctane are not stabilized by the benefit of endo-spiroconjugation, it is to be expected that the molecule will distort; it will maintain its principal S_4 axis, although it will distort towards a flattening out of the cyclooctane skeleton. Consequently, the local T_d symmetry of the inscribed tetrahedron will be lowered to one of D_{2d} symmetry and the scattering power at the endo-spiro center, $r_{spiro}(x,y,z)$, will be greatly diminished.

Construction of the model of the glitter lattice appealed to the geometry of the 1,4-cyclohexadiene molecule, information provided from diffraction studies [2]. The carbon–carbondouble bonds of the molecule are not conjugated and this appears in the bond alternation, the carbon–carbon single bonds being 1.51 Å and the carbon–carbon double bonds being 1.35 Å (in naphthalene the carbon–carbon bond lengths are approximately 1.40

Å, and in adamantane the carbon–carbon bond lengths are approximately 1.54 Å, respectively). Similarly, the bond angles are unsymmetrical in 1,4-cyclohexadiene with the C=C–C angles being 123° (ideal trigonal angle is 120°) and the adumbrated C–C–C angles are 114° (ideal tetrahedral angle is 109°28′).

Bond length and bond angle alternation present in the 1,4-cyclohexadiene molecule, and consequently the glitter lattice which is based upon it, distinguishes glitter from the metrically ideally regular sibling structures: graphite and diamond. See a later chapter for a further description of this metrical difference. Also discussed later, the standard topological indexes, polygonality and connectivity, distinguish glitter further as a topologically irregular (i.e., Wellsean) structure, while two of its siblings, graphite and diamond, are regular 2-dimensional and 3-dimensional structures, respectively; and the 3rd group of its siblings, the crystalline fullerenes, are semi-regular polyhedra (i.e., Archimedean).

Apparently quite by accident, the irregular metrical properties of the glitter lattice; the bond angle and bond length alternation, concomitant with its "low" topology (see a later chapter), result in the occurrence of nearly perfect tetrahedra, centered in space, which are about 2.5 Å on an edge. Such tetrahedra, being centered in space as opposed to being centered at an ideally tetrahedral C atom, as in diamond for example, provide the vehicle within which the 4-center p orbital interaction of endo-spiro-conjugation may occur. These tetrahedra possess nearly the full T_d symmetry that is evidently required for the optimum energetic stabilization associated with endo-spiroconjugation, refer to the earlier sections for a quantitative measure of this interaction energy.

Superimposed upon the unit cell of glitter; interpenetrating with it, which itself is outlined by the positions occupied by the 6 C atoms in $P4_2/mmc$ in glitter, are the set of points coincident with the spatial centers of the nearly perfect tetrahedra. These are the sites at which, through the effects of endo-spiroconjugation, are concentrated the π electron density from the fourcenters, through-space interaction. The latter array constitutes a primitive tetragonal lattice of points (P4/mmm) which is half the volume of the unit cell of glitter, the secondary lattice having its origin centered at (1/2,1/2,0) of the primary glitter lattice. Each of these sites may have a scattering power approaching two at low x-ray scattering angles,

Θ. In light of the T_d symmetry of these sites, and the more closely-spaced, regular array of such sites in the glitter lattice, compared to the situation in a molecular crystal of 1,3,5,7-tetramethylenecyclooctane discussed previously, it is to be expected that any weak reflections associated with x-ray scattering from the sites of $r_{spiro}(x,y,z)$ in glitter would be more evident than in a comparable diffraction pattern of a molecular crystal of 1,3,5,7-tetramethylenecyclooctane.

1.6 EXO- AND ENDO-SPIROCONJUGATION AND CHEMISTRY

Williams and Benson [10] report aspects of the chemistry of 1,3,5,7-tetramethylenecyclooctane, which provides some evidence for the proximity of π electrons within the cradle of the cyclooctane ring system. Effects on chemical reactivity due to exo-spiroconjugation have been referred to earlier [7]. It was hoped that evidence for the proximal interaction of adjacent, exocyclic methylene groups of the molecule would be evident through the formation of a transannular bond, in analogy with the cycloaddition reaction of tetracyanoethylene with bicycloheptadiene producing nortricyclene [11]. Cycloaddition of tetracyanoethylene with 1,3,5,7-tetramethylenecyclooctane falls within the class of symmetry-allowed, thermal s,s,s reactions; it is a $[_p2_s + _p2_s + _p2_s]$ cycloaddition reaction [12].

Tetracyanoethylene was chosen as the reagent to establish the interaction of pairs of opposite double bonds, the ring carbon p orbitals of which point towards each other along a line corresponding to one of the two orthogonal edges of the tetrahedron inscribed in the cradle of the cyclooctane skeleton, described earlier. An approximately equimolar reaction mixture of the two was prepared in tetrahydrofuran (THF) at room temperature. An exothermic reaction occurred within minutes, and after an hour the product was isolated, crystallized from acetonitrile and from spectroscopis analysis was determined to have a 3,3,4,4-tetracyano-8,11-dimethylenetricyclo[4.3.3.0]-dodecane structure. As the author's pointed out,"this appears to be the first example in which transannular bond formation can be directly attributed to the *proximity* of exocyclic methylene groups" [10].

1.7 DISTORTIONS OF THE GLITTER LATTICE

Based upon the crystal structures of diamond and graphite [9], in which the bond lengths and bond angles are uniform within each structure, respectively, a distortion was performed on the glitter lattice such that the bond lengths were all fixed at 1.46 Å (instead of alternating between 1.35 Å and 1.51 Å) and the bond angles were all fixed at 120° (except around the circumference of the tetrahedral C atoms, which is constrained to about 105°). The distortion to a more metrically regular structure destabilized the lattice by 0.80 eV/C atom relative to the metrically irregular structure.

Evidently, as the metrical properties of the lattice are modified to be more uniform, the tetrahedral sites that are centered in space, associated with the 4 center p orbital interaction of endo-spiroconjugation, concomitantly undergo a lowering of their local symmetry from T_d to D_{2d}. This is consistent with the model of endo-spiroconjugation presented here, in which the ideality of the tetrahedron yields optimum stabilization from the effect of endo-spiroconjugation.

1.8 SYNTHESIS OF GLITTER

Glitter is a 3-,4-connected net, so it is to be expected that a synthesis would be possible at conditions of temperature (T) and pressure (P) dictated by the phase boundary between graphite and diamond. However, the recent literature on C reflects the importance of kinetically driven syntheses, especially at conditions of T and P other than those along the graphite-diamond phase boundary. These include syntheses with vaporous carbon and hydrocarbon precursors in deposition apparatus [13]. A thermodynamic synthesis, in the vein of the original synthesis of diamond from graphite, could possibly be achieved in the presence of a suitable catalyst, such as a metallic solvent. Conventional high T-high P apparatus could be used for such a synthesis [14].

Identification of phases produced in such synthetic programs would principally include the use of diffractometry and spectroscopy. From the details of the electronic band structure of glitter reported herein, it should be possible to assign π–π^* electronic transitions through the analysis of the UV-visible absorption spectra, photelectron spectroscopy should elucidate features of the highest occupied crystal orbital [15]. With regard to diffraction evidence, theoretical diffraction patterns could be calculated for various versions of the crystal structure of glitter and these theoretical diffraction lines could be searched for in a synthesis experiment such as one employing a diamond anvil synthesis cell (DAC) and synchrotron radiation.

1.9 TOPOLOGICAL INDEXES OF CARBON ALLOTROPES AND GLITTER

Wells' seminal work on the structures of 2- and 3-dimensional nets and polyhedra was organized through the use of topological labels called Schlaefli symbols [3]. In such a scheme, there are two indexes, the polygonality represented by n and the connectivity represented by p. Each polyhedron, plane tessellation and 3-dimensional net has a Schlaefli symbol (n, p) to label its location in the space of topological structures in which Wells worked [16]. Interestingly, the Schlaefli symbols are rigorously determined for the convex polyhedra through a relation due to Euler.

Euler's work on the convex polyhedra resulted in an equation marking the origin of the discipline of topology, shown in one form as Eq. (1) [17].

$$V - E + F = 2 \tag{1}$$

Here, V, E and F are the number of vertices, edges and faces, respectively, in the convex polyhedron [18]. Geometrical arguments can be used to transform Eq. (1) into forms relating V, E and F; the primary topological indexes, to n, the polygonality and p, the connectivity; the latter being secondary topological indexes.

Each edge in a convex polyhedron is shared by two faces, therefore, nF is the same as 2E. Each edge has two vertices, therefore pV is the same as 2E. By substitution, Euler's equation now reads:

$$2E/n - E + 2E/p = 2 \tag{2}$$

Rearrangement of Eq. (2) into Eq. (3) shows that the values n, p and E must be positive integers for the expression to have validity for polyhedra [3]:

$$\frac{1}{n} - \frac{1}{2} + \frac{1}{p} = \frac{1}{E} \tag{3}$$

Further restrictions are imposed on the values of n and p in order to determine the solutions to Eq. (3). In order for the number of edges, E, to be positive the values of n and p must be less than 6, and in fact n and p must be greater than two because of the impossibility of faces of zero area or spikes in the convex polyhedron.

Substituting each of the nine combinations of 3, 4 and 5 into Eq. (3), there are only five finite, rational solutions. The five are well known and are shown as the tetrahedron, t, the octahedron, o, the icosahedron, i, the cube, c, and the dodecahedron, d, in the Table 1.

TABLE 1 Schlaefli Symbols for Regular Structures.

	p						
n	3	4	5	6	7	8	...
3	t	o	i	(3, 6)	(3, 7)	(3, 8)	
4	c	(4,4)	(4, 5)	(4, 6)	(4, 7)	(4, 8)	
5	d	(5,4)	(5, 5)	(5, 6)	(5, 7)	(5, 8)	
6	(6,3)	(6,4)	(6, 5)	(6, 6)	(6, 7)	(6, 8)	
7	(7,3)	(7,4)	(7, 5)	(7, 6)	(7, 7)	(7, 8)	
8	(8,3)	(8,4)	(8, 5)	(8, 6)	(8, 7)	(8, 8)	

Of the other four non-solutions, from the square net (4, 4) substitution into Eq. (3) indicates the number of edges, E, in this structure is infinite. Evidently, in (4, 4) the Euler equation reveals a transition to a 2-dimensional extended structure [19]. In (4, 5), (5, 4) and (5, 5) the non-solutions have values of –5, –5, and –10, respectively. Another transition has occurred, to a still higher dimensionality, specifically to 3-dimensional structures. (5, 5) corresponds to a 3-dimensional network that Wells first discovered in the 1960's [3]. For a sensible result, in which the number of edges, E, is positive, it is clear that a modification of the classical Euler equation, one applicable for 3-dimensional networks, is necessary [20].

The structure of the graphite net is well known, its Schlaefli symbol is (6, 3), the polygonality is six and the connectivity is three. Its place in the topological subspace comprised of regular structures, regular meaning the Schlaefli symbols n and p are integers, is to the right of the convex polyhedron (5, 3), which is the pentagonal dodecahedron, and to the left of (7, 3). Interestingly, (7, 3) represents a group of at least 4 unique, regular nets which Wells discovered [3]. To date, none of these nets in the family (7, 3) has yet been discovered to correspond to an actual crystal structure.

Regular structures are adjoined to their left or right, or adjacent to them vertically, by the semi-regular structures, semi-regular meaning that one of the Schlaefli indexes n and p is fractional and the other is an integer. One such semi-regular with contemporary importance is the Archimedean polyhedron called the truncated icosahedron. The Schlaefli symbol for this mathematical object, and for the Buckminsterfullerene molecule that is patterned after it, is given by ($5^{2/3}$, 3). It is intermediate in topology between the pentagonal dodecahedron (5, 3) and the graphite net (6, 3). Buckminsterfullerene is unique among the allotropes of C in that it is semi-regular.

Classical structures include the regular and Archimedean polyhedra; Archimedean meaning only n, the polygonality, is fractional. In the 19th century, Catalan identified the reciprocal polyhedra to the Archimedean polyhedra where n and p were exchanged with each other. Catalan polyhedra can be constructed from their reciprocals by joining the midpoints of each face of the Archimedean polyhedron to each other [21]. This reciprocal nature occurs also in two dimensions, one can easily visualize the reciprocals of (4, 4), (3, 6) and (6, 3), for example; but is apparently absent in 3-dimensional nets.

Beyond the semi-regular structures lie the Wellsean structures, those in which both the polygonality n and the connectivity p are fractional, in the Schlaefli symbol (n, p) [22]. These structures would lie diagonally, to the left or right and displaced vertically, from the entries into the table of Schlaefli symbols for the regular polyhedra, plane nets and 3-dimensional nets. Wells discovered many Wellsean structures, specifically classifying them as possessing fractional polygonality and fractional connectivity,in the course of his exploration of the space of 3-,4-connected nets [22].

It appears that the first actual Wellsean structure was identified, though not topologically indexed as such, by W.L. Bragg and W.H. Zachariasen, from their determination of the crystal structure of the silicate mineral called phenacite, Be_2SiO_4 [23]. This was an especially difficult structure determination because of the rhombohedral symmetry (R3) and the large size of the unit cell (six molecules of Be_2SiO_4 in each unit cell). In W.L. Bragg's description he reports, "It is difficult to give a clear figure of the structure, because the unit cell is large and a pattern with rhombohedral symmetry is harder to depict than one based upon rectangular axes. The principles of the structure are very simple, however, and are readily traced in a model. It is formed of linked tetrahedra, with Si and Be at their centers. Each O of the independent SiO_4 groups also forms part of two neighboring tetrahedra around Be atoms. Thus each Si is surrounded by 4 O atoms, and each Be by 4 O atoms, and each O is linked to 2 Be atoms and one Si atom at the corners of an equilateral triangle" [24].

With the identification of this structure type, which Bragg pointed out could be simplified by replacing the Be and Si atoms by one atom type, to a binary compound of formula A_3B_4 with a smaller unit cell with hexagonal symmetry ($P6_3/m$), entry had been made into a topologically new class of crystal structures. In such a hexagonal lattice, there are four 3-connected vertices and three 4-connected vertices in the asymmetric unit of structure, and there are six 6-sided polygons and four 8-sided polygons in the asymmetric unit of structure [9]. The Schlaefli symbol for the phenacite structure is therefore ($6^{4/5}$, 3.4285....). In 1939 and 1940, Juza and Hahn [25] synthesized and reported the crystal structure of Ge_3N_4, patterned on the structure of phenacite in the manner described by Bragg [26, 27]. About 17 years later, the synthesis and crystal structure of another polymorph of Ge_3N_4, with nearly an identical density, and the corresponding syntheses of the isomorphous α- and β-Si_3N_4 polymorphs was reported in part by several groups including Hardie and Jack [26] and Ruddleson and Popper [27].

Motivated by the reports of the synthesis of the IV–V nitrides, 1st principles calculations were performed on a hypothetical carbon nitride phase, C_3N_4, patterned on the β-Si_3N_4 structure type [28]. A semi-empirical formula for bulk modulus developed out of this work, predicted the latter nitride would be of hardness comparable to diamond [28]. Inspection of the glitter structure, a 3-,4-connected net with 6- and 8-sided polygons in its structural pattern, indicates the Schlaefli symbol (7, $3^{1/3}$). The polygonality is an admixture of 6's and 8's, which nonetheless has the integer polygonality 7. It is a Wellsean structure. Its polygonality is higher that in the diamond structures (6, 4), and the graphene tessellation (6, 3), and the molecule Buckminsterfullerene ($5^{2/3}$, 3).

In a separate vein, Waser and McClanahan reported [29] the 1st transition metal-containing crystal structure of a 3-,4-connected net, the Pt_3O_4 structure type, in 1951. As it turns out, this is a semi-regular structure (a Catalan network) with the Schlaefli symbol (8, 3.4285….). Here, as in the structures patterned on phenacite, the connectivity appears as a continued fraction. It might be expected that the symmetry of the structure would be low, on the basis of the low topology. Despite the irrationality of the connectivity, the Pt_3O_4 lattice possesses cubic space group symmetry (Pm3n). The IV–V nitrides are of hexagonal symmetry ($P6_3/m$). Topology (regularity) and symmetry are to some degree independent properties of structures. Even so, the well-known structures of graphite and diamond, for example, independent of their high symmetry, possess high topology.

In 1965, Wells [3] reported two new semi-regular (Catalan) 3-,4-connected nets; 2 unique nets possessing the same Schlaefli symbol (8, 3.5714….). Their topological identity is not distinguishable completely by their Schlaefli symbol, as in other cases, and a distinct topological identity is provided through the use of further labels based upon the number of polygons common to each link and vertex, respectively. In fact one net is comprised of square planar connectivity (space group I4/mmm) and the other contains tetrahedral connectivity (space group I4m2). Both of these nets possessed trigonal planar vertices bonded with four-connected vertices. Moreover, a 3rd 3-,4-connected net was reported in this paper, this net with the Schlaefli symbol ($8^{1/2}$, $3^{1/3}$). It possesses orthorhombic symmetry (Pmmm). With this work, Wells began his exploration of the 3-,4-connected networks and began identifying Wellsean networks.

Physically, the polygonality is approximately related to the openness of the structure. The connectivity index is approximately related to the closedness of the structure. For example, the glitter structure has a connectivity index of $3^{1/3}$, it is clearly intermediate between graphite at 3, and diamond at 4. From the connectivity index it can be identified topologically as a hybrid of graphite and diamond. Compare to the Buckminsterfullerene molecule, with a polygonality of $5^{2/3}$, it being intermediate between the pentagonal dodecahedron (5, 3) and the graphene tessellation (6, 3). Another useful topological index obtained from the polygonality and the connectivity, is formed by taking the ratio of n and p:

$$l=\frac{n}{p} \tag{4}$$

where l is a measure of the form of the structure as related to its average polygon size per unit average connectivity. It is a topological index useful in one sense for identifying similarities between regular, semi-regular and Wellsean structures.

Along the principal diagonal of Table 1 are structures in which the polygonality is the same as the connectivity, the topological index l is therefore unity. Such structures have the very highest topology. To the left of the principal diagonal lies the subspace of structures with l indexes less than one, to the right of the principal diagonal lie structures with indexes l greater than one. As a general index of topological relatedness, it is very interesting to note graphite and glitter both possess l indexes close to 2. It is hoped that this indicial similarity can be elucidated in a future communication about graphite and glitter.

Consideration of the work on 3-,4-connected nets initiated with Bragg and Zacharaisen in 1930 on phenacite; followed by Juza and Hahn in 1940 and Ruddleson and Popper, and Jack and Hardie in 1957, on IV-V nitrides; then the theoretical work of A.F. Wells on the 3-,4-connected networks begun in 1965; and most recently the attempts by the group led by Cohen to synthesize C_3N_4 in the 1990's; there is a clear progression towards the synthesis of 3-,4-connected nets of the 2nd period elements. Quite apart from this development, though apparently converging with it, is the descendency of synthetic allotropes of C from those C nets possessing a high topology, to new forms of C which break to lower topologies, the principal example of which is the self assembly of the Buckminsterfullerene molecule [13].

Isomorphous variants of the parent lattice of C atoms, are the III-IV series of compounds which could adopt the glitter structure. The fully metallic band profile of the hypothetical B_2C phase patterned on the glitter structure, has been briefly described [1]. Alternatively, the adjacency of trigonal centers across faces in the unit cell of glitter, suggests a denser structure in which trigonal planar points are transformed into trigonal bipyramidal points. Such a lattice would appear to be the first 4-,5-connected net, and a good model for exploration of 3rd period IV-V structures which have access to expanded octet hybridization (sp^3d) about the Group V elements, for example phases such as SiP_2.

1.10 THE FOUNDATIONS OF A SPIRO QUANTUM CHEMISTRY

With the foregoing analysis of spiroconjugation presented with respect to the 1-dimensional polyspiroquinoid polymer and the 3-dimensional glitter structure, the foundations have been laid for the elucidation of a new type of quantum chemistry, called spiro quantum chemistry. Spiro quantum chemistry complements the historically developed quantum chemistry of linear polyenes, the annulenes, polyhexes, 2D graphene sheets and the fullerenes. Fragments of the polyspiroquinoid polymer, called the spiro[n]quinoids, constitute the spiro quantum chemistry analogs to the linear polyenes. If one joins the spiro[n]quinoids at their ends, one forms the cyclospiro[n]quinoids. Thus the cyclospiro[n] quinoids constitute the spiro quantum chemistry analog to the annulenes of ordinary quantum chemistry. As they possess analogous nodal properties, in the spiro wave functions, to the ordinary π and π^* wave functions of the linear polyenes and the cyclic annulenes, it may be possible to identify an analog of the "4n+2" rule of aromaticity in the cyclospiro[n]quinoids.

In two dimensions one has the analog of the graphene sheet reported as the spirographene sheet in Ref. [1]. It's band structure shows it to be semi-metallic, in analogy with the ordinary graphene sheet. Fragments of the graphene sheet are termed poly[m,n]hexes, where the indexes m and n indicate the length and width of the polyhex fragment. In analogy to the poly[m,n]hexes there are the spiro[m,n]hexes, where the indexes m and n indicate the numbers of spiro nodes running lengthwise and widthwise in the spiro[m,n]hex fragment. From the band structure of the parent spirographene sheet, it is apparent that the spiro[m,n]hexes may have important electronic or optical properties. One could indeed wrap the spirographene sheet back onto itself and create spirographene cylinders analogous to the graphene cylinders seen in nanotubes.

In fully three dimensions one has the synthetic metal glitter, evidently there is no 3-dimensional analog to glitter in ordinary quantum chemistry, which extends to 2D graphene sheets and no further. Glitter, as a synthetic metal, is already interesting from the point of view of its undoped metallic status. That its metallic state can be so tightly linked to the endo-spiro-conjugation in it is also remarkable. One could envision fragments of the glitter lattice, analogous to the recently reported diamondoid hydrocarbon fragments of the diamond lattice, called the glitter[m,n,o]enes in which the indexes indicate the numbers of spiro nodes in the fragment along its length, width and height.

Clearly there is a rich area of research accessible to organic chemists, theorists, materials scientists and biologists in spiro quantum chemistry, and the results in this paper are only the beginning. Each class of spiro analogs to the ordinary hydrocarbons of quantum chemistry presents its own, unique synthetic challenges to the organic chemistry community. And certainly there is a wealth of problems for theorists pointed out by this paper. The abundant literature on spiroconjugation, alluded to in Ref. [7], suggests its importance in chemistry, biology, materials science and fundamental theory. It is hoped that chemists may be interested enough by the prospects of this manuscript to take up the cause of elucidating the fundamental features of spiro quantum chemistry.

KEYWORDS

- **π combination**
- **glitter**
- **polyspiroquinoid**
- **spiro**
- **spiroconjugation**

REFERENCES

1. Bucknum, M. J.; Hoffmann, R. *J. Am. Chem. Soc.*, **1994,** *116*, 11456.
2. (a) Carreira, L.A.; Carter, R.O.;Durig, J. R. *J. Chem. Phys.*, **1973,***59*, 812. (b) H. Oberhammer, S.H. Bauer, *J. Am. Chem. Soc.*, **1969,** *91*, 10. The structural information is based upon electron diffraction data of the gaseous 1,4-cyclohexadiene molecule. Carreiraet al. report the ring in a planar equilibrium conformation.
3. Wells, A.F. The Geometrical Basis of Crystal Chemistry: (a) Part 1, *ActaCryst.*,**1954,** *7*, 535. (b) Part 2, *ActaCryst.*,**1954,** *7*, 545. (c) Part 3, *ActaCryst.*,**1954,** *7*, 842. (d) Part 4, *ActaCryst.*,**1954,** *7*, 849. (e) Part 5, *ActaCryst.*,**1955,** *8*, 32. (f) Part 6, *ActaCryst.*,**1956,** *9*, 23. (g) Part 7, A.F. Wells, R.R. Sharpe, *ActaCryst.*,**1963,** *16*, 857. (h) Part 8, *ActaCryst.*,**1965,** *18*, 894. (i) Part 9, *ActaCryst.*,**1968,** *B24*, 50. (j) Part 10, *ActaCryst.*,**1969,** *B25*, 1711. (k) Part 11, *ActaCryst.*,**1972,** *B28*, 711. (l) Part 12, *ActaCryst.*,**1976,** *B32*, 2619. (m) In:*Three Dimensional Nets and Polyhedra*, 1st edition, Wiley and Sons: New York, 1977. (n) In:*Further Studies of Three-dimensional Nets*, American Crystallographic Association, Monograph #8, 1st edition, ACA Press, 1979.
4. Merz, K.M.;Hoffmann, R.;Balaban, A.T. *J. Am. Chem. Soc.*, **1987,** *109*, 6742.
5. (a) Hoffmann, R.;Lipscomb, W.N. *J. Chem. Phys.*, **1962,** *37*, 2872. (b) Whangbo, M-H.;Hoffmann, R.*J. Am. Chem. Soc.*, **1978,** *100*, 6093. (c) Whangbo, M-H.;Hoffmann, R.; Woodward, R.B.*Proceedings of the Royal Soc.*, **1979,** *A366*, 23.
6. (a) Simmons, H.E.;Fukunaga, T.*J. Am. Chem. Soc.*, **1967,** *89*, 5208. (b) Hoffmann, R.; Imamura, A.; Zeiss, G.*J. Am. Chem. Soc.*, **1967,** *89*, 5215.
7. (a) Galasso, V. *Chem. Phys.*, **1991,** *153*, 13. (b) Smolinski, S.;Balazy, M.;Iwamura, H.; Sugawara, T.; Kawada, Y.; Iwamura, M.*Bulletin Chem. Soc. Jpn.*, **1982,** *55*, 1106. (c) Duerr, H.;Gleiter, R.*AngewandteChemie International Edition in English*, **1978,** *17*, 559. (d) Simkin,B.Y.;Makarov, S.P.; Furmanova, N.G.; Karaev, K.S.; Minkin, V.I.*Chem. Heterocyclic Compounds*, **1978,** *1*, 948. (e) Foos, J.S. "Synthesis and Reactivity of Spiro4.4]nonatetraene and Spiro4.4]nona-1,3,6-triene," PhD thesis, CornellUniversity, 1974. (f) C. Batich, E. Hielbronner, E. Rommel, M.F. Semmelhack, J.S. Foos, *J. Am. Chem. Soc.*, **1974,** *96*, 7662. (g) R. Sustmann, R. Schubert, *AngewandteChemie International Edition in English*, **1972,** *11*, 840. (h) H. Duerr, H. Kober, *Justus Liebigs Ann. Chem.*,**1970,***740*, 74. (i) E.W. Garbisch, Jr., R.F. Sprecher, *J. Am. Chem. Soc.*, **1966,** *88*, 3433. (j) P.E. Eaton, R.A. Hudson, *J. Am. Chem. Soc.*, **1965,** *87*, 2769.
8. Minkin, V.I.;Glukhovtsev, M.N.;Simkin, B.Y. In:*Aromaticity and Antiaromaticity: Electronic and Structural Aspects*, 1st edition, Wiley and Sons, NY, 1994.
9. Wells, A.F. In *Structural Inorganic Chemistry*, 4th edition; Oxford University Press: Oxford, UK, 1984.
10. Williams, J.K.;Benson, R.E. *J. Am. Chem. Soc.*, **1962,** *84*, 1257.
11. Benson, R.E.;Lindsey, Jr., R.V. *J. Am. Chem. Soc.*, **1959,** *81*, 4247.
12. Woodward, R.B.;Hoffmann, R.In:*The Conservation of Orbital Symmetry*, 1st edition, VerlagChemie, GmbH, Weinheim/Bergstrasse, 1970.
13. (a) Donohue, J. *The Structure of the Elements,*, 1st edition, Wiley and Sons, NY, 1974. (b) Kraetschmer, W.;Huffmann, H. *Nature*, **1990,** *347*, 354. (c) H. Kroto, J.R. Heath, S.C. O'Brien, R.F. Curl, R.E. Smalley, *Nature*, **1985,** *318*, 162. (d) E. Osawa, *Kagaku*,

1970, *25*, 85. (e) F.P. Bundy, J.S. Kasper, *J. Chem. Phys.*, **1967,** *46*, 3437. (f) R.B. Aust, H.G. Drickamer, *Science*, **1963,** *140*, 817. (g) L. Zeger, E.Kaxiras, *Phys. Rev. Lett.*,**1993,** *70*, 2929. (h) D.A. Muller, Y. Zhou, R. Raj,,J. Silcox, *Nature*, **1993,** *366*, 725. (i) H. Hiura, T.W. Ebbensen, J. Fujita, K. Tanigaki, T. Takada, *Nature*, **1994,** *367*, 148. (j) C.L. Renschler, J. Pouch, D.M. Cox, Eds.;*Novel Forms of Carbon,* Materials Research Society (MRS) Symposium Proceedings, *270*, Materials Research Society (MRS), Pittsburgh, 1992. (k) F. Moshary, N.H. Chen, I.F. Silvera, C.A. Brown, H.C. Dorn, M.S. de Vries, D.S. Bethune, *Phys. Rev. Lett.*,**1992,** *69*, 466. (l) S. Iijima, *Nature*, **1991,** *354*, 56. (m) W. Utsumi, T. Yagi, *Science*,**1991,** *252*, 1542. (n) H. Hirai, K. Kondo, *Science*, **1991,** *253*, 772. (o) K.E. Spear, A.W. Phelps, W.B. White, *J. Mater. Res.*, **1990,** *5*, 2277. (p) A.V. Baitin, A.A. Lebedev, S.V. Romanenko, V.N. Senchenko, M.A. Scheindlin, *High Temperatures-High Pressures*, **1990,** *21*, 157. (q) J.C. Angus, C.C. Hayman, *Science*, **1988,** *241*, 913. (r) V.M. Melnitchenko, Y.N. Nikulin, A.M. Sladkov, *Carbon*, **1985,** *23*, 3. (s) I.V. Stankevich, M.V. Nikerov, D.A. Bochvar, *Russian Chem. Rev.s*, **1984,** *53*, 640. (t) A.G. Whittaker, E.J. Watts, R.S. Lewis, E. Anders, *Science*, *209*, 1512, **1980,.** (u) A.G. Whittaker, B. Tooper, *J. Am. Ceramic Soc.*, **1974,** *57*, 443.(v) D.A. Bochvar, E.G. Galpern, *Dokl. Akad. Nauk. SSSR*, **1973,** *209*, 612.(w) A. El Gorsey, G. Donnay, *Science*, **1968,** *161*, 363. (x) S. Ergun, *Carbon*, **1968,** *6*, 141. (y) H. Drickamer, *Science*, **1967,** *156*, 1183. (z) F.P. Bundy, *J. Chem. Phys.*, **1963,** *38*, 631.

14. Hazen, R.M. In:*The New Alchemists*, 1st edition, Times Books-Random House: New York, 1994.
15. Atkins, P.W. In:*Physical Chemistry*, 4th edition, W.H. Freeman and Company: New York, 1990.
16. More descriptive topological indexes are sometimes used in place of Schlaefli symbols because of the need to further identify nets with the same Schlaefli symbol (n, p). For example the graphene tessellation with the Wells point symbol 6^3, has the index (6, 3), which is identical to that of the tessellation of squares, hexagons and octagons with the Well point symbol 4.6.8. Such coincidental patterns are known as topoisomers.
17. *Elementadoctrinaesolidorum and Demonstratiumnonannularum in signiumproprietatumquibus solid heddrisplanisinclusasuntpraedita*, L. Euler, included in the Proceedings of the St. PetersburgAcademy, 1758.
18. Explicit Euler equations for 3-, 4- and 5-connected polyhedra elegantly show the existence of the 5 regular convex polyhedra and the semi-regular Archimedean and Catalan polyhedra.
19. For 1-dimension, n and p can be viewed as each being equal to 2. (2, 2) has a number of edges, E, equal to 2. For n-sided polygons, the Schlaefli symbol is (n, 2) and the number of edges, E, is properly equal to n.
20. Wells, A.F. The Geometrical Basis of Crystal Chemistry: Part 7, Wells, A.F.;Sharpe, R.R. *ActaCryst.*,**1963,** *16*, 857. This paper reports on a modification of the Euler relation for application to 3-dimensional polyhedra. Such polyhedra are infinite in extent and are represented by the Schlaefli symbols (3, p) where p is greater than 6, and other Schlaefli symbols, some of which are identical to those of the ordinary 3-dimensional nets.
21. Wenninger, M. In:*Dual Models*, 1st edition, Cambridge University Press: Cambridge, UK, 1983. And the references therein.

22. Wells, A.F. In:*Further Studies of Three-dimensional Nets*, American Crystallographic Association (ACA) Monograph #8, 1st edition, ACA Press, Pittsburgh, 1979. And the references therein. Much previously unpublished material, concerned mainly with the vast topological subspace of 3-,4-connected networks, was reported in this publication. Scores of previously unknown structures, perhaps numbering more than 100, were reported. The explicit identification of structures with both fractional polygonality and fractional connectivity was made, anticipating later work.
23. Bragg, W.L.;Zachariasen, W.H. *ZeitschriftfuerKristolograffie*, **1930,** *72*, 518.
24. Bragg, W.L. In:*Atomic Structure of Minerals*, 1st edition, Cornell University Press: Ithaca, NY, 1937.
25. (a) Juza, R.;Hahn, H. *Naturwissenschaften*, **1939,** *27*, 32. (b) Juza, R.;Hahn, H. *ZeitschriftfuerAnorganischeChemie*, **1940,** *244*, 125.
26. Hardie, D.;Jack, K.H.*Nature*, **1957,** *180*, 332.
27. Ruddleson, S.N.;Popper, P. *ActaCryst.*,**1958,** *11*, 465.
28. Cohen, M.L. *Phys. Rev. B*, **1985,** *32*, 7988.
29. Waser, J.; McClanahan, E.D.*J. Chem. Phys.*, **1951,** *19*, 413.

CHAPTER 2

SYNTHESIS OF GLITTER

MICHAEL J. BUCKNUM and EDUARDO A. CASTRO

CONTENTS

SUMMARY OF SYNTHESIS OF GLITTER

Structural analysis on carbonaceous samples produced from high temperature-pressure conditions by Palatnik et al., in 1984 indicate the existence of a metallic allotrope of carbon with a diffraction pattern closely matching that of cubic diamond. A structure proposed for this phase by Konyashin et al., and others suggests that the four interstitial carbon atoms occupying ½ the tetrahedral holes in the ordinary cubic diamond lattice are vacant in the new structure. This leads to a transformation of the Fd3m, ordinary cubic diamond structure-type, to a simple face centered cubic carbon lattice in space group Fm3m, with a lattice parameter of 3.56 Å, identical with that of cubic diamond. The new structure supports the diffraction evidence accumulated for this so-called "n-diamond" phase, but does not hold up to a first principles total energy optimization at the DFT level of theory for the fcc lattice, which reports the Konyashin et al. structure to be unstable. Here we report an alternative tetragonal carbon structural-type, which we have called "glitter," that explains the observed diffraction pattern of n-diamond reasonably well, and that is stabilized by extensive spiroconjugation in three dimensions leading to a metallic status for the carbonaceous structural-type.

2.1 DISCOVERY OF METALLIC CARBON

In 2001, Konyashin et al. [1] reported the observation of a metallic modification of carbon from techniques of high temperature-pressure synthesis. These results confirmed earlier studies by Palatnik et al. [2] in 1984. In their work, the scientists synthesized a crystalline carbon material that reproduced the diffraction pattern of diamond, but in addition, there were the presence of symmetry forbidden diamond reflections contained in the diffraction evidence.

Gogotsi et al. [3], also with work reported in 2001, used a reaction-based method centered upon the carbide derived carbon (CDC) strategy developed in their group, and synthesized bulk polycrystalline samples of carbon, which by interrogation with diffraction techniques were seen to produce Bragg reflections consistent with carbon diamond, but in addition the samples were seen to exhibit some symmetry forbidden reflections of diamond, like the (200) reflection of cubic diamond at 1.78 Å. This work was reminiscent of that described by Konyashin et al. [1] and Palatnik et al. [2], in that symmetry forbidden reflections of diamond were consistently observed, along with the usual diamond polytype reflections, in the diffraction analysis of their carbonaceous materials, as produced in this instance from the CDC strategy.

Hirai et al. [4], in research first reported in 1991, catalogued similar results from studies of shock-compressed graphite using a rapid cooling technique. They observed carbon diamond polytype reflections in their heated, shock-compressed carbon samples, but in addition they recorded the (200), (222) and (420) symmetry forbidden Bragg reflections of cubic diamond in their diffraction analysis.

Collectively, the researchers call this diamond-like material "n-diamond" (or γ-carbon) after the resemblance of its diffraction signature to ordinary carbon diamond. Like the other studies on n-diamond, the work of Hirai et al. demonstrated the metallic nature of the carbon materials they produced. Their report will be used as the basis for a comparison of the calculated diffraction pattern of a novel, hypothetical tetragonal allotrope of carbon called "glitter," that has been theoretically characterized by Bucknum et al. previously [5-7], to that of their n-diamond samples. Finally, Bursill et al., have recorded similar diffraction observations for carbon ion-implanted carbon samples crystallized in a quartz matrix [8].

A structure for the modified diamond form in these studies, the so-called "n-diamond," has been proposed by Palatnik et al. as early as 1984. It would be physically unrealistic to ascribe the symmetry forbidden reflections observed in these studies to ordinary cubic diamond, and in fact they are proposed to originate from a material consisting of a simple face centered cubic (fcc) cell of carbon atoms, in space group Fm3m. Furthermore, in order to explain the diffraction evidence they obtained, the researchers were constrained to define the fcc n-diamond lattice of carbon as possessing the same unit cell edge as does ordinary cubic diamond (Fd3m) at 3.56 Å. The proposed structure of n-diamond and its relationship to the structure of ordinary cubic diamond are shown in Fig. 1.

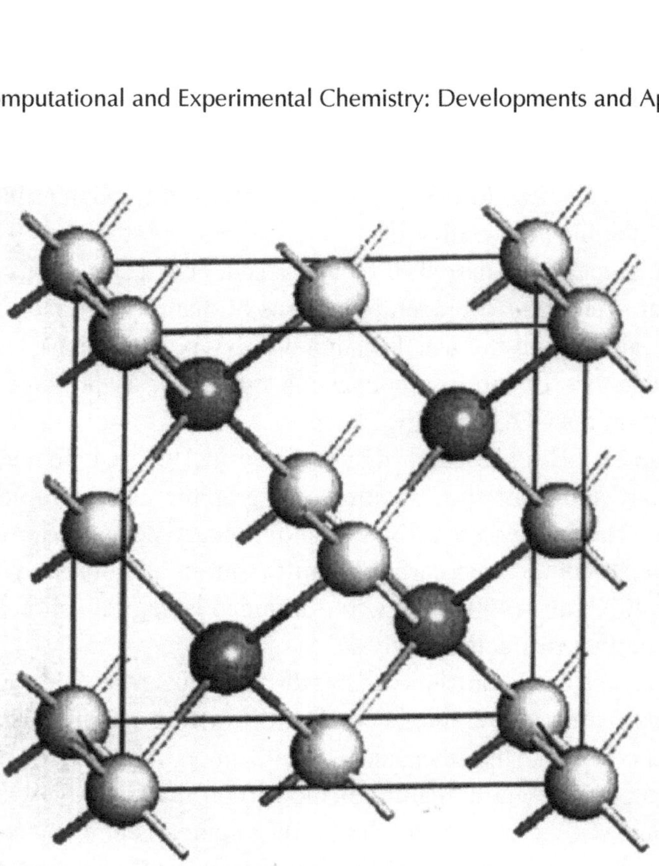

FIGURE 1 The relationship between Fd3m cubic diamondand an Fm-3m cell.

In this proposal, note that the n-diamond structure has a density exactly ½ that of the ordinary cubic diamond lattice, as there are ½ as many C atoms contained in the same cubic cell. The density of their proposed n-diamond structure is therefore 1.78 g/cm^3 (the x-ray density of diamond is 3.56 g/cm^3).

The justification for this face centered cubic (fcc) structure of n-diamond is simply that in the Fm3m space group, where the restrictions of symmetry forbidden reflections present in the Fd3m lattice of carbon are lifted, the resultant diffraction patterns observed to contain symmetry forbidden reflections of Fd3m cubic diamond are readily explained.

However, this proposed structure of n-diamond by Palatnik et al. and others seems unlikely to be the true structure from the point of view of elementary chemical bonding theory, as the carbon atoms in the fcc lattice are no longer in the conventional tetrahedral coordination of covalent carbon–carbon single bonding that is ubiquitous in C, as observed in ordinary organic chemistry for an example [9]. In the fcc lattice, the C atoms are forced to have a higher coordination number than four, at distances that are unrealistic for carbon–carbon covalent bonding. In addition, the idea that an allotrope of carbon with ½ the density of cubic diamond could be metallic, is counter-intuitive from a physical standpoint. One would reasonably expect an increase in density from cubic diamond, as corresponding to the onset of metallization in a novel modification of carbon [10].

With the counter-intuitive nature of the proposed model for the structure of n-diamond in mind, the research group of Pickard et al.reported in 2001 [11], in theoretical studies at the state-of-the-art density functional level of theory (DFT) [12], that an Fm3m structure of carbon would be unstable, with the constraint of a cubic lattice parameter of 3.56 Å. They showed that such an fcc C lattice would have an energy-minimized optimum lattice parameter of about 3.08 Å, resulting in a diffraction pattern that would not correspond to the observed evidence for n-diamond. This result from DFT calculations seems to completely rule out the Fm3m structure as an explanation for the observed diffraction pattern of n-diamond. It does not, however, explain the nature of the metallic status observed in these crystalline carbon materials [1–4], nor the unusual diffraction patterns observed for them by several leading research groups around the world, as referenced earlier.

2.2 THE GLITTER MODEL OF METALLIC CARBON

Bucknum et al., using a strategy based upon a mixture of sound chemical intuition and semi-empirical molecular orbital calculations (EHMO), to characterize band structures of potential C allotropes [13–16], reasoned that a potential allotrope of carbon might be constructed from a structural basis constituted by a 1,4-cyclohexadieneoid motif. The structural pattern produced by the 1,4-cyclohexadieneoid motif was suggested to them as a potential allotropic structure of C, strictly based upon an analogy with the diamond and graphite polytypes of carbon, which are built upon cyclohexaneoid and benzenoid motifs, respectively. In some sense, such a structure would represent a form of carbon intermediate between graphite and diamond in the degree of unsaturation in the lattice, and in other properties including the coordination number.

Therefore, in 1994 Bucknum et al. reported the prediction of a tetragonal modification of carbon, in space group $P4_2/mmc$ (#131), which contains six atoms in the unit cell and has a density of about 3.08 g/cm^3. This calculated density is based upon the optimized lattice parameters for the tetragonal cell given as a = 2.560 Å and **c** = 5.925 Å. These lattice parameters have been optimized using a computational procedure based upon Density Functional Theory (DFT) [12], the computational procedure is known as CASTEP and it was developed at CambridgeUniversity by Payne et al. [17]. The optimization calculations were performed in CASTEP using a functional based upon the local density approximation (LDA). The earlier optimization calculations on the fcc carbon cell by Pickard et al. were performed with this code as well [11, 17]. Thus Table 1 provides the crystallographic coordinates of the glitter unit cell from the CASTEP-DFT optimized geometry of the glitter lattice.

TABLE 1 DFT-CASTEP Crystallographic Coordinates of the Glitter Unit Cell.

Atom#	X	Y	Z	a	c
1	0	0	0	2.564Å	5.928Å
2	1.2820	0	0.8168		
3	1.2820	0	2.1468		
4	0	0	2.9640		
5	0	1.2820	3.7808		
6	0	1.2820	5.1108		

This tetragonal structure-type was built entirely upon a 1,4-cyclohexadieneoid motif [18–20], as shown in Fig. 2, and constituted a 3-,4-connected network of C, related to many other such 3-,4-connected structural-types that were derived and constructed by A.F. Wells in the period from the early 1950's to the mid 1980's. The structure was given the name "glitter" based upon the metallic nature of the bonding in it, produced from extensive p_σ and p_{spiro} interactions within the network [5–9].

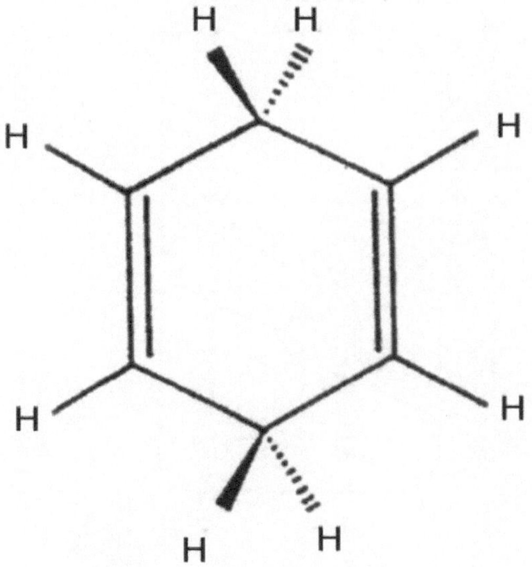

FIGURE 2 The 1,4-cyclohexadiene model molecule.

In their initial report on the glitter structure, Bucknum et al.showed the material was alternatively built upon a motif of 3-,4-connected 1-dimensional polymer chains of C, with the adjacent 1-dimensional chains of such substructures coordinated to each other in an orthogonal fashion. The resulting glitter allotrope, in the tetragonal space group $P4_2/mmc$, was thus shown to be intermediate in constitution to the polyethylene chains of the diamond lattice, in which adjacent chains are joined orthogonally (and with the latter structure in space group Fd3m), and the polyacetylene chains of the Wells-constructed 3-connected net known as (10, 3)-b, in which adjacent polyacetylene chains are joined orthogonally (and with the latter structure in space group $I4_1/amd$) [21]. In addition, it was seen that glitter could be derived from the known mineral structure of Cooperite [22] by replacing the square planar vertices in the Cooperite structural-type with the trigonal planar atom pairs of glitter. Such a transformation is described as a topological isomorphism. A drawing of the glitter unit of pattern is shown in Figs. 3 and 4 shows a view of an array of several unit cells of glitter.

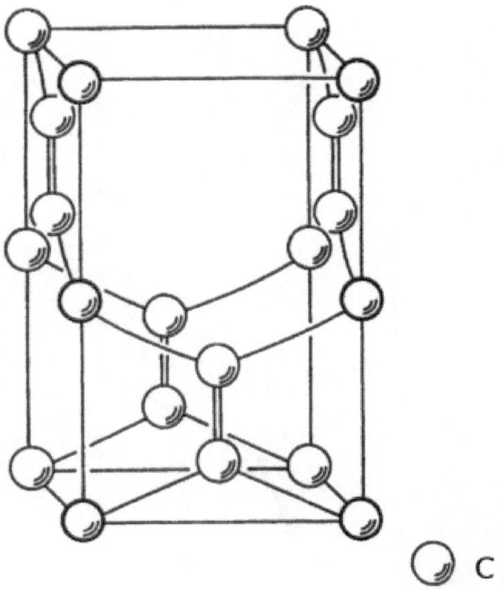

FIGURE 3 The unit of pattern of glitter.

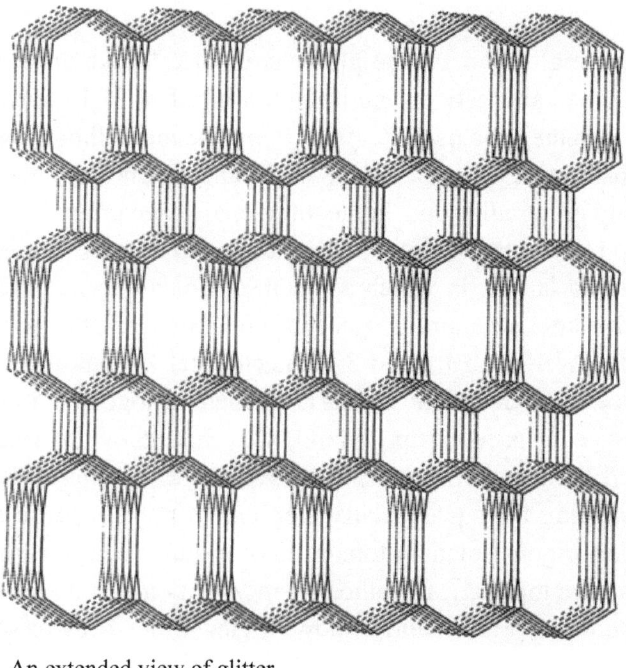

FIGURE 4 An extended view of glitter.

By calculating a theoretical diffraction pattern for the glitter lattice, based upon, for example, the DFT optimized set of lattice parameters given earlieras a = 2.560 Å and c = 5.925 Å, one can show a good fit to the diffraction pattern of n-diamond, as reported, for example, by Hirai et al. in 1991 [4]. The close match between the calculated Bragg reflections for the glitter lattice and those reflections measured for the n-diamond material, suggest the central message of this monograph. Perhaps the structure of n-diamond and the structure of glitter are one and the same. A tabulation of the associated diffraction data is given in Table 2 [4, 23]. Note that the mean of the deviations of the 10 matched Bragg reflections in Table 2 is given as x(mean) = 0.001983 nm/reflection (0.01983 Å/reflection) and the standard deviation of the corresponding dataset for the 10 matched Bragg reflections is given as s_x = 0.002363 nm/reflection (0.02363 Å/reflection). Therefore, all the data fit within $3s_x$ of x(mean) (0.009072 nm/reflection) with the glitter (003)-to-n-diamond (111) pair of matched reflections having the largest deviation at 0.008500 nm (0.08500 Å). For reference a high resolution electron micrograph (HREM) of a representative 25 nm diameter nanocrystalline droplet of n-diamond is shown in Fig. 5 [27].

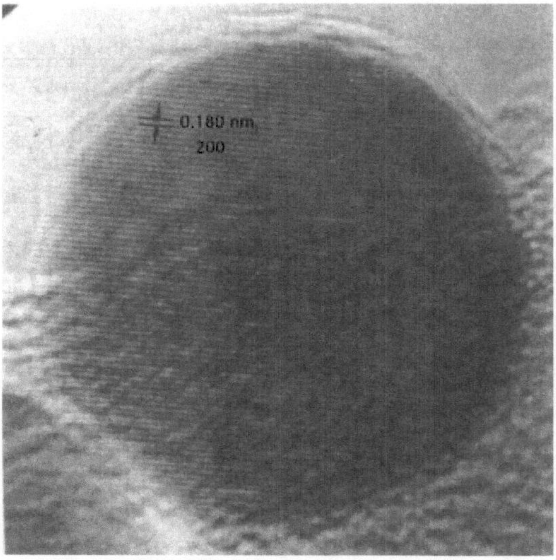

FIGURE 5 High Resolution Electron Micrograph (HREM) of 25 nm diameter droplet of representative Nanocrystalline n-diamond (lattice image indexed to a cubic diamond model, see Ref. [7]).

With the data in Table 2, which, as stated earlier, consist of 10 n-diamond reflections, as well as the calculated tetragonal glitter and ordinary cubic diamond reflections, it can be seen that the set of data is entirely consistent with that of Fd3m cubic diamond, except that 3 of the reflections, the (200) at 1.780 Å, the (222) at 1.040 Å, and the (420) at 0.7960 Å, are symmetry forbidden cubic diamond reflections in the analysis. Quite in contrast, one can consider the data from the theoretical diffraction pattern of glitter, where it can be seen that all 10 reflections of n-diamond are reproduced in the glitter spectrum, with only the (003) glitter reflection, at 1.975 Å, being symmetry forbidden. The fit between glitter and the n-diamond data is thus seen to be a closer fit than that of cubic Fd3m diamond, especially given that the lattice parameters giving rise to this glitter diffraction pattern have been optimized for the lattice at the DFT level of theory [12, 17].

TABLE 2 Observed diffraction data of n-diamond compared to theoretical diffraction data of $P4_2/mmc$ tetragonal glitter and Fd3m cubic diamond.

Calculated glitter reflections a = 0.2560 nm, c = 0.5925 nm		**n-Diamond reflections* d-spacing, nm**	**Cubic diamond reflections** d-spacing, nm**
(hkl)	**d-spacing, nm**		
100	0.2560		
001	0.5925		
110	0.1810	0.1780 (200)	"forbidden"
101	0.2350		
111	0.1731		
200	0.1280		
002	0.2963		
102	0.1937		
120	0.1145		
201	0.1251	0.1260 (220)	0.1261 (220)
211	0.1124		
221	0.08947	0.08980 (400)	0.08916 (400)

TABLE 2 *(Continued)*

Calculated glitter reflections a = 0.2560 nm, c = 0.5925 nm		n-Diamond reflections* d-spacing, nm	Cubic diamond reflections** d-spacing, nm
(hkl)	**d-spacing, nm**		
212	0.1068	0.1040 (222)	"forbidden"
222	0.08656		
300	0.08533		
003	0.1975	0.2060 (111)	0.2060 (111)
103	0.1564		
130	0.08095	0.07960 (420)	"forbidden"
301	0.08446		
311	0.08021		
331	0.06003		
313	0.07491		
333	0.05771		
203	0.1074	0.1070 (311)	0.10750 (311)
302	0.08200	0.08180 (331)	0.08182 (331)
320	0.07100	0.07260 (422)	0.07281 (422)
223	0.08228		
232	0.06905	0.06830 (511)	0.06864 (511)
332	0.05912		
323	0.06682		
321	0.07050		
312	0.07809		
213	0.09905		
104	0.1282		
401	0.06363		

*H. Hirai and K. Kondo, Science, 253, 772 (1991).
**JCPDS6, Card # 675.

Yet a second piece of evidence supporting the glitter model as an explanation of the structure of n-diamond, is the data supplied from a state of the art, density functional theory (DFT) calculation of the band structure of the corresponding tetragonal lattice [24]. In fact, the structure is shown to be a good metal at the DFT level of theory, with the π* band dipping down into the occupied bands of glitter at symmetry point M in the reciprocal space of the material [5–7]. Metallic status for glitter is consistent with the observations of Palatnik et al., and others, mentioned earlier, with regard to the electrical conductivity of n-diamond [1–4]. The calculated band structure of glitter, with the lattice parameters close to those given earlier, is shown in Fig. 6.

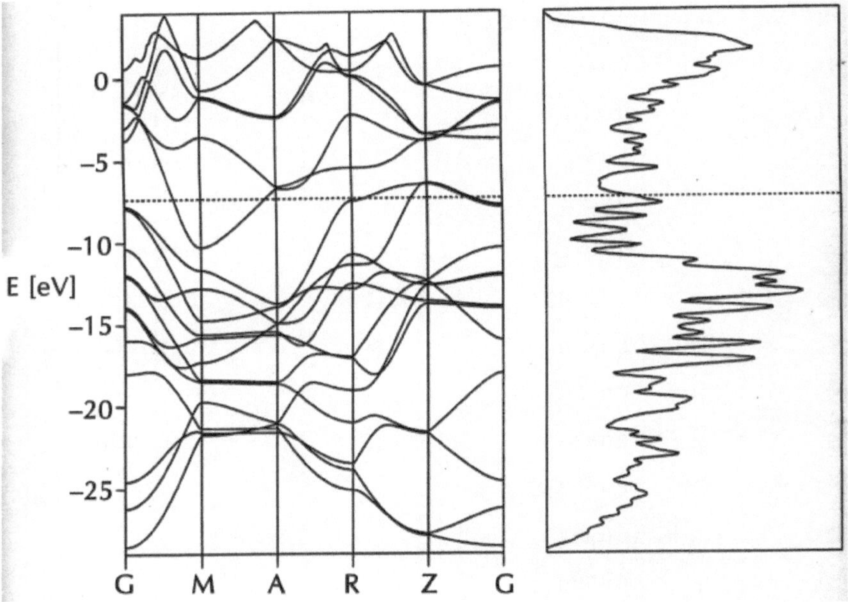

FIGURE 6 Calculated band structure and density of states (DOS) of tetragonal glitter at DFT level of theory.

2.3 THEORETICAL STABILITY OF GLITTER

First principles calculations of the stability of the glitter lattice have been performed with the CASTEP computational method described earlier. It has been shown that under the local density approximation (LDA) the DFT-based CASTEP method calculates a stability of 0.511 eV/C atom above the corresponding stability of graphite, which is the zero on the energy scale [17]. As a comparison, fullerene is calculated to be 0.3 eV/C atom less stable than graphite by the same method. Although this is a fairly large energy above that of graphite, if the glitter structure was synthesized it would persist, as kinetic routes to its decomposition to graphite or diamond have high activation barriers due to the reconstructive nature of such a phase transition [19]. These considerations are based upon the unique 3-,4-connected bonding in the structure.

In addition to the kinetic parameters, the spiroconjugation effects reported for it previously [25], lend great support for the relative stability of glitter as a structure of carbon. Conjugation and resonance are known to be important themes in the explanation of the reactions of organic chemistry, and in explaining the structural stability of graphite and the fullerenes [9, 22]. In the graphene sheet, one is constrained to conjugation in the 2D plane. With the fullerenes, one is constrained to conjugation on a 2D spherical surface. From the MO calculations done on glitter in the 1990's [5–7], it is clear that another type of p-orbital conjugation, called spiroconjugation, is present in the glitter lattice in fully 3D. It constitutes the first fully 3-dimensional conjugated organic system that has ever been conceived in the chemical literature [5–7].

In direct analogy with the conjugation present in the graphene sheet, as shown in Fig. 7 [22], there are a set of resonance structures that can be written over the glitter unit cell to indicate its relative stability from spiroconjugation,this is illustrated in Fig. 8.

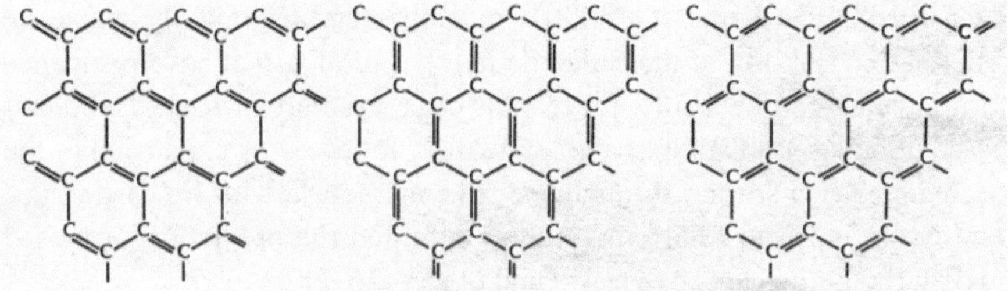

FIGURE 7 Resonance structures of graphite in 2D.

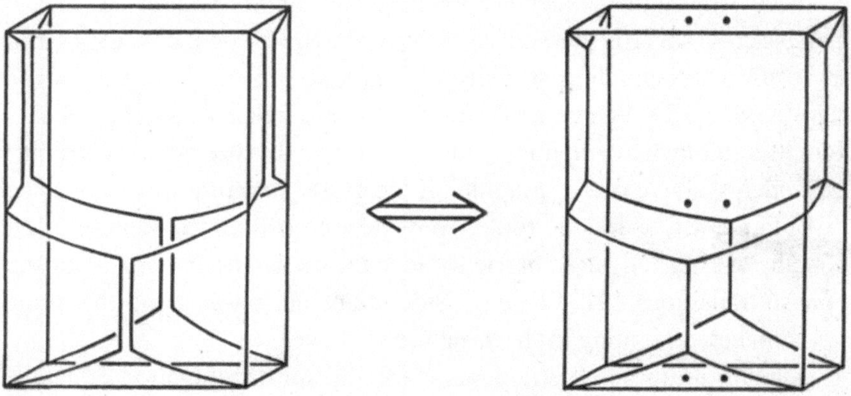

FIGURE 8 Resonance structures of glitter in 3D.

It is clear from these resonance structures in 3D, that the electrical conductivity in glitter will be highly anisotropic, with conduction electrons, as π electrons carried by the overlapping π and π^* bands in glitter's electronic structure, moving in pairs in a direction parallel to the **c**-axis of the unit cell. It is indeed a very unusual, organic metal with a correspondingly unusual π electron pair conductivity mechanism, and potentially a unique type of stability due to the organic chemical phenomenon of conjugation, as spiroconjugation, exhibited in 3D in the electronic structure of glitter [5–7].

In comparison, one can consider the various members of the fullerene family, the parent Buckminsterfullerene molecule C_{60}, although calculated to be 0.3 eV/C atom less stable than graphite [17], was predicted by Osawa in 1970 [26, 27] to be a uniquely stable molecule based upon the resonance structures that could be drawn over the C_{60} skelton. It wasn't until 15 years later in 1985 that Kroto et al. [28] accidently stumbled onto the existence and relative stability of the fullerene family (all of which have resonance forms which can be written over their cage structures) from a synthesis based upon laser ablation of a graphite disk, followed by expansion of the resultant carbon species through a supersonic nozzle with He carrier gas. The work of Kroto et al., therefore, confirmed the prediction of closed shell carbon molecules by Osawa and others [26–28].

It can be seen from this lesson that conjugation and resonance are a hallmark of stability in molecules and extended structures of carbon. One should therefore take seriously the proposal of glitter as an extended structure of carbon, even in addition to the close match between its calculated diffraction pattern as compared to that of the experimentally observed n-diamond diffraction data, because it, like its sibling structures graphite and the fullerenes, has a unique stability due to the spiroconjugation present in fully 3D within the structure [5–7].

The symmetry adapted π and π* crystal orbitals of glitter, at symmetry point M of the Brillouin zone of the tetragonal material, have been shown previously in Fig. 11, and are seen to support the picture of resonance in 3D shown in Fig. 8. However, this comparison should be looked upon with the caveat that the glitter crystal orbitals are derived from a molecular orbital (MO) perspective, while the valence bond (VB) method is employed in arriving at the resonance structures of glitter shown in Fig. 7 [5–7].

2.4 CONCLUSIONS

It is quite evident from the data comparison shown in Table 2, in which the mean deviation over 10 matched Bragg reflections between glitter and n-diamond is less than 0.02 Å/reflection, that glitter is a plausible model for n-diamond. Of course, there are many proposals for carbon allotropes that have been published [29–55], some of these are chemically intuitive, like the original fullerene proposal by Osawa [26], while other models for carbon allotropes, like the fcc structure of carbon proposed for n-diamond, are counterintuitive. In what way can we eliminate these myriad other possible carbon allotropes as structures of n-diamond?

Clearly some of these hypothetical carbon allotropes have diffraction patterns which show reflections that may overlap with those observed for n-diamond. However, the possibility that the complete set of 10 n-diamond reflections observed by Hirai et al. [4] should be entirely reproduced by another hypothetical structure of carbon, in addition to the glitter structure, is remote. The unlikelihood of such a set of 10 coincidental overlaps is even more remote, when one includes in the analysis that the proposed carbon structure should be chemically intuitive, as well as unique in its electronic structure from the point of view of conjugation and resonance effects, as well as being metallic in its band profile.

Glitter is special in that it satisfies all of these requirements. It reproduces the n-diamond diffraction spectrum reasonably well, it is a chemically intuitive structure based upon reasonable bond parameters (the carbon–carbon single bonds in this optimization are 1.5202 Å and the carbon–carbon double bonds are 1.3300 Å, while the C–C=C angle is 122.5 degrees and the C–C–C bond angle is 115 degrees), it is comprised of C atoms in archetypal tetrahedral and trigonal connection motifs [38], the trigonal carbon atoms are grouped into parallel pi bonded pairs rather than existing as separate radical centers or being in orthogonally coordinated pairs, it has the unique property of conjugation and resonance, and it is a metallic carbon structure.

Lastly one must consider the myriad of reflections calculated for glitter but not observed in the n-diamond analysis. Obviously, from considerations of both powder diffraction and Laue diffraction, all Bragg(Laue) reflections possible from a given material are rarely observed in practice. There are many reasons for diminished diffracted X-ray intensity, the most important of which is symmetry [56]. In the analysis given in Table 2, the glitter reflection (003) at 1.975 Å is calculated to be forbidden by symmetry [23]. Therefore, its potential match to the n-diamond reflection at 2.060 Å is in question.

It is possible that by extending the degree of reciprocal space that is sampled in an n-diamond diffraction experiment, and lengthening the exposure time over which n-diamond diffraction data is collected, one might see some additional reflections as predicted by the data given in Table 2. As n-diamond is a phase produced only intermittently, and therefore n-diamond has not been exhaustively explored to date, it is not surprising that a complete diffraction pattern has not been obtained yet. Of course, for the reasons mentioned above, the glitter model is still a very plausible structure for n-diamond, and those reflections predicted for it in Table 2 should be taken seriously by investigators in the field of carbon materials research.

As to its utility, polycrystalline glitter will be potentially useful as an abrading material for machining applications and, in the form of a single crystal, as a material that exceeds even cubic diamond in its degree of hardness [57–60]. That the material is an organic metal is yet another technological bonus of this structure. Potentially it could be used in situations where strength and conductivity are both required characteristics. The use of glitter as a single crystal anvil material in an opposed anvil high-pressure device, analogous to the well-known diamond anvil cell (DAC) [61] is obvious. Overall, it seems that there is quite a lot of technological promise in the glitter structure with its metallic and superhard characteristics. Finally, we should mention the synthetic work on glitter and its isovalent analog silicon dicarbide being carried out in Romania contemporaneously with the publication of this manuscript, indeed Stamatin et al., have shown that these glitter phases can be readily synthesized from Novolac resin (a Bakelite resin) using a cross-linking agent like HMTA (hexamethylene tetramine), which is appropriate to cross-linking in a spiroconjugated fashion [62].

KEYWORDS

- **Bragg reflections**
- **face centered cubic**
- **n-diamond**
- **spiroconjugation**

REFERENCES

1. Konyashin, I.; Zern, A.; Mayer, J. *Diamond and Related Mater.*, **2001,** *10*, 99.
2. Palatnik, L.S.; Guseva, M.B.; Babaev, V.G.; Savchenko, N.F.; Fal'ko, L.L. *Soviet Phys., JETP*, **1984,** *60*, 520.
3. Gogotsi, Y.;Weiz, S.; Ersoy, D.A.; McNallan, M.J. *Nature (London)*, **2001,** *411*, 283.
4. Hirai, H.; Kondo, K.*Science*, **1991,** *253*, 772.
5. Bucknum, M.J.; Castro, E.A.*J. Math. Chem.*, **2004,** *36(4)*, 381.
6. Bucknum, M.J.; Ienco, A.; Castro, E.A.*J. Mol. Struct.:Theochem.*, **2005,** *716*, 73.
7. Bucknum, M.J.; Castro, E.A.*J. Math. Chem.*, **2012,** *50(5),* 1034.
8. Bursill, L. et al., *Int. J. Modern Phys. B*, **2001,** *15*, 3107.
9. McMurry, J.E. In: *Organic Chemistry*, 6th edition, Brooks-Cole: New York, 2004.

10. Although as is reported here, a heretofore hypothetical metallic modification of carbon with a density of about 3.1–3.2 g/cm^3, called glitter, has been described by Bucknumet al. in 1994. See Ref. [1] of Chapter 1.
11. Pickard, C.J.; Milman, V.; Winkler, B.*Diamond Related Mater.*,**2001,** *10*, 2225.
12. Hohnenberg, P.; Kohn, W.*Phys. Rev. B*, **1964,** *136*, 864.
13. Hoffmann, R.*J. Chem. Phys.*, **1963,** *39*, 1397.
14. Hoffmann, R.*J. Chem. Phys.*, **1964,** *40*, 2745.
15. Hoffmann, R.*J. Chem. Phys.*, **1964,** *40*, 2474.
16. Hoffmann, R.*J. Chem. Phys.*, **1964,** *40*, 2480.
17. CASTEP (Cambridge Serial Total Energy Package) was created and developed by Professor M.C. Payne and others in the 1980's. It is a computational procedure based upon the density functional theory (DFT) referenced in 41. Professor M.C. Payne and his colleagues are located in the Theoretical Condensed Matter group in the Cavendish laboratory at CambridgeUniversity. Dr. Chris J. Pickard kindly performed and provided the calculations on the glitter structure to the authors. The details of the CASTEP program are described in *Rev. Mod. Phys.*, **1992,** *64*, 1045. For the present implementation of CASTEP, used to optimize the structural parameters of glitter, ultrasoftpseudopotentials were employed, the basis set had an energy cutoff of 400 eV and k-point sampling was done with a 10x10x4 mesh.
18. Carreira, L.A.; Carter, R.O.; Durig, J.R.*J. Chem. Phys.*, **1973,** *59*, 812.
19. Oberhammer, H.; Bauer, S.H. *J. Am. Chem. Soc.*, **1969,** *91*, 10.
20. Dallinga, G.; Toneman, L.H. *J. Mol. Struct.*,**1967,** *1*, 117.
21. Hoffmann, R.; Hughbanks, T.; Kertesz, M.; Bird, P.H.*J. Am. Chem. Soc.*, **1983,** *105*, 4831.
22. Pauling, L. In: *The Nature of the Chem. Bond*, 3rd edition, Cornell University Press: Ithaca, New York, 1960.
23. Theoretical diffraction data for the glitter lattice, including intensity calculations, with lattice parameters for the tetragonal cell of $a = 2.530$ Å and $c = 5.980$ Å, have been reported previously. See M.J. Bucknum, E.A. Castro, *J. Mol. Model.*, **2005,** *12(1)*, 111.
24. The Amsterdam Density Functional program (ADF2002.02) was employed in the calculations of the glitter band structure and DOS. Details of this program package can be found in the following reference, G. teVelde, F.M. Bickelhaupt, E.J. Baerends, C.F. Guerra, S.J.A. van Gisbergen, J.G. Snijders and T. Ziegler, *J. Comput. Chem.*, **2001,** *22*, 931.
25. These were thoughts made in a personal communication with the one of the authors (M.J.B.) by Professor Roald Hoffmann at Cornell University.
26. Osawa, E. *Kagaku*, **1970,** *25*, 85.
27. Jones, D.E.H. *New Scientist*, **1966,** *32*, 245.
28. Kroto, H.W.; Heath, S.C. O'Brien, R.F. Curl, R.E. Smalley, *Nature(London)*, **1985,** *318*, 162.
29. Bucknum, M.J.; Castro, E.A. *J. Math. Chem.*, **2006,** *39(3/4)*, 611.
30. Balaban, A.T.; Klein, D.J.; Folden, C.A. *Chem. Phys. Lett.*, **1994,** *217*, 266.
31. Baughman, R.H.; Galvao, D.S. *Chem. Phys. Lett.*, **1993,***211*, 110.
32. Baughman, R.H.; Galvao, D.S. *Nature (London)*, **1993,** *365*, 735.
33. Tanaka, K.; Okahara, K.; Okada, M.; Yamabe, T. *Chem. Phys. Lett.*,**1992,** *191*, 469.
34. Karfunkel, H.R.; Dressler, R. *J. Am. Chem. Soc.*, **1992,** *114*, 2285.

35. Diederich, F.; Rubin, Y.*Angew. Chem., Int. Ed. Engl.*,**1992,** *31*, 1101.
36. Boercker, D. *Phys. Rev. B*, **1991,** *44*, 11592.
37. Liu, A.Y.; Cohen, M.L.; Hass, K.C.; Tamor, M.A.*Phys. Rev. B*, **1991,** *43*, 6742.
38. Mailhot, C.; McMahan, A.K.*Phys. Rev. B*, **1991,** *44*, 11578.
39. Laqua, G.; Musso, H.; Boland, R. Ahlrichs, *J. Am. Chem. Soc.*,**1990,** *112*, 7391.
40. Tamor, M.A.; Hass, K.C. *J. Mater.Res.*,**1990,** *5*, 2273.
41. Johnston, R.L.; Hoffmann, R. *J. Am. Chem. Soc.*, **1989,** *111*, 810.
42. Baughman, R.H.; Eckhardt, H.; Kertesz, M. *J. Chem. Phys.*, **1987,** *87*, 6687.
43. Robertson, J.; O'Reilly, E.P. *Phys. Rev. B*, **1987,** *35*, 2946.
44. Wen, X.D.; Cahill, T.J. Roald Hoffmann, *Chem. Eur. J.*, **2010,** *16*, 6555.
45. Balaban, A.T. *Comput. Math. Appl.*, **1987,** *17*, 397.
46. Stein, S.E.; Brown, R.L. *J. Am. Chem. Soc.*, **1987,** *109*, 3721.
47. Burdett, J.K.; Lee, S.*J. Am. Chem. Soc.*, **1985,** *107*, 3050.
48. Biswas, R.; Martin, R.M.; Needs, R.J.;Neilsen, O.H.*Phys. Rev. B*, **1984,** *30*, 3210.
49. Kostadinov, L. N.; Dobrev, D. D.*Phys. Status Solidi A*, **1988,** *109*, K85.
50. Kertesz, M.; Hoffmann, R. *J. SolidState Chem.*, **1980,** *54*, 313.
51. Davidson, R.A. *Theor. Chim. Acta*, **1981,** *58*, 193.
52. Hoffmann, R.; Eisenstein, O.; Balaban, A.T. *Proc. Natl. Acad. Sci. U.S.A.*, **1980,** *77*, 5588.
53. Balaban, A.T.; Rentia, C.C.; Ciupitu, E.*Rev. Roum. Chim.*, **1968,** *231*, 1233.
54. Kakinoki, J. *Acta Crystallogr.*, **1965,** *18*, 578.
55. Riley, H.L. *J. Chim. Phys.*, **1950,** *47*, 565.
56. Sands, D.E. *Introduction to Crystallography*, 1st edition, Dover Publications, Inc. Mineola, NY, 1993.
57. Bucknum, M.J.; Castro, E.A.*J. Math. Chem.*, **2006,** *40(4),* 341.
58. Bucknum, M.J.; Castro, E.A.*J. Math. Chem.*, **2005,***38(4)*, 27.
59. Bucknum, M.J.; Castro, E.A. *J. Mol. Model.*,**2005,** *12*, 111.
60. Bin Wen, et al., *Diamond and Related Mater.*, **2008,** *17(7–10)*, Special Issue SI, 1353.
61. Hazen, R.M. In: *The Diamond Makers*, 1st Ed.;, Cambridge University Press: Cambridge, U.K., 1999.
62. Stamatin, I.; Dumitru, A.;Bucknum, M.J.; Ciupina, V.;Prodan, G.*Mol. Crystals and Liquid Crystals*, **2004,** *417*, 167.

CHAPTER 3

CLASSIFICATION OF CARBON ALLOTROPES AND GRAPHS

MICHAEL J. BUCKNUM and EDUARDO A. CASTRO

CONTENTS

SUMMARY

In this chapter, we describe the tenets of a chemical topology of crystalline matter and certain associated rational approximations to the transcendental mathematical constants φ, e and π, that arise out of considerations of both: (1) the Euler relation for the division of the sphere into vertices, V, faces, F, and edges, E, and: (2) its simple algebraic transformation into the so-called Schläfli relation, which is an equivalent mathematical statement for the polyhedra, in terms of parameters known as the polygonality, defined as $n = 2E/F$, and the connectivivty, defined as $p = 2E/V$. It is thus the transformation to the Schläfli relation from the Euler relation, in particular, that enables one to move from a simple heuristic mapping of the polyhedra in the space of V, F and E, into a corresponding heuristic mapping into Schläfli-space, the space circumscribed by the parameters of n and p. It is also true, that this latter transformation equation, the Schläfli relation, applies only directly to the polyhedra, again, with their corresponding Schläfli symbols (n, p), but as a bonus, there is a direct 1-to-1 mapping result for the polyhedra, that can be seen to also be extendable to the tessellations in 2-dimensions, and the networks in 3-dimensions, in terms of coordinates in a 2-dimensional Cartesian grid, represented as the Schläfli symbols (n, p), as discussed earlier, which do not involve rigorous solutions to the Schläfli relation. For while one could never identify the triplet set of integers (V, F, E) for the tessellations and networks, that would fit as a rational solution within the Euler relation, it is in fact possible for one to identify the corresponding values of the ordered pair (n, p) for any tessellation or network. The identification of the Schläfli symbol (n, p) for the tessellations and networks emerges from the formulation of its so-called Well's point symbol, through the proper translation of that Well's point symbol into an equivalent and unambiguous Schläfli symbol (n, p) for a given tessellation or network, as has been shown by Bucknum et al. previously. What we report in this communication, are the computations of some, certain Schläfli symbols (n, p) for the so-called Waserite (also called platinate, Pt_3O_4, a 3-,4-connected cubic pattern), Moravia (A_3B_8, a 3-,8-connected cubic pattern) and Kentuckia (ABC_2, a 4-,6-,8-connected tetragonal pattern) networks, and some topological descriptors of other relevant structures. It is thus seen, that the computations of the polygonality and connectivity indexes, n and p, that are found as a consequence of identifying the Schläfli symbols for these relatively simple networks, lead to simple and direct connections to certain rational approximations to the transcendental mathematical constants φ, e and π, that, to the author's knowledge, have not been identified previously. Such rational approximations lead to elementary and straightforward methods to estimate these mathematical constants to an accuracy of better than 99 parts in 100.

3.1 A CHEMICAL TOPOLOGY SCHEME

Bucknum [1] in work first described in 1997, outlined a general scheme for the systematic classification and mapping of the polyhedra, 2-dimensional tessellations and 3-dimensional networks in a self-consistent topological space for these structures. This general scheme begins with a consideration of the Euler relation [2] for the polyhedra, shown as Eq. (1), which was first proposed in 1758 to the Russian Academy by Euler, and was, in fact, the point of departure for Euler into a new area of mathematics thereafter known explicitly as topology.

$$V - E + F = 2 \tag{1}$$

Equation (1) stipulates that for any of the innumerable polyhedra, the combination of the number of vertices, V, minus the number of edges, E, plus the number of faces, F, resulting from any such division of the sphere, will invariantly be that number 2, known as the Euler characteristic of the sphere. The variables known as V, E and F are topological properties of the polyhedra, or, in other words, they are invariants of the polyhedra under any kind of geometrical distortions. It is from this simple Eulerian relation, that we can develop a systematic and, indeed otherwise rigorous, mapping of the various, innumerable structures that present themselves, in levels of approximation, as models for the structure of the real material world within the domain of that area of science known as crystallography.

About a century after Euler's relation for the polyhedra was first proposed, as described earlier, the German mathematician Schläfli introduced a simple algebraic transformation of Euler's relation, for various purposes of understanding the relation better, and adopting it more effectively in proofs [3]. Thus, Schläfli introduced two new topological variables, like V, E and F before them, that were derived from them. Schläfli, therefore, defined the so-called "polygonality," hereafter represented by n, of a polyhedron as the averaged number of sides, or edges, circumscribing the faces of a polyhedron. He conveniently defined such a polygonality, as n = 2E/F, where, in this instance one can see that because each edge E straddles two faces, F, the definition is rigorous. Similarly, Schläfli introduced the topological parameter called the "connectivity," hereafter represented by p, of a polyhedron as the averaged number of sides, or edges, terminating at each vertex of a polyhedron. He conveniently defined such a connectivity, as p = 2E/V, where, in this instance one can see that because each edge E terminates at two vertices, V, the definition is rigorous.

From these definitions of n and p as topological parameters of the polyhedra, Schläfli was able to show quite straightforwardly, by algebraic substitution, that a further relation exists among the polyhedra in terms of their Schläfli symbols (n, p) [3]. It is from this equation, the Schläfli relation, shown as Eq. (2), that one can see that not only do the polyhedra rigorously obey 6, but it is also true that their indices as (n, p), that serve as solutions to 6, in addition lead to a convenient 2-dimensional grid, or Schläfli space, over which the various polyhedra can be unambiguously mapped, as has been explained by Wells in his important 1977 monograph on the subject [3, Chapter 1].

$$\frac{1}{n} - \frac{1}{2} + \frac{1}{p} = \frac{1}{E} \qquad (2)$$

Figure 1 due to Wells [3, Chapter 1], illustrates the application of this type of Schläfli mapping for the regular Platonic polyhedra, where one sees that the point that belongs to the origin of this mapping, is indeed given by the Schläfli symbol (n, p) = (3, 3). The symbol (3, 3) represents the Platonic solid known as the tetrahedron, or by the symbol "t" in the map, known since Antiquity by the Greeks. Thus, as it is cast as the origin of this mapping

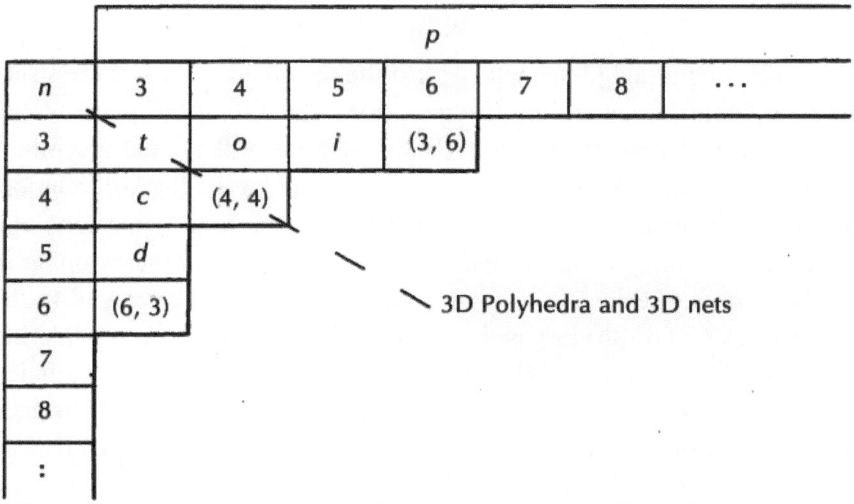

FIGURE 1 Topology mapping of the Platonic polyhedra due to Wells.

of polyhedra, it is apparently, the only self-dual polyhedron. Similarly (4, 3) is the Platonic solid of the Greeks known as the cube, or “c” in the map, (5, 3) is the Platonic solid of the Greeks known as the dodecahedron, or “d” in the map, (3, 4) is the Platonic solid of the Greeks known as the octahedron, or “o” in the map, and (3, 5) is the Platonic solid of the Greeks known as the icosahedron, or “i” in the map.

Although the mapping of the Platonic polyhedra, shown in Fig. 1, indeed involves only those polyhedra in which the ordered pairs (n, p) are integers, it is readily transparent that one could *magnify* the map to include those polyhedra in which the polygonality “n” is fractional, these are the so-called Archimedean polyhedra, discovered by Archimedes in Ancient Greece [4]. In addition, the map could be *magnified* to include those polyhedra in which the connectivity “p” is fractional, these are the so-called Catalan polyhedra, discovered in Europe in the 19th century [5]. And finally, the so-called Wellsean polyhedra, discovered in the 21st century [6], in which both indexes, “n” and “p,” are fractional, could be mapped in this Schläfli space of the polyhedra, without any loss of mathematical rigor.

3.2 MAPPING OF TESSELLATIONS AND NETWORKS

Wells also suggested [3, Chapter 1] that the tessellations, which are structures or tilings extended into 2-dimensions and filling the plane, could have nominal labels attached to them, in the form of the Schläfli symbols (n, p), that while not leading to rational solutions of the Schläfli relation, were, still it seems, rigorously defined from inspection of the topology of these elementary tessellations. Thus in order to *extend* the mapping in Schläfli space, the space of (n, p), some loss of rigor with regard to the Schläfli relation for the polyhedra had to be introduced, when considering the patterns known as tessellations. Wells, therefore included, explicity, the mapping of the square grid, given by the Schläfli symbol (4, 4), and the honeycomb grid, given by (6, 3), as well as the closest-packed tessellation, given by (3, 6), in his topology mapping, shown in Fig. 1, along with the Platonic polyhedra. He thereby *extended* the mapping to the tessellations, and later he implied that such a mapping could be *extended* to include the 3-dimensional (3D) networks as well, with a concomitant further loss of mathematical rigor, in that the values assigned as (n, p) to the various tessellations and networks were not rigorous solutions to Eq. (2).

It is also true that Wells [3, Chapter 1], perfectly well introduced a systematic and rigorous coding of the topology of tessellations and networks he worked with, which is now called the Wells point symbol notation, and that this was a simple coding scheme over the circuitry and valences, about the vertices, in the unit of pattern of the tessellations and networks. The Wells point symbol notation was, however, nonetheless an important development for the rigorous mathematical basis it put the tessellations and networks on, formally, as quasi-solutions (n, p) for the Schläfli relation shown as Eq. (2).

Therefore, in a generic case of the Wells point symbol notation, one could have such a symbol for an a-, b-connected, binary network, given as $(A^a)_X(B^b)_Y$, such that the exponents "a" and "b" nominally represent the valences of the two vertices in the tessellation or network of interest, and the bases "A" and "B" give the respective relative polygon sizes (circuit sizes) in the tessellation or network, while the parameters "X" and "Y" describe the binary stoichiometry of the network. In this generic case, we see that "X/Y" represents the number of structural components, identified by their topology character as (A^a), to the number of structural components identified by their respective topology as (B^b), that occur in this characteristic ratio in the structure, as specified by the unit of pattern.

Despite his invention of this elegant notation, Wells, for some odd reason, never explicitly showed how to *translate* the language of the Wells point symbol $(A^a)_X(B^b)_Y$, rigorously into a Schläfli symbol (n, p), as was later shown elsewhere. Thus Bucknum et al. in 2004 [7], showed that the translation of the Wells point symbol into an, otherwise, from the perspective of Eq. (2), rigorous set of values (n, p) for the purpose of mapping tessellations and networks, was achievable if one used the following straightforward, simple formulas (applicable in this case for a generic Wellsean, binary stoichiometry structure) for the ordered pair (n, p), which can later be employed in the mapping of the structure, as is shown by Eq. (3).

$$n = (a \cdot A \cdot X + b \cdot B \cdot Y)/(a \cdot X + b \cdot Y) \quad (3a)$$

$$p = (a \cdot X + b \cdot Y)/(X + Y) \quad (3b)$$

It should, of course, be noted that one has to proceed with care in using Eqs. (3a) and (3b), in this topology analysis of structures, by carefully normalizing the circuitry traced around the p-connected vertices of the structure (paying careful attention to the parameters "a" and "b"), by rigorously translating the circuitry of a given structure, into a vertex connectivity for the network of interest, by employing a vertex translation table like that shown in Table 1.

TABLE 1 Vertex Connectivity, p, as a Function of Circuit Number.

Name	Vertex connectivity	Circuit number
Trigonal planar	3	3
Square planar	4	4
Tetrahedral	4	6
Trigonal bipyramidal	5	9
Square pyramidal	5	10
Octahedral	6	12
Cube centered	8	24
Anti-cube centered	8	28
Closest packed	12	60

Therefore, from the use of Eq. (3), for a binary stoichiometry, Wellsean net with topology $(A^a)_X(B^b)_Y$, or some other homologous translation formulas, as shown, for example, by Bucknum et al.for various elementary structural cases [7], it becomes, evidently, rigorously possible to *precisely* map the topology of any structure, including of course the polyhedra, but *extendable* to the vast body of the known tessellations and networks, that have been discovered and characterized crystallographically, otherwise by their symmetry character, now by their topological character in the form of a mapping in an *extended* Schläfli space, as is shown in Fig. 2.

n \ p	3	4	5	6	7	8	...
3	t	o	i	(3, 6)	(3,7)	(3, 8)	
4	c	(4, 4)	(4, 5)	(4, 6)	(4,7)	(4, 8)	
5	d	(5, 4)	(5, 5)	(5, 6)	(5,7)	(5, 8)	
6	(6, 3)	(6, 4)	(6, 5)	(6, 6)	(6,7)	(6, 8)	
7	(7, 3)	(7, 4)	(7, 5)	(7, 6)	(7,7)	(7, 8)	
8	(8, 3)	(8, 4)	(8, 5)	(8, 6)	(8,7)	(8, 8)	
⋮							

FIGURE 2 Extended Schläfli space of the Platonic Structures.

3.3 SURVEY OF MAPPED PATTERNS AND CARBONS ALLOTROPES

As Eqs. (1) and (2), and Figs. 1 and 2 explicitly reveal, it is the Platonic solids that form the basis of this mapping formulation of structures described in this paper. These forms, as shown in Fig. 3 with their appropriate polyhedral face symbolism [8], as discovered in Ancient Greece from the application of pure thought, were implicated later on in Plato's Timaeus, as

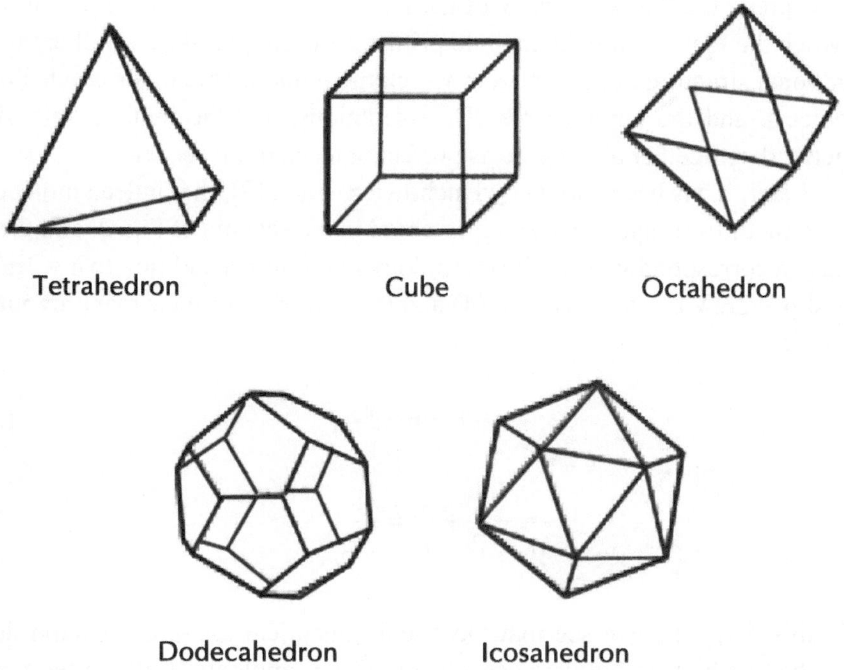

FIGURE 3 The Platonic polyhedra, with their corresponding polyhedral face symbols and Wells point symbols, comprised of the tetrahedron (3^4 and 3^3), the octahedron ($3^8$and 3^4), the icosahedron (3^{20} and 3^5), the cube (4^6 and 4^3), and the dodecahedron (5^{12} and 5^3).

the building blocks of Nature [9]. With the advent of modern crystallographic techniques by the Bragg's in the 20th century [10], we have come to learn that the structure of matter does indeed often take on various vestiges of these eternal objects. And so they have come to be important in modern structural chemistry as elucidated by Pauling [11] and others.

The polyhedra, thus forming the basis of the topology map of structures in Fig. 2, and also rigorously obeying the topology relations shown as Eqs. (1) and (2), are positioned uniquely in this construction to support the vast space of tessellations and networks that, as we have seen in the preceding chapter, can be mapped, rigorously, in Fig. 2 by the identification and proper translation of their Well's point symbols, as described earlier, into ordered pairs as Schläfli symbols (n, p). Plato's great work, Timaeus, thus predicted the ascendancy of the material world into perfect forms, in which the Platonic polyhedra hold primacy and support the overall organizational structure of matter, from which the innumerable other polyhedral objects, and the innumerable 2D tessellations, and the innumerable 3D networks, together all emerge as perfect objects, in this scheme.

Later, it has been shown by Duchowicz et al. [12], that indeed molecular structures (graphs) can be represented in the scheme of Fig. 2, and they have a corresponding set of two topology relations (in addition to n = 2E/F and p = 2E/V), shown as Eqs. (4) and (5), that govern their mapping into Fig. 2.

$$V - E + F = 1 \tag{4}$$

$$\frac{1}{n} - \frac{1}{2} + \frac{1}{p} = \frac{1}{2E} \tag{5}$$

In this way, one can see that the overall chemical topology scheme described here, is a complete description of the topology of all matter constituting the material world. This communication will not treat the specific applications of Eqs. (4) and (5), but it will be left to the reader to refer to those applications suggested in the literature [12].

Moving from the polyhedra and molecular fragments, as described earlier, one can then map the regular tessellations as shown in Fig.4, with the honeycomb tessellation, given by the Wells point symbol notation as 6^3, and translated into the mapping symbol or Schläfli symbol as (n, p) = (6, 3), and the square grid, given by the Wells point symbol notation as 4^4, and translated into the mapping symbol or Schläfli symbol as (n, p) = (4, 4), and finally, the third regular tessellation, which thus outlines the space of Fig. 4 in terms of the tessellations, as the closest-packed grid, given by the Wells point symbol notation as 3^6, and translated into the mapping symbol or Schläfli symbol as (n, p) = (6, 3).

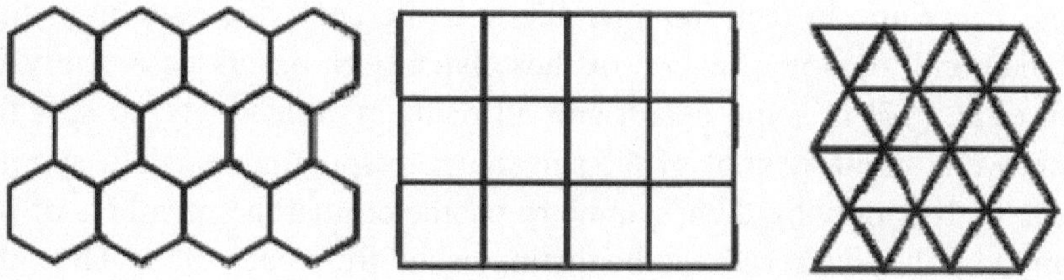

FIGURE 4 The Platonic tessellations, with their corresponding Wells point symbols, given as the closest-packed grid (3^6), the square grid (4^4) and the honeycomb grid (6^3).

One can, of course, insert all manner of hybrid tessellations, among these threeregular ones, and generate innumerable Archimedean, Catalan and Wellsean tessellations. Some hybrid tessellations of the square grid-honeycomb grid pair have been analyzed topologically by Bucknum et al. [13]. There are, of course, an infinity of such structural tessellations, and they indeed fill the space in the neighborhood of the borderline between the polyhedra and the tessellations, on the one hand, and the tessellations and the networks in 3D, on the other hand. It is also true that Wells and others [3, Chapter 1], have identified tessellations of the plane comprised of 5-gons and 7-gons, and their have been tessellations of 4-gons and 6-gons that admit unstrained 8-gons, and there are many more tessellations proposed, some of which have been taken as models of various C allotropes [14–16], which essentially can possess any n-gons in their pattern, provided that the restraint of being regular n-gons is relaxed. And it is thus true, that all of this infinity of tessellations can be mapped, rigorously, by the methods in Eqs. (3a) and (3b) outlined earlier.

Finally, the map in Fig. 2 outlines the 3D networks, and a prominent member is, of course, the diamond lattice given by the Wells point symbol 6^6, which is translated [17] into the Schläfli symbol (6, 4). By examination of Fig. 2, one can see that the diamond network, given the Schläfli symbol (6, 4), is situated just across the borderline from the 2-dimensional honeycomb tessellation given by (6, 3) in the map. One member of the diamond network topology, is in a cubic symmetry space group of

Fd-3m, space group #227, one of the highest symmetry space group patterns. There are, in fact, innumerable possible polytypic patterns within the diamond topology, several of these have been discussed recently by Wen et al. [17] in some detail, and all of them collectively possess the same Wells point symbol of 6^6, and the corresponding Schläfli symbol of (6, 4). It is only by their symmetry character, that the members of the diamond polytypic series can be distinguished from each other. Thus, the simplest cubic diamond polytype, known as 3C, is shown in Fig. 5.

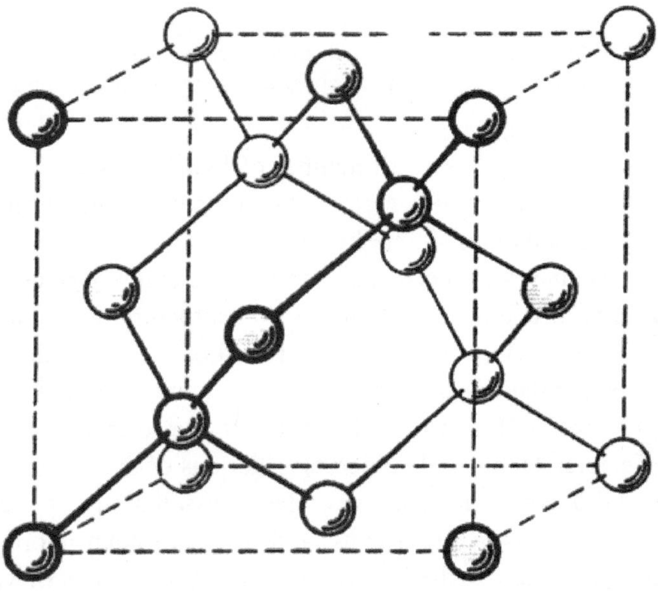

FIGURE 5 Cubic diamond (3C) polytype, with the Wells point symbol (6^6), lying in symmetry space group (Fd-3m).

Thus in the diamond network, which corresponds to the Platonic (integer) topology of the Platonic polyhedra, one can readily trace the uniform 6-gon, puckered circuitry of the network connected together by all 4-connected, tetrahedral vertices. Diamond's topology classifies the network as a regular, Platonic structure-type.

Next, we move to the space between (6, 3) and (6, 4), seemingly between 2D and 3D forms, and investigate what potential structures might emerge along this boundary area. Such an examination turns up two distinct families of Catalan networks, that together possess the Catalan Wells point symbol $(6^6)_x(6^3)_y$. One can see, through this Wells point symbol notation, that we are describing hybrid structures of the honeycomb tessellation, the so-called graphene grid, 6^3, and the diamond network, 6^6. The notation "y/x" specifies the stoichiometry of the net, in terms of the ratio of 3-connected, trigonal planar vertices, to 4-connected, tetrahedral vertices in the hybrid structure [18].

Thus one example of such a class of hybrid "graphene-diamond" structures is shown in Fig. 6, and these forms are known, by their hybrid topology, as the "graphite-diamond hybrids." They come in infinite series,' in

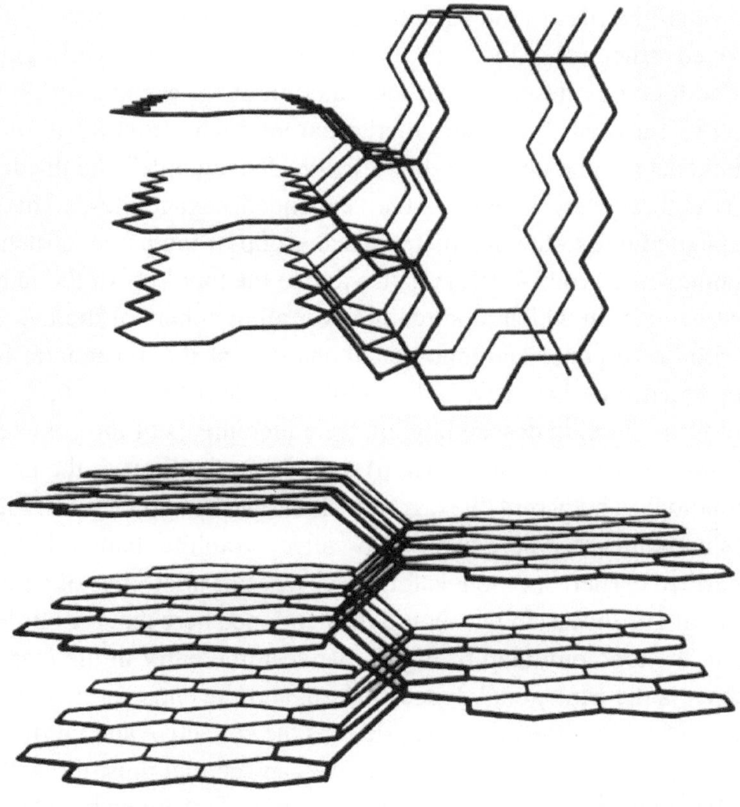

FIGURE 6 Representatives of the infinite families of ortho- and para-graphite-diamond hybrid structures, with the collective Wells point symbol $(6^6)_x(6^3)_y$, of orthorhombic symmetry (Pmmm).

each of two varieties that are known as the ortho-form, and the para-form, these have been elegantly described by Balaban et al. in 1994 [46, Chapter 2]. Collectively, as a family, they possess the Schläfli symbol of $(6, 3^{(x/(x+y))})$, where the parameters "x" and "y" have the stoichiometric significance ascribed to them in the preceding paragraph. These structures, as a family, occupy the border-line area between (6, 3) and (6, 4) in the Schläfli map in Fig. 6.

Yet another family of "graphite-diamond" hybrid structures to be considered, with the Wells point symbol $(6^6)_x(6^3)_y$, and the corresponding Schläfli symbol given by $(6, 3^{(x/(x+y))})$, is the family of structures described first by Karfunkel et al. [34, Chapter 2] in 1992, as being built from the barrelene hydrocarbon molecular fragment, and extended by the insertion of benzene-like tiles to the parent framework, to generate many infinities of derived structures, which all, collectively, possess the hybrid graphite-diamond topology described earlier. Later, in 2006, Bucknum [29, Chapter 2, 18] clarified the details of the parent such structure derived by Karfunkel et al., and he called this structure "hexagonite" and the derived, such structures were known as the "expanded hexagonites." This name was assigned due to the symmetry space group of the parent structure, in P6/mmm, space group #191, and also due to the topology of the family of such structures, in which all circuitry over all members of the family, are comprised of 6-gons. The topological analysis of the hexagonite family, suggests that they begin with the Schläfli symbol $(n, p) = (6, 3^{2/5})$, and extend from there, in descending, discrete increments of the connectivity, p, towards their termination at (n, p) = (6, 3), in the limit of the graphene grid topology. The parent "hexagonite" is shown in two views in Fig. 7.

It should be noted here that the so-called "graphite-diamond hybrids," as described earlier, are models for novel types of allotropes of carbon. In the vein of this discussion, it should be mentioned here that the border-line space in the topology map of Fig. 2, between the entry in the map of (5, 3), which is the pentagonal dodecahedron, and the entry in the map (6, 3) which is, of course, the graphene grid, lies the collective and infinite space of the fullerene C structures [19]. So, we can see that thus, the collective Schläfli symbol for the family of fullerenes is, in fact, given by $(5^{(x/(x+y))}, 3)$, where, in this instance, "x" is the number of hexagons in the polyhedron, and "y" is the number of pentagons in the polyhedron [8]. The Schläfli

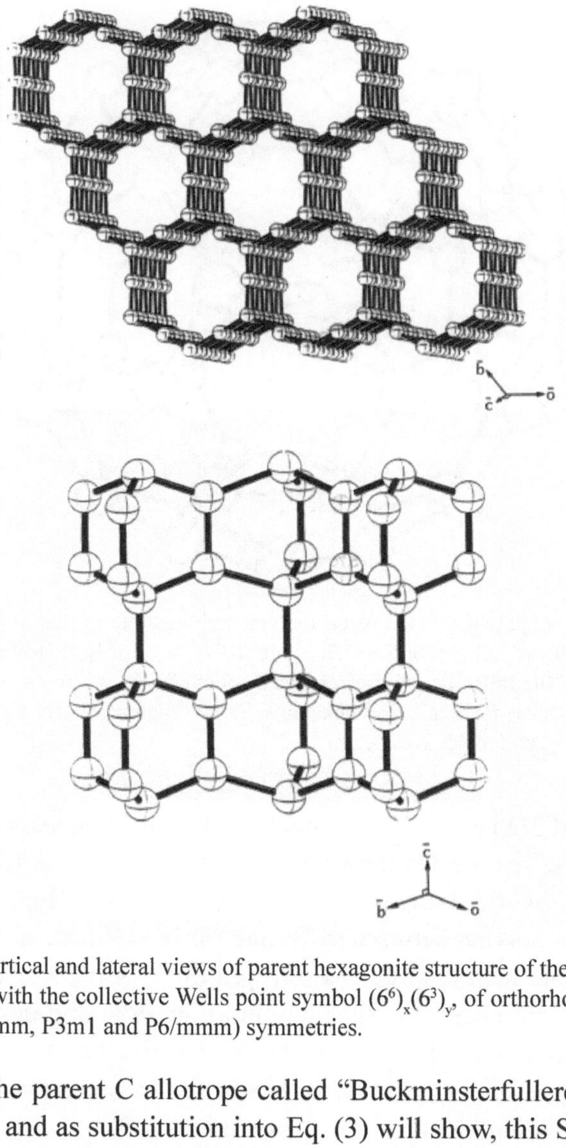

FIGURE 7 Vertical and lateral views of parent hexagonite structure of the infinite family of hexgonites, with the collective Wells point symbol $(6^6)_x(6^3)_y$, of orthorhombic-trigonal-hexagonal (Pmmm, P3m1 and P6/mmm) symmetries.

symbol for the parent C allotrope called "Buckminsterfullerene" (C_{60}) is ($5^{5/8}$, 3) [19], and as substitution into Eq. (3) will show, this Schläfli symbol rigorously describes the Buckminsterfullerene polyhedron fully. Figure 8 shows a view of this polyhedron of icosahedral symmetry, and it is clear from this view that the polyhedron is uniformly 3-connected, (as the Schläfli symbol reveals, the fullerenes are Archimedean) and comprised entirely of 5-gon and 6-gon circuitry.

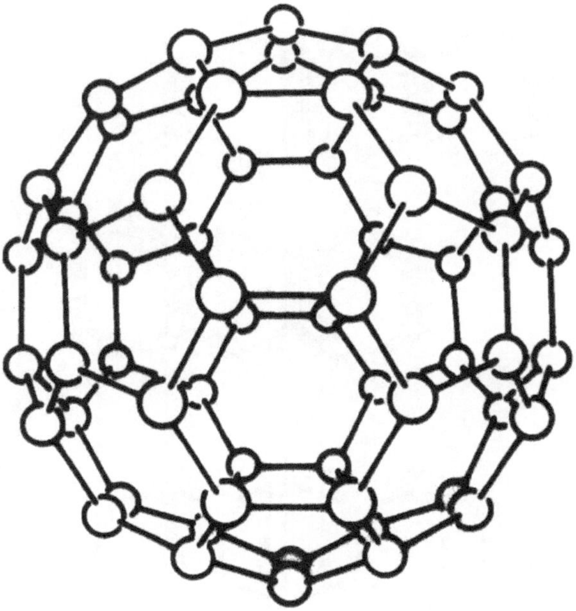

FIGURE 8 The parent Buckminsterfullerene polyhedron, of the infinite family of fullerenes, with the collective Schläfli symbol for the family of fullerenes given by ($5^{(y/(x+y))}$, 3), and the fullerene polyhedral face symbol of 5^x6^y, where "x" is the number of pentagons, and "y" is the number of hexagons, in the fullerene, where such structures are of icosahedral (I_h) and lower symmetry.

Yet a final 3D network to be mentioned in this connection, which is a model of a 3-,4-connected network of C, as opposed to a straight "graphite-diamond" hybrid, lying between (6, 3) and (6, 4) of Fig. 2, or a fullerene polyhedron, lying between (5, 3) and (6, 3) of Fig. 2, as in the preceding discussions with respect to allotropes of C; is the so-called "glitter" network of C proposed by Bucknum et al. in 1994 and described in the previous Parts I and II [1, Chapter 1]. As described earlier in Parts I and II, this structure can be envisioned as being constructed from a 1,4-cyclohexadiene building block, and it is shown in Fig. 9.

As can be seen in Fig. 9, the Wells point symbol for this Wellsean network is given by (6^28^4) $(6^28)_2$, and derived from this, is its Schläfli symbol of $(7, 3^{1/3})$ [7]. It is comprised of 6-gons admixed with 8-gons, in its topology, and an admixture of two trigonal vertices for every one tetrahedral vertex in its connection pattern. This particular C network has been important from the perspective of the 3-dimensional (3D) resonance structures, which can be drawn over it, see Fig. 23 shown previously in Chapter 2. There have been some favorable indications that its synthesis has been achieved [20–21].

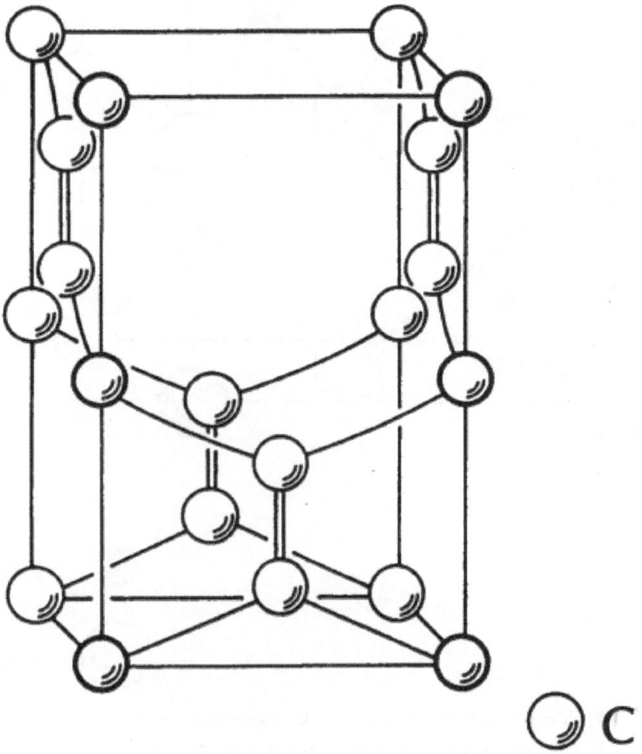

FIGURE 9 Tetragonal glitter network of carbon, with Wells point symbol given by $(6^2 8^4)(6^2 8)_2$, and of space group symmetry ($P4_2/mmc$).

Other inorganic networks that are of interest [22], that are not C allotropes, or models of C allotropes, would include the Archimedean Cooperite network, the structure of the minerals PtS and PdO [7], shown in Fig. 10, which is a 4-connected network comprised of an equal mixture of tetrahedral and square planar vertices, both of which are distorted in their geometries.

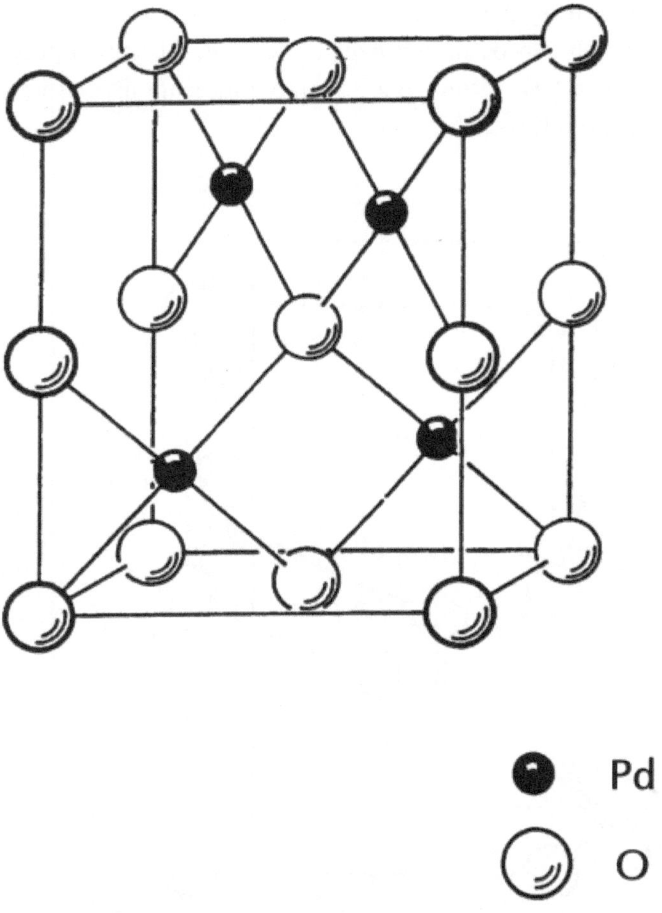

FIGURE 10 Archimedean Cooperite network as structure of PdO and PtS, with the Wells point symbol of $(4^2 8^4)(4^2 8^2)$, and of space group symmetry ($P4_2/mmc$).

As Fig. 10 indicates, taking both tetrahedral and square planar vertices as equally 4-connected, one can thus assign an Archimedean topology to this network, with a Wells point symbol of $(4^2 8^4)(4^2 8^2)$, and a Schläfli symbol of $(6^{2/5}, 4)$ [7]. In this case, the network nominally is binary, with two distinct types of connectivity, and an apparently Wellsean topology associated with this, but, in fact in this instance, we have a 4-connected network where the square planar vertices (with 4 independent circuits) are viewed as equivalent in topology (if distorted) versions of the tetrahedral vertices (with 6 independent vertices).

Yet another inorganic network includesthe Catalan fluorite network [11], the structure of a number of mineral fluorides including CaF_2, shown in Fig. 11.

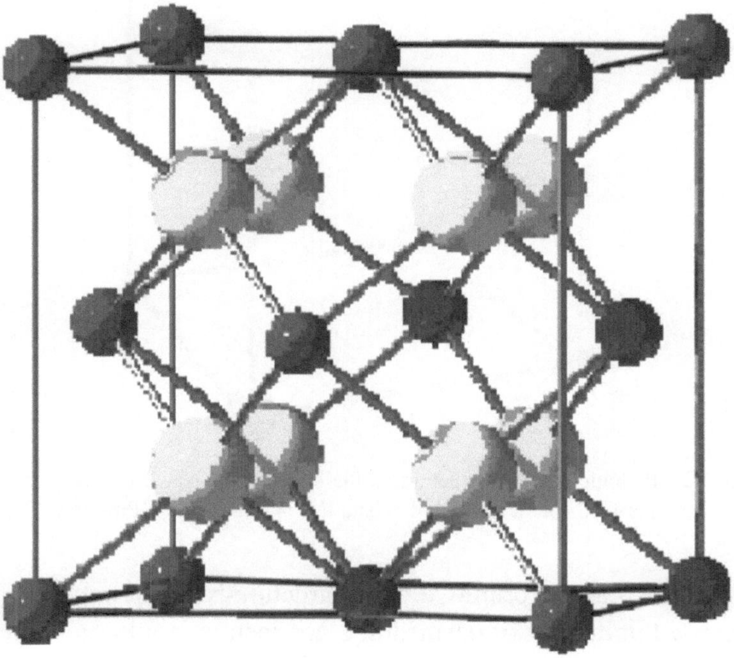

FIGURE 11 Catalan fluorite structure as the structure of CaF_2, this network can be represented by the Wells point symbol $(4^{24})(4^6)_2$, and lies in space group (Fm-3m).

This densely connected network is comprised of 4-connected tetrahedral vertices, and 8-connected cube-centered vertices, which are connected to each other through a uniform set of 4-gons. The topology of this network can be represented by the Wells point symbol $(4^{24})(4^6)_2$, and this can be translated into a Catalan Schläfli symbol of $(4, 5^{1/3})$. It can be mapped in Fig. 2 just beyond the entry (4, 5).

Still another inorganic structure-type we can provide the topology of, is the rocksalt (or primitive cubic) lattice [11], which is the structure of a number of inorganic alkali metal halides and alkaline earth chalcogenides. The rocksalt lattice is shown in Fig. 12.

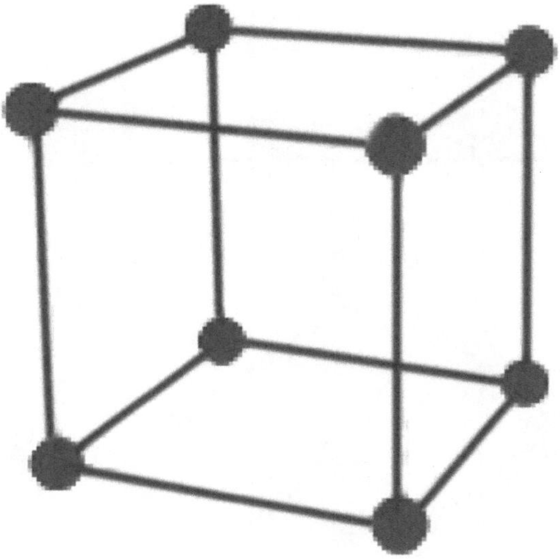

FIGURE 12 Platonic rocksalt structure as the structure of NaCl, this network can be represented by the Wells point symbol (4^{12}), and lies in space group (Pm-3m).

Also known as the primitive cubic structure-type, the rocksalt lattice is comprised uniformly of octahedral 6-connection in all 4-gon circuitry. The Wells point symbol for the network is given as 4^{12}, and this can be translated into a Schläfli symbol for the network of (4, 6) [7]. Because this rocksalt network is of a Platonic (regular, integer) topology, it can readily be seen where it maps in Fig. 2. It represents, structurally, an extension into 3-dimensions (3D) of the square grid 4^4, or (4, 4) which extends in 2D, through a layering, in exact register, of other square grids onto a parent square grid, and their interconnection through perpendicular interlayer bonding through the respective vertices.

As a final inorganic structural-type that we can analyze here topologically in this chapter, we have the so-called body-centered cubic (bcc) structure of CsCl [11], and a number of inorganic structures including alkali metal halides and alkaline earth chalcogenides and other materials. The bcc structure-type is shown here in Fig. 13.

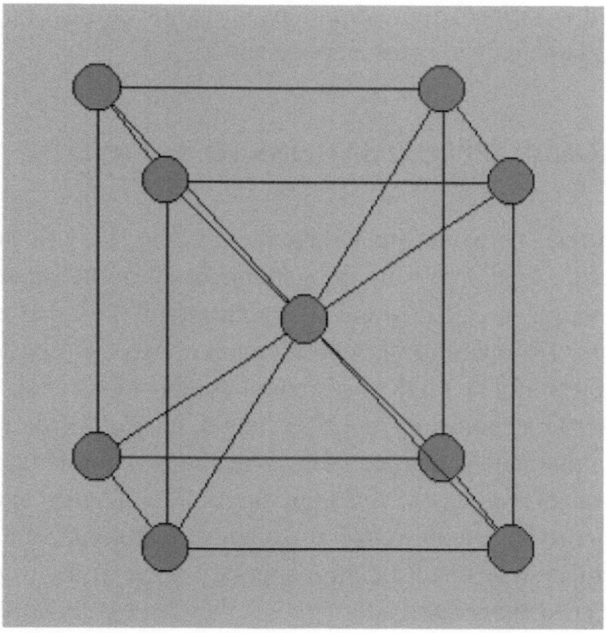

FIGURE 13 Platonic body-centered cubic (b.c.c.) structure as the structure of CsCl, this network can be represented by the Wells point symbol (4^{24}), and lies in space group (Im-3m).

It is uniformly comprised of 8-connected, cube-centered vertices that are mutually interconnected by all 4-gon circuits in the net. The topology of the bcc lattice can be specified by 4^{24}, or by the corresponding Platonic Schläfli symbol of (4, 8) [7].

There are, of course, innumerable other network structures that possess 3-dimensional geometries, one only has to look at the exhaustive works of O'Keeffe et al., to discern the scope of this field [22]. In the present discussion, we move, in the next chapter, into the area of the topological analysis of some networks in 3-dimensions that exhibit rather odd Schläfli symbols (n, p). These networks, the oldest of which was identified only in 1951 (Waserite) [29, Chapter 1], and the others of which were identified only in 1988 (Kentuckia) [23], and 2005 (Moravia) [24], suggest from the computation of their respective Schläfli symbols (n, p), by the methods outlined in Eqs.(3a) and (3b), that such numbers represent not only topological parameters of these networks, but they coincidently and fortuitously can be construed as rational approximations, in various instances, to the transcendental mathematical constants φ, e and π.

3.4 RATIONAL APPROXIMATIONS TO Φ, E AND P

Certain material networks, including the $CaCuO_2$ [25] (Kentuckia network) structure-type, that is the progenitor for all of the superconducting cuprates, and the so-called Moravia structure-type [26], that is the prototype structure for a number of coordination network (also called metal-organic frameworks or MOF's) structural compositions [24], and the so-called Waserite structure-type [29, Chapter 1], that is the structure of the anionic, platinate sublattice of the ionic conducting lattice, known as sodium platinate ($NaPt_3O_4$), have here, through topological analysis, been shown, either to have a network polygonality, n, (Kentuckia network) the value of which serves as a rational approximation to the transcendental mathematical constant π, or, alternatively, they have a network connectivity, p, the value of which serves as a rational approximation to the product of the transcendental mathematical constants e and φ (Moravia network), or e and π (Waserite network).

The Kentuckia structure-type, proposed by Bucknum et al. in 2005 [25], is the pattern adopted by the high temperature superconducting cuprate composition; it is $CaCuO_2$ [23] that adopts this pattern, which is the progenitor to all the superconducting cuprates discovered so far. This tetragonal structure-type lies in space group P4/mmm, #123 and is shown in Fig. 14.

It can be seen from Fig. 14 that this tetragonal oxide, bears a relation to the cubic perovskite ($BaTiO_3$, [11]) structure-type which lies in the cubic space group Im-3m, not shown here, in which, by removal of an axial pair of oxygen vertices, one can generate the Kentuckia structure-type from the perovskite structure-type. However, it should be pointed out that the perovskite structure-type is a 6-,12-connected network, in which the transition metal titanium and chalcogenide oxygen centers, attain octahedral 6-coordination, while the alkaline earth barium cation bears a closest packed coordination sphere of 12. Whereas in the Kentuckia structure-type, the lattice, by great contrast as can be seen in Fig. 14, bears an oxygen vertex with a square planar, 4-connected coordination, and the transition metal copper vertex bears an octahedral, 6-connected coordination, while the alkaline earth calcium cation is in cube-centered, 8-fold coordination. It is a ternary, 4-,6-,8-connected tetragonal structural pattern, as is described in Fig. 14.

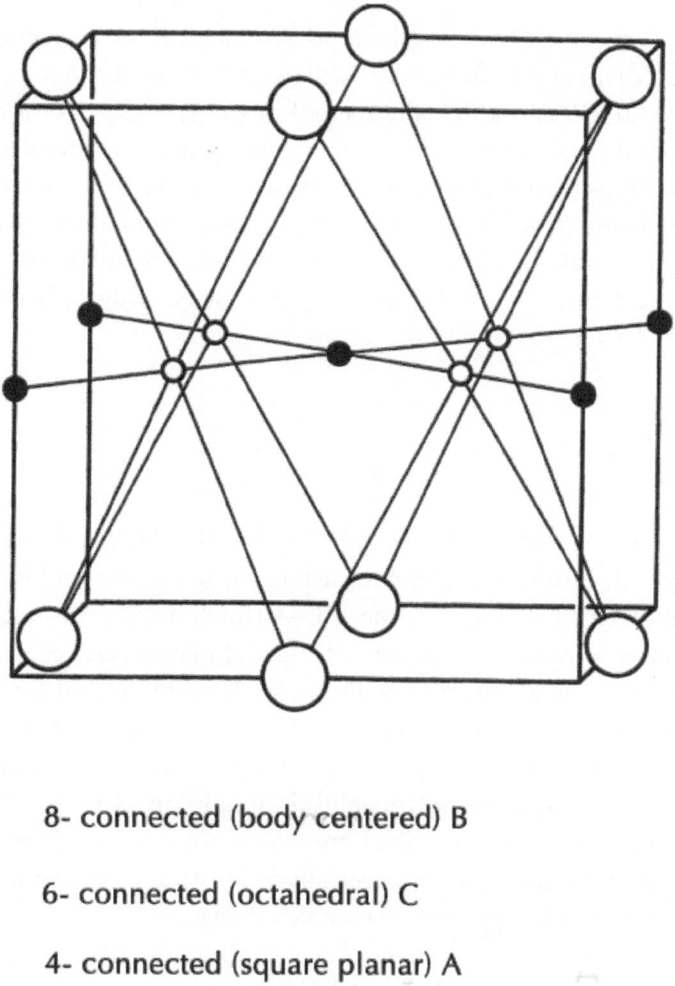

FIGURE 14 Wellsean Kentuckia (ABC_2) structure-type as the structure of the superconducting cuprate salt $CaCuO_2$, this network can be represented by the Wells point symbol $(4^4)(4^{12}6^{12})(4^{12})_2$, and lies in space group (P4/mmm).

Thus the connectivity in Kentuckia, is that of a ternary 4-,6-,8-connected structural-type, while the connectivity in perovskite appears to be that of a binary 6-,12-connected network topology. It appears that all the circuitry in perovskite is comprised of 4-gons with, perhaps, the stoichiometry AB_4, in which the vertex "A" is in 12-connected, closest-packed topology, and the vertex "B" is 6-connected, octahedral coordination. This leads to a Wells point symbol for perovskite of $(4^{60})(4^{12})_4$ and a Schläfli symbol of (n, p) = (4, $7^{1/5}$). Where, as a reference, the hexagonal closest packed (hcp) and cubic closest packed (ccp) networks, have the Wells point symbol 3^{60}, and a Schläfli symbol of (n, p) = (3, 12). In contrast, we see that the polygonality in the Kentuckia pattern, as revealed in Fig. 15, is a composite of 4-gons and 6-gons, with the Wells point symbol for the lattice being $(4^4)(4^{12}6^{12})(4^{12})_2$. As Eq. (6) reveals, the polygonality, n, for this structural-type, bears an odd resemblance to the transcendental mathematical constant π [27], occurring as it does, within 1% of exactly the value of $\sqrt{2}\cdot\pi$.

$$n = (40\cdot 4 + 12\cdot 6)/52 \tag{6a}$$

$$n = \sqrt{2}\cdot p \tag{6b}$$

Turning to the Waserite network [29, Chapter 1], which is shown to be a relatively simple, binary 3-,4-connected network topology in Fig. 15.

As is revealed in Fig. 15, the 3-to-4 stoichiometry of 4-connected square planar vertices to 3-connected trigonal planar vertices, present in the Waserite topology, together with the Wells point symbol for this network as $(8^4)_3(8^3)_4$, thus demonstrates that this simple structure is indeed a binary, Catalan network comprised of all 8-gon circuitry. Therefore it is apparent, readily, that the polygonality is simply given by n = 8, in this pattern. But as has been described previously for this so-called Waserite network [29, Chapter 1], the connectivity index of it, as shown in Eq. (7), suggests that its topology is more complex than meets the eye.

$$p = (3\cdot 4 + 4\cdot 3)/7 \tag{7a}$$

$$p = (2/5)e\cdot\pi \tag{7b}$$

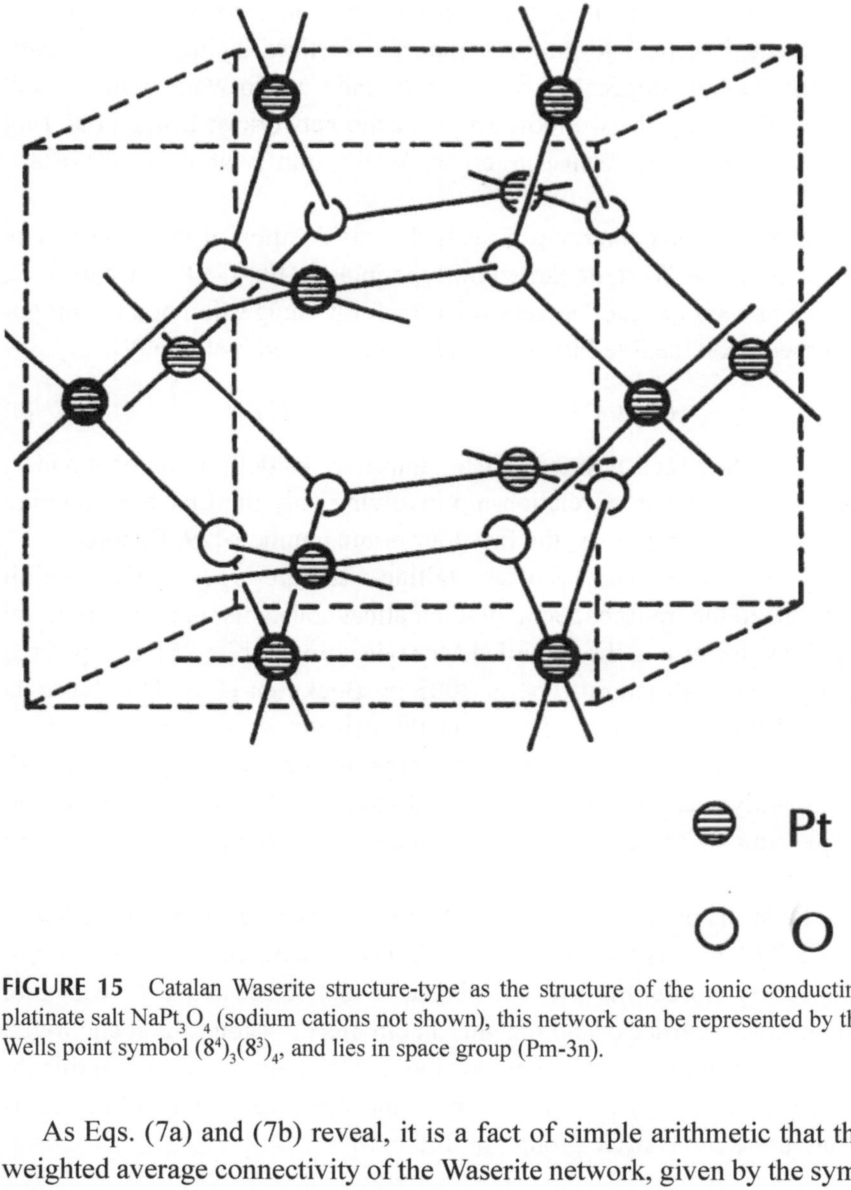

FIGURE 15 Catalan Waserite structure-type as the structure of the ionic conducting platinate salt $NaPt_3O_4$ (sodium cations not shown), this network can be represented by the Wells point symbol $(8^4)_3(8^3)_4$, and lies in space group (Pm-3n).

As Eqs. (7a) and (7b) reveal, it is a fact of simple arithmetic that the weighted average connectivity of the Waserite network, given by the symbol p, is in fact equal, to better than 99 parts in 100, to (2/5)e·π. Here π is, the familiar ratio of a circle's circumference to its diameter [27], and e is the natural base of exponentials [28]. These numbers are transcendental, as they are infinite, non-repeating fractions [27, 28]. An identical relation will also hold for the other structures patterned on a stoichiometry of four 3-connected vertices-to-three 4-connected vertices including, for example, the rhombic dodecahedron (given by the Catalan Wells point symbol as $(4^4)_6(4^3)_8$), and the well-known phenacite network of Bragg et al. (not shown, given by the Wellsean ternary Wells point symbol $(8^3)(6^3)_3(6^38^3)_3$ [29]).

Other relations emerging from such consideration of the connectivity index, p, in the Waserite structural-type include Eq. (1) [29, Chapter 1], known hereafter as the Timaeus relation, for its suggestion of a Cosmogony based upon the five Platonic solids as enunciated by Plato [9].

$$(1)\,(2.3333333\ldots\ldots\ldots\ldots)\cdot e\cdot \pi = (4)\cdot(5) \tag{8}$$

Equation (8) suggests the ultimate simplicity of definitions of e and π, through an elementary relationship involving only the first five counting numbers, or alternatively, the first four prime numbers [29, Chapter 1].

Finally, in this survey of crystalline structure-types, which exhibit relations to the transcendental mathematical constants, in their structural topology, we turn to the so-called Moravia network [24, 26], first posited as a potential structural-type in 2005 by Bucknum et al. This Moravia network has, in fact, turned out to be the structure adopted by several coordination networks known as metal-organic-frameworks (MOF's) [26]. It is readily seen to be a Wellsean, 3-,8-connected network upon careful inspection of the drawing for valences, and tracing of circuitry in Fig. 16.

The Wells point symbol for the Moravia structural-type is encoded as $(4^46^88^{12})_3(4^3)_8$, it is thus a complex, Wellsean network composed of two connection motifs, the trigonal planar, 3-connected, and cube-centered, 8-connected, vertices, held together in circuits of 4-gon, 6-gon and 8-gon sizes. The complex Well's point symbol for the network, belies in this instance, the relatively high symmetry of the structure, in which Moravia is lying in the cubic space group Pm-3m, #221.

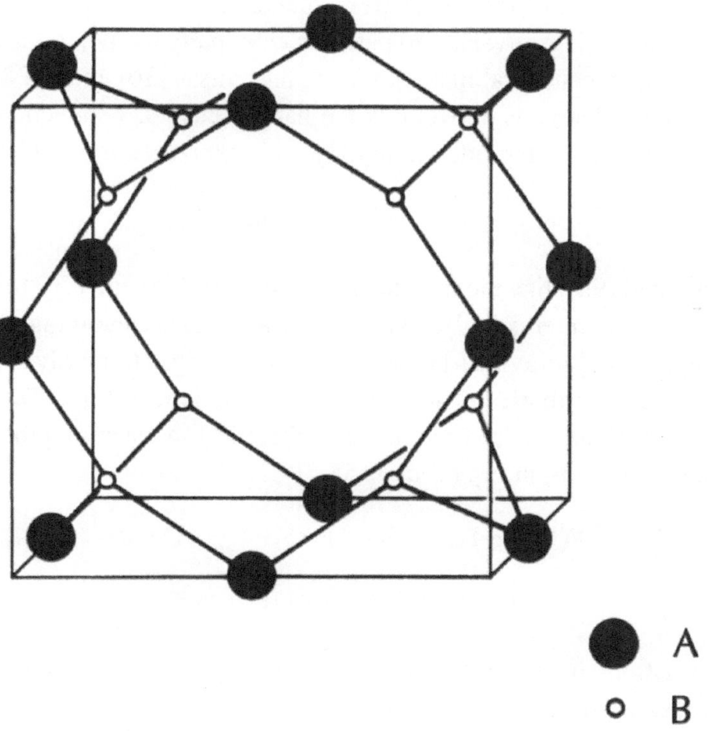

FIGURE 16 Wellsean Moravia structure-type as the structure of several coordination networks (metal organic frameworks, MOF's), this network can be represented by the Wells point symbol $(4^4 6^8 8^{12})_3(4^3)_8$, and lies in space group (Pm-3m).

If we take the point symbol for the network, and analyze it according to Eqs. (3a)–(3b) described earlier, the Well's point symbol translation formulas, one obtains the result that the weighted average polygonality for the network is indeed given by n = 6. The Waserite net is thus pseudo-Catalan, with an integer polygonality of 6, which is nonetheless the result of averaging over 4-, 6- and 8-gons in its structural pattern. Upon calculating the connectivity index, p, for Moravia, however, we get the result shown in Eqs. (9a) and (9b).

$$p = (3 \cdot 8 + 8 \cdot 3)/11 \tag{9a}$$

$$p = e \cdot \varphi \tag{9b}$$

Thus it is seen that the connectivity index, p, of Moravia, as a 3-,8-connected network, is equivalent, to better than 99 parts in 100, to the product of the two transcendental mathematical constants φ [30] and e [28], given simply as φ·e. Here, as above, e is the natural base of exponentials [28], and φ the well-known golden ratio [30], as is expressed in Eq. (10).

$$\varphi = (\sqrt{5} + 1)/2 \tag{10}$$

Eqs. (9a) and (9b), like the transformation of Eqs. (7a) and (7b) to Eq. (8), can be factored, interestingly, so that the relation of φ·e evolving out of the topology of the Moravia structure, as shown in Fig. 16, involves the first six Fibonacci numbers, F(1-to-6) (given, on the left, in Eq. (11a) as 1, 1, 2, 3, 5, and 8), and these are related to the 10th Fibonacci number, F(10) (given, on the right, in Eq. (11b) as 55).

$$F(1)\cdot F(2)\cdot F(3)\cdot F(4)\cdot F(5)\cdot F(6) = (e)\cdot(\varphi)\cdot F(10) \tag{11a}$$

$$(1)\cdot(1)\cdot(2)\cdot(3)\cdot(5)\cdot(8) = (e)\cdot(\varphi)\cdot(55) \tag{11b}$$

These relations between the topology of these structures in this chapter, as is revealed by the computation and mapping of their corresponding Schläfli symbols (n, p), and the transcendental mathematical constants φ, e and π, that can be thus correlated to their structural character, suggest the mathematical, and potentially scientific, richness that such structures may lead to.

3.5 CONCLUSIONS

In this chapter we have reviewed the basic tenets of a chemical topology scheme, one that can be applied to classify and effectively map the innumerable polyhedra, tessellations and networks, based upon a simple computation of their Schläfli symbols (n, p), from translation of their corresponding Wells point symbols. A restriction pointed up by this work, is that all structures in such a chemical topology scheme must, indeed, be *simply connected.* The phrase *simply connected* means that all edges, E, in a structure, be it a polyhedron, tessellation or network, must terminate at distinct vertices, V, in the network, where such edges, E, are known as proper edges. A second condition on a net being *simply connected,* is that all faces, F, in the structure should be bounded by proper edges, E, as defined in the preceding sentence. It is not clear, at this juncture, what form a systematic chemical topology would take on for the *non-simply connected* structures. As there are innumerable *non-simply connected* structures, to accompany the infinite number of *simply connected* structures in Schläfli space (the space of the Schläfli symbols (n, p)), it would seem that a topological analysis of these complex structures would be desirable and necessary to get a more complete handling of the chemical topology of crystal chemistry.

The brunt of this paper has been dedicated to a survey of some of the more prominent (well-known or obvious) organic and inorganic structure-types. Organic structures included some well-known C allotropes, like the regular, graphene grid and the regular, diamond network, both forms of C known since Antiquity. And, also, more modern C forms were surveyed, like the 3-,4-connected, Catalan graphite-diamond hybrids [46, Chapter 2], the 3-,4-connected, Catalan hexagonite lattices [29, 50, Chapter 2] and the Wellsean, 3-,4-connected glitter C form [1, Chapter 1], for which there is currently some evidences of their syntheses from the growth of C nanocrystals [20, 21]. Inorganic structures included in this survey, were the 4-,8-connected, Catalan fluorite lattice, the 4-connected, Archimedean Cooperite lattice, the 8-connected, regular CsCl, body-centered cubic structure-type, and the 6-connected, regular rocksalt structure-type [11]. Finally, in this survey, lattices which admitted connections in their topology to the transcendental numbers included the 3-,8-connected, Wellsean Moravia net, discovered in 2005 [24, 26] (related to φ and e, through the connectivity), the 4-,6-,8-connected, Wellsean Kentuckia (cuprate structure-type) net, discovered in 1988 [23, 25] (related to π, through the polygonality), and finally the 3-,4-connected, Catalan Waserite net (platinate structure-type), discovered in 1951 [29, Chapter 1].

The occurrence of relations to the transcendental numbers of mathematics, in the computations of the topology character of some of these networks, is indeed a mysterious outcome. It is not clear whether such relations could imply that the topology of these lattices, like the Kentuckia lattice, in which $n = \sqrt{2}\cdot\pi$, could indeed be equated to some type of ordering parameter for the lattice, such that by the introduction of systematic defects in the connectivity, p (or thereby the polygonality, n) over the bulk lattice, might lead to a corrected value of n, that asymptotically approaches the true value of π, and that, that might have some bearing on bulk properties of the Kentuckia network, like the critical superconducting transition temperature, T_c, in the cuprate composition $CaCuO_2$ [25]. Such considerations as these, open up new avenues of explorations for solid state scientists based upon the intrinsic topology character of such networks as these.

KEYWORDS

- **Euler relation**
- **graphene-diamond**
- **hexagonite**
- **Platonic polyhedral**
- **Schläfli relation**
- **Timaeus relation**

REFERENCES

1. Bucknum, M.J. *Carbon*, **1997,** *35(1)*, 1.
2. Leonhard Euler, In: *Element adoctrinaesolidorum et Demonstration onnularuminsigniumpropriet atumquibus solid aheddrisplanisinclusasuntpraedita, Proceedings of the St. Petersburg Academy*, St. Petersburg, Russia 1758.
3. Burckhardt, J.J. Ib *Der mathematische Nachlass von Ludwig Schläfli, 1814–1895, in der SchweizerischenLandesbibliothek*, 1st edition, Bern, 1942.
4. Gardner, M. In: *Archimedes: Mathematician and Inventor*, 1st edition, Macmillan Publishing: New York, NY, 1965.
5. Frederico, P.J. In: *Descartes on Polyhedra: A Study of the De Solidorum Elementis*, 1st edition, Springer-Verlag: New York, NY, 1982.
6. Peters, I. *Science News*, **2001,** *160(25/26)*, 396.
7. Bucknum, M.J.; Castro, E.A.*(MATCH) Commun. Math. Comp. Chem.*, **2005,** *54*, 89.
8. It appears that the 5 Platonic polyhedra obey Eqs. (1) and (2) of the text, if one specifies their topology by a Well's point symbol, given by A^a, in which n = A and p = a, and there is no relation between this polyhedral face symbol, A^a, and the computation of V, E and F in the Euler model of Eq. (1). It is also the case, that the 5 Platonic polyhedra

obey Eqs. (1) and (2) of the text, if one specifies their topology by a polyhedral face symbol, given as A^b, in which (b·A) = 2E, b = F, and V is identified through inspection of the polyhedron, then n = 2E/F and p = 2E/V, by definition. For the Archimedean polyhedra it appears that it is only possible to specify (n, p), for insertion into Eq. (2), by encoding a polyhedral face symbol, given as $A^aB^bC^c$......., where a is the number of A-gons, in a ratio with b B-gons etc., for the polyhedron, and thus, in which (a·A + b·B + c·C + ...) = 2E, (a + b + c + ...) = F, and V is identified through inspection of the polyhedron, and where finally then, n = 2E/F and p = 2E/V, by definition. While for the Catalan and Wellseanpolyhedra, by contrast, it appears that it is only possible to specify (n, p), for insertion into Eq. (2), by encoding a Wells point symbol, given as $(A^a)_x(A^b)_y(A^c)_z$.........., where a is the number of A-gons meeting at a, and b is the number of A-gons meeting at b etc., and x, y, z etc. specify the stoichiometry of the polyhedron, for the Catalan or Wellsean polyhedron of interest, in which (a·A + b·A + c·A + ...) = E, (a·x + b·y + c·z + ...) = 4F and (x + y + z + ...) = V, where finally then, n = 2E/F and p = 2E/V by definition, as throughout.

9. Cooper, J.M., Ed.; In:*Plato: Complete Works*, 1st edition, Hackett Publishing Company: Indianapolis, IN, 1997.
10. Bragg, W.L. In:*The Development of X-ray Analysis*, 1st edition, Dover Publications, Inc.: Mineola, NY, 1975.
11. Pauling, L. In: *General Chem.*, 4th edition, Dover Publications, Inc.: Mineola, NY, 1988.
12. Duchowicz, P.; Bucknum, M.J.; Castro, E.A. *J. Math. Chem.*, **2007,** *41(2)*, 193.
13. (a) Bucknum, M.J.; Castro, E.A.*(MATCH) Commun. Math. Comp. Chem.*, **2006,** *55*, 57. (b) Bucknum, M.J.; Castro, E.A. *Solid State Sci.*, **2008,** *10(9)*, 1245.
14. Fahy, S.; Louie, S. G.*Phys. Rev. B,* **1987,** *36*, 3373.
15. Zhu, H.; Balaban, A.T.; Klein, D.J.; Zivkovic,T.P. *J. Chem. Phys.*, **1994,** *101*, 5281.
16. Crespi, V.H.; Benedict, L.X.; Cohen, M.L.; Louie, S.G. *Phys. Rev. B*, **1996,** *53*, R13303.
17. Wen, B.; Zhao, J.; Si, D.; Bucknum, M.J.; Li, T. *Diamond Related Mater.*, **2008,** *17(3)*, 356.
18. Bucknum, M.J.; Castro, E.A. *J. Chem. Theory and Computation*, **2006,** *2(3)*, 775.
19. Baggott, J. In: *Perfect Symmetry: The Accidental Discovery of Buckminsterfullerene*, 1st edition, Oxford University Press: Oxford, U.K., 1996.
20. Bucknum, M.J.; Castro, E.A. *J.oretical and Comput. Chem.*, **2006,** *5(2)*, 175.
21. Bucknum, M.J.; Castro, E.A. *Mol. Phys.*, **2005,** *103(20)*, 2707.
22. (a) M. O'Keeffe, B.G. Hyde, *Crystal Struct.s I. Patterns and Symmetry*, 1st edition, Mineralogical Society of America (M.S.A.), Washington, D.C., 1996. (b) M. O'Keeffe, M.A. Peskov, S.J. Ramsden, O.M. Yaghi, *Acc. Chem. Res.* **2008,** *41*, 1782.
23. Siegrist, T.; Zahurak, S.M.; Murphy, D.W.; Roth, R.S. *Nature*, **1988,** *334*, 231.
24. Dincã, M.; Dailly, A.; Liu, Y.; Brown, C.M.; Neumann, D.A.; Long, J.R. *J. Am. Chem. Soc.*, **2006,** *128*, 16876.
25. Bucknum, M.J.; Castro, E.A. *Russian J. General Chem.*, **2006,** *76(2)*, 265.
26. Bucknum, M.J.; Castro, E.A.*Central European J. Chem. (CEJC)*, **2005,** *3(1)*, 169.
27. (a) Beckmann, P. In: *A History of* π, 1st edition, The Golem Press, New York, NY, 1971. (b) Blatner, D. In: *The Joy of Ø*, 1st edition, Walker Publishing Company, Inc., USA, 1997.

28. Maor, E. In: *The Story of a Number*, 1st edition, Princeton University Press: Princeton, NJ, 1994.
29. Bucknum, M.J.; Bin Wen, E.A. Castro, *J. Math. Chem.*, **2010,** *48,* 816.
30. Livio, M. In: *The Golden Ratio: The Story of Ø the World's most Astonishing Number*, 1st edition, Broadway Books: New York, NY, 2002.

CHAPTER 4

DYNAMIC ELASTICITY OF CARBON

MICHAEL J. BUCKNUM and EDUARDO A. CASTRO

CONTENTS

SUMMARY

The present report is an account of the generalization of the dynamic elasticity theory earlier proposed by Bucknum et al. and applied to the cubic diamond and tetragonal glitter lattices. It describes a theory of elasticity in which the elasticity moduli are based upon the microscopic constants of the various structure-types. Such microscopic constants include the force constants of the chemical bonds in the unit of pattern of the material, its associated lattice parameters, and the elastic chemical bond deformation parameters of the material. In developing the outward features of the dynamic elasticity model, it is shown that an integral over the force density in the unit cell of a given material; where the force is modeled based upon the elastic deformation forces of the chemical bonds in the unit of pattern of the material, and the volume is written as a function of the deformations taking place inside the unit cell of the material; generates the terms for calculating its modulus of elasticity at pressure, in components, that are directed along the principal axes of the unit cell. Several potential solutions to the problem of super-hardness are discussed and illustrated.

4.1 DYNAMIC ELASTICITY

Bucknum et al. previously reported the calculated bulk modulus at pressure of the tetragonal glitter [59, Chapter 2] and cubic diamond [57, Chapter 2] structure types by applying a dynamic elasticity theory to these structures. In this work, the bulk modulus at pressure of a given material, B, was calculated starting from the zero-pressure bulk modulus, B_0, which can be approximated for crystalline covalent materials according to a formula due to Cohen [1]. Corrections to the zero-pressure bulk modulus were subsequently calculated on the basis of the potential elastic chemical bond deformations taking place inside the given unit cell in response to an applied stress. In particular, the theory relied upon the projections of these elastic chemical bond deformation forces, inside the unit cell, as stresses across the principal crystallographic planes of the given lattice. From this analysis, it was shown that tetragonal glitter attains a higher bulk modulus at pressure than does cubic diamond, along its **c**-axis, and therefore, tetragonal glitter appears to be a superior material to cubic diamond, in terms of the axial stresses tolerated against its basal plane in response to an applied mechanical stress [57, 59, Chapter 2].

In what follows, an attempt is made to generalize the results of these two earlier studies on cubic diamond and tetragonal glitter, by assuming a generic force density integral to model stress in a material. By carrying out the machinations of the integration of such a force density function, it results in the generation of a power series in the attendant strain on the unit cell. The terms of the power series can be identified with physically realistic corrections to the zero-pressure, static bulk modulus. These are the result of the manifestation of the elastic chemical bond deformation forces existent in the elastically deformed unit cell of the covalent material studied, as stresses inside the unit cell.

4.2 HISTORY OF ELASTICITY

Historically, the field of elasticity has been focused on the elastic properties of bulk materials, without reference to any crystalline structure [2]. One of the first such phenomenological moduli of elasticity discovered was that of a modulus of elasticity with respect to length deformation [3]. The well-known Young's modulus, Y, is defined in Eq. (1):

$$\text{stress} = Y\left(\frac{\Delta l}{l}\right) \tag{1}$$

In this expression, the stress applied as a force-per-unit-area of the material studied, is measured against the strain produced in the material, where the strain is measured as the incremental deformation, Dl/l, of the length of the specimen.

A curve relating stress to strain, in the bulk material, is plotted, and from the slope of this curve one obtains the Young's modulus for the material. A typical plot for obtaining the Young's modulus is shown in Fig. 1:

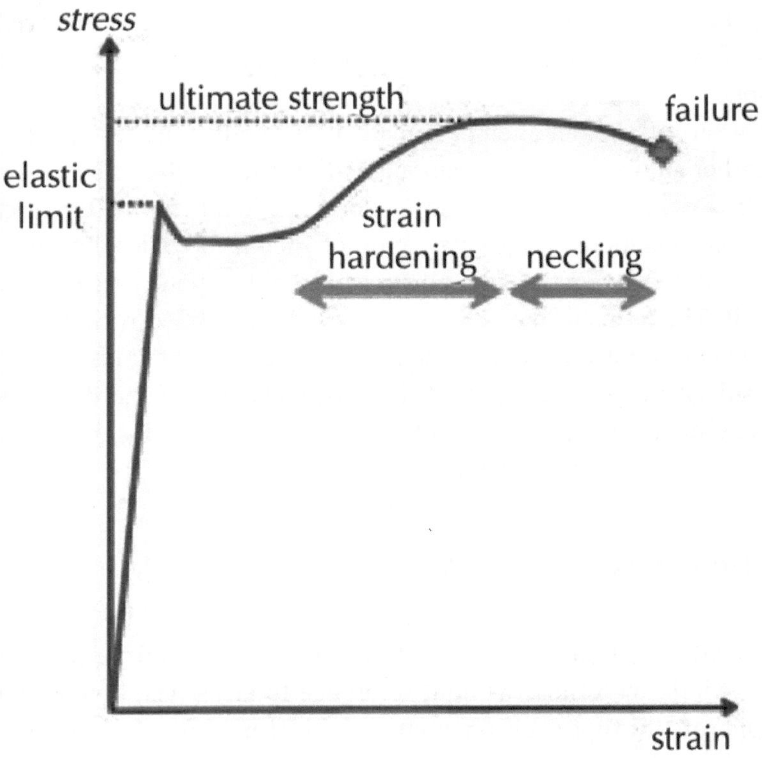

FIGURE 1 Stress-strain Curve for Determining Moduli of Elasticity.

From Fig. 1, it is noticed that the regions of elastic deformation and plastic deformation. In order to determine the modulus of elasticity one must make measurements in the region of elastic deformation of the material. The study of plastic deformation is an important field in its own right and will not be discussed further here.

Some typical values of the Young's modulus are shown in Table 1. Note the Young's modulus represents the slope of the stress-strain curve for elastic length deformation of a bulk material. This modulus of elasticity, as can be seen from Table 1, can be considerable, reaching a value in excess of 1.8 TPa for carbon nanotubes [4]:

TABLE 1 Selected Values of the Young's Modulus in Various Materials.

Material	Young's Modulus, Y, GPa	Reference
Carbon nanotube	1800	[4]
Steel	200	[6]
Aluminum	70	[6]
Glass	65	[6]
Concrete	30	[6]
Wood	13	[6]
Polystyrene	3	[6]

An important point should be addressed here with regard to the Young's modulus, Y, of a bulk material. It together with a measure of the ratio of the longitudinal strain to the lateral strain in a stressed bulk material, the so-called Poisson's ratio, σ, can be used to calculate, in principle, all the elastic constants of a bulk material, including the bulk modulus, B, to be discussed in Table 1 [5].

The idea of modulus of elasticity was generalized to three dimensions, later on, with the introduction of the notion of the bulk modulus, B [6]. The bulk modulus is also known as the volume modulus of elasticity, it is defined in Eq. (2):

$$\text{stress} = \mathrm{B}\left(\frac{\Delta V}{V}\right) \tag{2}$$

In this expression, the stress applied over the volume of the material is measured against the strain produced in the material. This is incremental deformation, DV/V, of the volume of the specimen. An analogous stress-strain curve for the bulk material, similar to Fig. 1, but based upon incremental volume deformation, DV/V, is used to determine the bulk modulus of the material under consideration.

B is defined in Eq. (17) with respect to the incremental change of volume of the material, DV/V. In contrast, the bulk modulus of elasticity can be identified with respect to a bulk sample of an ideal gas. This can be accomplished by defining the bulk modulus in terms of the volume derivative of the internal pressure of a material, as shown in Eq. (3):

$$B = -V \left(\frac{\partial P}{\partial V} \right) \tag{3a}$$

$$B = -V \frac{\partial}{\partial V} \left(\frac{nRT}{V} \right) = P \tag{3b}$$

From Eq. (3b), we see that the bulk modulus of an ideal gas, B, is just equal to the pressure of the ideal gas, P. This gives some perspective on what the physical meaning of the bulk modulus is, and suggests its central importance among the various moduli of elasticity in determining properties like the relative strength of materials with respect to volume deformation. It can also be thought of as a measurement of the internal energy of a material divided by its respective volume, U/V, or an energy density [7]. Some typical values of the bulk modulus of materials are shown in Table 2. One can see that the bulk modulus of cubic diamond, at 435 GPa, represents the zenith in B_0 among real materials.

TABLE 2 Selected Values of the Bulk Modulus in Various Materials.

Material	Bulk Modulus, B, GPa	Reference
Cubic diamond	435	[1, Chapter 1]
Steel	160	[6]
Mercury	27	[6]
Glycerine	4.8	[6]
Water	2.2	[6]
Ethanol	0.90	[6]

4.3 THE CRYSTALLINE MODULUS OF ELASTICITY

In what follows here, an alternative definition of elastic modulus, applicable to crystalline materials (as opposed to macroscopic, bulk samples of material), is given in terms of a coefficient which has the dimensions of a force density, multiplied by an integral in the strain over the unit cell of the crystalline material, the strain being generated by elastic chemical bond deformation forces created by an applied mechanical stress. The generalized expression is shown in Eq. (4):

$$\text{elastic modulus} = \int \frac{F(r')}{V(r')}\,\mathrm{d}r' \tag{4a}$$

$$\text{elastic modulus} = \sum_{i=1}^{N}\left\{\int \frac{N_i k_i x_i'}{bc(a+d_i x_i')}dx_i' + \int \frac{N_i k_i y_i'}{ac(b+d_i y_i')}dy_i' + \int \frac{N_i k_i z_i'}{ab(c+d_i z_i')}dz_i'\right\} \tag{4b}$$

Equation (4a) is the generic statement of the new elasticity law [59, Chapter 2]. It states that elastic modulus is equal to an integral over the force density of a material undergoing elastic chemical bond deformations inside the unit of pattern of the material, in response to an applied stress.

Equation (4b) recasts the relation in terms of a crystalline material with orthogonal axes a, b and c, N bonds per unit cell, the length decrement along the crystallographic axes from elastic chemical bond deformation, given by, (a + dx′), (b + dy′) and (c + dz′), and the assumption of a harmonic potential for chemical bonds. Here x′, y′ and z′ are the elastic chemical bond deformations represented by r′cos α, r′cos β, and r′cos χ, respectively (where $r' = (r - r_e)$ is the actual deformation of the chemical bond considered from its equilibrium value, r_e). The given angular factors refer to the projections of those elastic chemical bond deformations with respect to the three orthogonal crystallographic axes a, b and c [8]. The factor k is just the force constant of the chemical bond considered, assuming a harmonic potential energy function between atom pairs in the crystal. Finally, the factor d refers to the number of elastic chemical bond deformations connected with the compressions of the respective crystallographic axes (and concomitantly the unit cell volume) a, b and c. This latter factor can be obtained from elementary trigonometric analysis applied to the lattice under consideration.

As an example of such a crystalline material, consider the cubic diamond lattice shown earlier in Fig. 5 from Chapter 3 of Part III [22, Chapter 2].Here a, b and c are the three orthogonal unit cell vectors of cubic diamond, N is the number of bonds per unit cell, which is given by

16, the force constant for the carbon–carbon single bonds in cubic diamond is given by 450 N/m [9], and the length decrements are given by the expressions (a + 4x′), (b + 4y′) and (c + 4z′), which can easily be justified on the basis of elementary trigonometric analysis [10]. Cubic diamond is a particularly simple lattice because of its high symmetry and the presence of only one type of chemical bond in the unit cell.

One can see in Eq. (4b), that the elastic modulus of the material is expressed in terms of the dimensions and microscopic parameters of the unit cell of the material. The terms in Eq. (4b) correspond to the compressions along each of its axes a, b and c. Therefore it is broken up into threecomponent Young's-like moduli of elasticity. Nonetheless, the three component length deformation moduli of elasticity have a composite value that is a measure of the bulk modulus, B, of the crystalline material at pressure, or the volume modulus of elasticity over the unit cell volume at pressure.

The motivation for the use of a force density integral to model elastic modulus has been given previously [59, Chapter 2] in the form of Feynman's analysis of static elasticity in crystalline materials [5]. In the Feynman analysis, the NaCl lattice is treated and the chemical bonds are equated to harmonic springs having a force constant of k. Using strain parameters taken over the entire unit cell, and harmonic potential energy functions based upon these strain parameters of the NaCl crystal lattice, Feynman expanded out and collected the resulting harmonic potential energy terms, and divided by the static volume of the crystal, given by a^3 (where a is the lattice parameter of the NaCl unit cell). From this analysis, term-by-term he equated the calculated coefficients in the energy density, U/V, with the given elastic constants of the lattice. Finally, by taking advantage of the high symmetry of the cubic NaCl structure-type, in space group Fm-3m, he determined the following general relation between the moduli of elasticity of the cubic crystal and its corresponding microscopic constants, see Eq. (5).

$$\text{elastic modulus } \alpha \frac{k}{a} \tag{5}$$

Here k is the force constant of the chemical bonds in the NaCl lattice, with the dimensions of N/m, and a is the respective lattice parameter in m. Note the ratio of these microscopic constants has the dimensions of N/m^2, or the dimensions of a pressure, therefore it is consistent with an elasticity modulus for a material [11].

Taking this result as a theoretical basis, it was conjectured by the author in [1, Chapter 1] that a dynamic elasticity term for a material, analogous to the static elastic constants of the Feynman analysis, could be gotten in the form of the elasticity expression shown in Eq. (6):

$$\text{elastic modulus } \alpha \frac{kr'}{a^2} \tag{6}$$

where r′ represents an elastic chemical bond deformation, and a^2 is the area of a crystalline plane normal to that elastic chemical bond deformation.

It was discovered in this work [59, Chapter 2] that a methodology of generating such dynamic elasticity terms, in their exact formulation on the basis of the unit of pattern of the respective crystalline material, was by defining the elastic modulus of a unit cell as an integral over the force density in the material. The force density is defined as the total force of elastic chemical bond deformations within the unit of pattern, divided by the corresponding volume function of the unit cell, as that unit cell volume is distorted from the initial volume by the elastic chemical bond deformations. This is generally accomplished by taking distortions of the unit cell along the threecomponent axes, a, b and c, in 1-dimension at a time.

4.4 THE STRAIN INTEGRAL

From Eq. (4b), one can see the elasticity expressions for crystalline materials take the generic form shown in Eq.(7):

$$\text{elastic modulus} = \frac{Nk}{ab} \int \frac{z'}{c+dz'} dz' \tag{7}$$

Where the collection of constants, given by Nk/ab, has the dimensions of a force density, and the integral over the strain parameters (unlike the strain, Dl/l, in Eq. (1)) has the dimensions of a length when it is integrated.

Clearly Eq. (1), as it is written for the Young's modulus, Y, has a close, one might say analogous, relationship to the elastic modulus relationship shown in Eq. (7). One could, in fact, rearrange Eq. (1) to make the connection more concrete, as is shown in Eq. (8):

$$\text{stress} = \left(\frac{Y}{l}\right)\Delta l \tag{8}$$

In such an expression, the term Y/l is derived from a set of constants characteristic of the bulk material. The term Y/l has the dimensions of a force density, while the strain parameter is modified to have the dimensions of a length, Dl. This is analogous to the term shown in Eq. (7), which is alternatively written down on the basis of the consideration of the microscopic constants of a crystalline material.

Therefore, in Eq. (7), a collection of crystallographic-molecular constants over the unit cell of the material considered, defines a characteristic microscopic force density constant for the unit cell of the material. This is multiplied by an integral over the strain of the unit cell, which nonetheless, upon integration, yields an infinite series of terms, each of which has the dimensions of a length. The product of the force density constant, and the strain integral over the unit cell, is therefore a measure of the dynamic elastic modulus in the material.

Starting from Eq. (7), we can derive the elastic modulus expression on the unit cell of a material in the following manner (where we treat only one dimension, z′, for clarity):

$$\text{elastic modulus} = \int \frac{Nkz'}{ab(c+dz')}dz' \tag{9a}$$

$$\text{elastic modulus} = \frac{Nk}{ab}\int \frac{z'}{(c+dz')}dz' \tag{9b}$$

$$\text{elastic modulus} = \text{B}_0 + \frac{Nk}{ab}\left\{\frac{z'}{d}+\frac{c}{d^2}\ln|c+dz'|\right\} \tag{9c}$$

Upon expanding the logarithm, we obtain:

$$\text{elastic modulus} = \text{B}_0 + \frac{N}{d}\frac{kz'}{ab}+\left\{\frac{Nk}{ab}\frac{c}{d^2}\left(\left(\frac{dz'}{c}\right)^1+\frac{1}{2}\left(\frac{dz'}{c}\right)^2+\frac{1}{3}\left(\frac{dz'}{c}\right)^3+\ldots\right)\right\} \tag{9d}$$

which reduces to the following power series in the attendant strain, z′/c.

$$\text{elastic modulus} = \mathrm{B_0} + \frac{2N}{d}\frac{kz'}{ab}\left(\frac{dz'}{c}\right)^0 + \frac{1}{2}\frac{N}{d}\frac{kz'}{ab}\left(\frac{dz'}{c}\right)^1 + \frac{1}{3}\frac{N}{d}\frac{kz'}{ab}\left(\frac{dz'}{c}\right)^2 + \cdots \quad (9e)$$

From the derivation given in Eq. (9) one can see, term-by-term, the moduli of elasticity of crystalline materials emerge, generated from the kernel function in the form of a strain integral multiplied by a force density constant. One can see that the constant of integration, B_0, can be equated to the static elastic constant of zero-pressure bulk modulus of the material. However, it is cast as a function of the crystallographic-molecular parameters of the crystal, k the force constant of the chemical bonds inside the unit cell, and the lattice parameter a of the unit cell. This is analogous to the case of the Feynman analysis of the static elastic constants of the rocksalt lattice, where the ratio k/a properly has the dimensions of a stress [12]:

$$\mathrm{B_0 = f(k/a)} \quad (10)$$

The 1st-order term in the power series is given by the elastic chemical bond deformation force, modeled as a harmonic, Hooke's law force, kz', divided by the area of the crystalline plane (hkl) normal to that deformation force. It is a term in force-divided- by-area, F/A, and represents the first dynamic contribution to the bulk modulus at pressure, B. Note that other chemical bond potentials, including a Morse potential [59, Chapter 2], can be substituted for a harmonic potential in order to arrive at a more accurate elastic modulus expression.

$$\text{elastic modulus } \alpha \; \frac{2N}{d}\left(\frac{kz'}{ab}\right) \quad (11)$$

The 2nd-order term in the power series is given in terms of the energies of elastic chemical bond deformations, modeled as Hooke's law springs, divided by the corresponding volume of the unit cell, U/V. Such higher order terms may be important at high degrees of strain associated with plastic deformation.

$$\text{elastic modulus } \alpha \; \frac{1}{2}\frac{N}{d}\frac{Kz'}{ab}\left(\frac{dz'}{c}\right)' \quad (12)$$

The 3rd-order term, and higher terms in the power series, correspond to harmonic potential energy densities, U/V, multiplied successively by strain parameters, z′/c, (y′/b and x′/a), to higher and higher powers.

$$\text{elastic modulus } \alpha \ \frac{1}{3}\frac{N}{d}\frac{Kz'}{ab}\left(\frac{dz'}{c}\right)^{2} \tag{13}$$

It is clear from the magnitudes of x′ = r′cos α (y′ = r′cos β and z′ = r′cos χ), the elastic chemical bond deformation parameters, and a (b and c), the lattice parameters of the unit cell of the material, that major contributions to the stress on a crystalline material generally cut off at the term in the harmonic energy density, U/V. The 2nd-order term in the power series. Therefore, the major contributions to the bulk modulus at pressure in crystalline materials, under conditions of elastic deformation, occur from the zero-th-order term, the static elastic modulus (given by B_0 = f(k/a)) and the 1st-order term in the strain integration, the force-over-area term (F/A α kz′/ab). The higher terms in the power series (U/V α kz'^2/2abc and higher terms) all converge to 0 as the strain parameter, x′/a (y′/b and z′/c), is successively raised to higher and higher powers. At high degrees of strain of the chemical bonds, as in the situation in plastic deformation, such higher order terms may become important.

4.5 REVIEW OF DYNAMIC ELASTICITY THEORY

The formulation of elasticity in covalent crystalline materials, according to the relation given in Eq. (4a) and (4b), provides a generalized approach to defining and describing the elasticity of these materials on a microscopic scale using microscopic constants of the structure.

The terms of 1st-order, in the strain integration power series, taking the form of force-over-area stresses, F/A, may be considered to be analogous to the Young's moduli in bulk materials, as defined by Eq. (8). Nonetheless, the sum of such 1st-order terms, in the physically meaningful form of stresses against each principal crystallographic face of the unit cell of the material (i.e., the (100), (010) and (001) faces), constitute, in composite, a measure of the dynamic bulk modulus of a crystalline material, B. Under elastic deformation, higher order terms in the strain integration power series all vanish, as the strain parameter, x′/a (y′/b and z′/c), approaches 0.

As all crystal systems (cubic, hexagonal, tetragonal, trigonal, orthorhombic, monoclinic and triclinic) can be modeled in terms of orthogonal axes, a, b and c. Even if this means lowering the symmetry of the unit cell. It thus becomes possible to calculate elastic moduli in any crystal system undergoing elastic deformation, from the formulation provided in Eq. (4) of this Part, with some modification for different types of chemical bonds (with pair potentials which may differ from harmonic, Hooke's law potentials) and for different orientations of those chemical bonds with respect to the crystallographic axes a, b and c as needed. One can model the dynamic elastic moduli of such materials exactly in terms of the elastic chemical bond deformation forces attendant on each of the three orthogonal crystallographic planes (100), (010) and (001).

The problem of the constant of integration in this analysis, the static zero-pressure bulk modulus of a crystalline material, becomes increasingly difficult to evaluate as the bonds in a crystal structure are distorted from orthogonality [5]. The prescription of dynamic elasticity doesn't provide a guide as to how one can get at the exact expressions for this constant of integration, as it comes up in each such elastic modulus analysis. Some important guiding principles for the formulation of B_0 have been provided by the lead of investigators such as Pauling et al. [13], Feynman [5], and Cohen [1], who have proposed approximations to the static moduli of elasticity, generally based upon considerations of the force constants in the unit cell of a material and the lattice parameters [14].

As an indication of the uniformity of the problem of identifying a modulus of elasticity in arbitrary crystalline materials, one can consider the work of Pauling and Waser [14] who identify the following Badger-like relation [15] between the force constant k of the bonds in the elements in their solid states, and the respective bond distances, d, in the elements:

$$k\ a\ d^{-3} \tag{14}$$

An attempt at calculating an explicit covalent-ionic B_0 from semi-empirical considerations, based upon calculations of electronic energy density, U/V, of the corresponding covalent-ionic chemical bonding in the materials, is that given by Cohen [1]:

$$B_0 = \left(\frac{1972 - 220(I)}{\langle d \rangle^{3.5}} \right) \left(\frac{\langle N_c \rangle}{4} \right) \tag{15}$$

Here, I is the ionicity in the material, N_c is the averaged coordination number, and d is the averaged bong length, inside the unit cell. Fortunately, the Cohen formula extends its accuracy to all covalent-ionic materials studied so far with tetrahedral bond structures (the so-called diamond-like solids) and is generalized to substances with coordination numbers different from 4. So one can use the formula to accurately model the zero-pressure elasticity of the diamond lattice in space group Fd3m, and related covalent-ionic materials like hexagonal β-C_3N_4 in space group $P6_3/m$, shown in Fig. 2 [16], and tetragonal glitter in space group $P4_2/mmc$, shown earlier in Fig. 9 from Chapter 3 of Part III [1, Chapter 1].

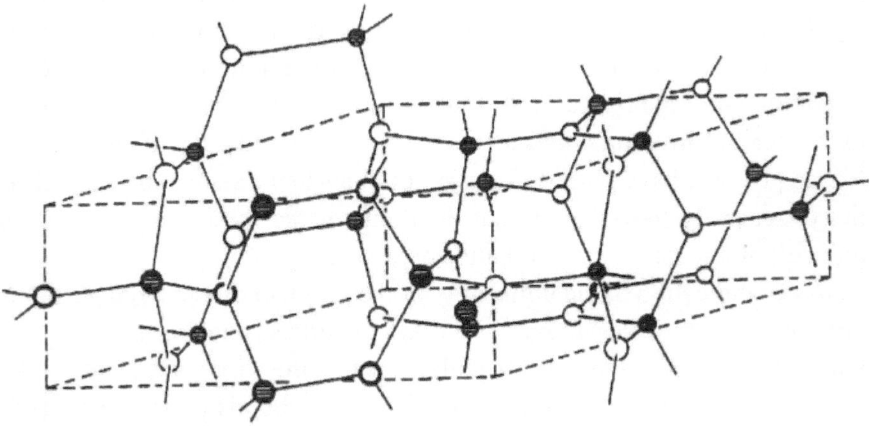

FIGURE 2 Hexagonal β-C_3N_4 Lattice, Space Group $P6_3/m$.

4.6 POTENTIAL AND REAL SUPERHARD MATERIALS

From Eq. (15) we can identify those covalent-ionic materials that lie at the zenith of hardness, according to the identification of the zero-pressure bulk modulus, B_0, as a measure of a material's hardness. Table 3 summarizes the calculations of B_0 by Eq. (15) for candidate covalent-ionic crystalline materials including, cubic diamond, hexagonal diamond in space group $P6_3/mmc$, hexagonal β-C_3N_4 [16], hexagonal hexagonite in space group P6/mmm, shown in Fig. 3 [18, Chapter 3], and tetragonal glitter shown in Fig. 9 from Chapter 3 [1, Chapter 1].

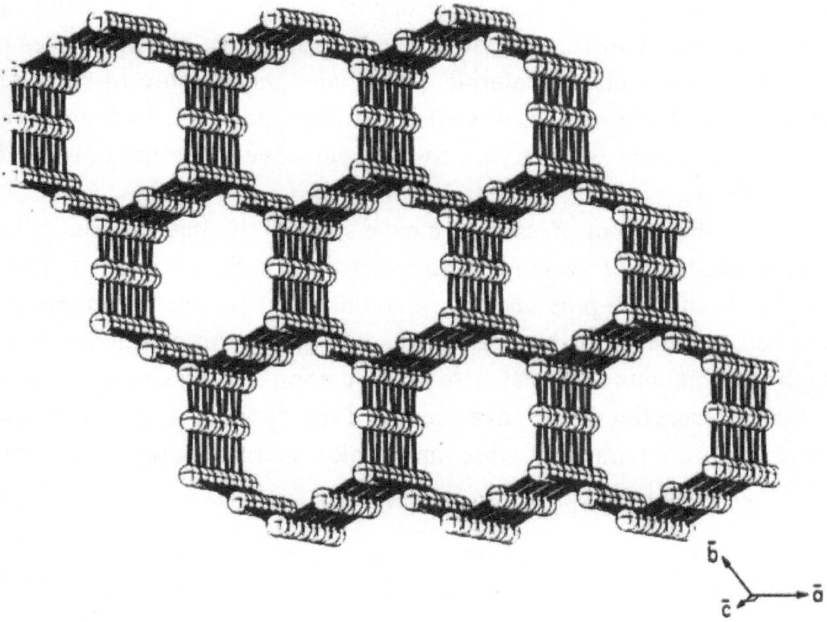

FIGURE 3 Hexagonal Hexagonite Lattice, Space Group P6/mmm.

TABLE 3 Calculated Zero-pressure Bulk Moduli of Superhard Materials According to the Cohen Prescription.

Material	B_0 in GPa	Reference
Cubic diamond	435	[57, Chapter 2]
Hexagonal diamond	435	[1]
β-C_3N_4	427	[7]
Hexagonite	427	[18, Chapter 3]
Glitter	440	[59, Chapter 2]

The constant of the strain integration, the zero-pressure bulk modulus, seems to reach a barrier at about 440 GPa, with all the calculated zero-pressure bulk moduli clustered not far from this value, with glitter being at the top ($B_0 = 440$ GPa).

Carbon allotropes, whether real, as in the case of cubic and hexagonal diamond, or hypothetical, as in the case of hexagonal hexagonite and tetragonal glitter, comprise many of the hardest possible materials. Even hexagonal β-C_3N_4 is in large degree comprised of C. Earlier on, it has been seen that carbon nanotubes have the very highest values of the Young's modulus of materials [4], Y, another measure of stiffness. Judging from these results, it is clear that one should be looking in the direction of carbon when trying to discover novel superhard materials [1, Chapter 3].

Beyond the zero of pressure, we have some preliminary results of the dynamic elasticity of some of these materials [57, 59, Chapter 2]. When factoring in the zero-pressure bulk modulus, B_0, assuming harmonic potential energy functions between atom pairs [59] and with modest bond length deformations of about 0.1 Å, on the carbon–carbon single, and double bonds respectively, we have identified the dynamic lateral and axial elastic moduli in tetragonal glitter and cubic diamond shown in Eqs. (16) and (17), respectively:

$$B_{glitter} = B_0 + \left(\frac{2k_2r_2'\cos 0°}{a^2} + \frac{4k_1r_1'\cos 57°}{\mathbf{a}^2}\right)_{axial} + \left(\frac{4k_1r_1'\cos 33°}{ac} + \frac{4k_1r_1'\cos 33°}{ab}\right)_{lateral} \tag{16}$$

$$B_{diamond} = B_0 + \left(\frac{8kr'\cos 54.73°}{bc} + \frac{8kr'\cos 54.73°}{\mathbf{ac}}\right)_{lateral} + \left(\frac{8kr'\cos 54.73°}{ab}\right)_{axial} \tag{17}$$

The results of the calculations on these lattices are shown in Table 4. From the data in Table 4, it can be seen that tetragonal glitter has a 33% higher bulk modulus at pressure, directed along its **c**-axis, B_c, at 893 GPa [59, Chapter 2], than does cubic diamond at 600 GPa [57, Chapter 2].

TABLE 4 Axial Correction Stresses and Total Bulk Moduli Parallel to the **c**-axis for Cubic Diamond and Tetragonal Glitter at a 0.1 Angstrom Chemical Bond Deformation.

Material	Axial Correction Stress	B_c, GPa
Diamond	$B_c \alpha B_0 + \left(\frac{8kr' \cos 54.73°}{ab}\right)_{axial}$	600
Glitter	$B_c \alpha B_0 + \left(\frac{2k_2 r_2' \cos 0°}{a^2} + \frac{4k_1 r_1' \cos 57°}{a^2}\right)_{axial}$	893

4.7 CONCLUSIONS

Evidently, the solution of the integral over the force density, shown earlier in Eq. (4), provides the key to solving the problem of the ultimate strength of materials, as measured from their elastic moduli. Solution of Eq. (4) shows that maximizing parameters such as the force constant of the chemical bonds in the unit of pattern of the given material, or alternatively minimizing the lattice parameters for a given material, are both directions one could go in to maximize the dynamic bulk modulus, B. The cubic diamond, hexagonal diamond, hexagonal β-C_3N_4, hexagonal hexagonite and tetragonal glitter structure-types, clearly provide some of the most obvious solutions to the problem of the ultimate hardness of materials.

Cubic diamond is a dense material ($\rho = 3.56$ g/cm^3) with strong, covalent carbon–carbonsingle bonds with an associated force constant, k, of 450 N/m [9] and a relatively small lattice parameter of 3.55 Å [22, Chapter 2]. Cubic diamond is the very hardest material known. As an alternative, one can consider tetragonal glitter ($\rho = 3.15$ g/cm^3) with its stronger, covalent carbon–carbon double bonds with their associated force constant, k, of 960 N/m [9] and smaller basal plane lattice parameters of 2.53 Å [59, Chapter 2]. Because of the geometry of the tetragonal glitter structure, the Hooke's law forces, attendant on all thecarbon–carbon bonds in glitter, will have components directed across the structure's basal plane. Such a situation leads to a maximization of the 1st-order term in the strain integration power series, shown in Eq. (9), compared to the situation in the other structures like cubic diamond.

It therefore appears that tetragonal glitter, although of a comparable stiffness to cubic diamond in the lateral bc- and ac-planes (with an approximate modulus of 540 GPa along [100] and [010] directions), will be considerably stiffer than cubic diamond across the basal plane of the lattice (i.e., along the [001] lattice direction in tetragonal glitter the modulus of elasticity is 893 GPa). When one considers the constraints associated with creating an extended 3D structure, that has nominal structural stability along all three Cartesian directions (i.e., a, b and c), with stronger, multiple covalent bonds directed across a relatively small crystallographic face (like (001) in glitter), and that is a chemically intuitive structure, it appears that tetragonal glitter is perhaps a unique structure-type with the potential for achieving the absolute zenith in dynamic bulk modulus, B, of any such conceivable structure drawn from the Periodic Table of the elements. Certainly it appears that tetragonal glitter is the hardest potential covalent material from this analysis. It is not clear whether a structure drawn from the metallic elements in the Periodic Table could ever achieve a greater dynamic bulk modulus, B, than carbon-based glitter. The metallic potential is different from the Hooke's law (or Morse's law) potential of covalent bonding. It appears to be the case, from experience, that the metallic potential has lateral contributions to it in addition to the usual axial pair potential of the Hooke's law force. The metallic bonds tend to be slippery and appear not to be good candidates for maximizing out the dynamic bulk modulus, B, as it has been worked out here from solution of Eq. (4).

If one can roughly equate the bulk modulus of a material (especially its bulk modulus of elasticity with respect to the crystallographic faces, of a well-oriented single crystal, in contact with a sample in a high pressure anvil device like the diamond anvil cell (DAC) [14, Chapter 1]) to the maximum accessible static pressure, it can attain, on a sample of material. One could then boldly state that the glitter lattice, which is evidently Nature's hardest material, seems to put a limit on the highest accessible static pressures attainable in the laboratory at somewhere between 900 to 1100 GPa (in the neighborhood of 10 Mb). The higher figure is obtained from analysis of the carbon–carbon bonds in glitter as Morse's law springs [59, Chapter 2]. The 10 Mb regime represents a doubling of the record set at near 5 Mb by Mao et al. in studies of metallic hydrogen [14, Chapter 1]. It would appear that difficult issues, such as the potential high-pressure metallization of hydrogen, could possibly be resolved by an opposed anvil device comprised of carefully oriented tetragonal glitter single crystals. This is in addition to the phenomenal amount of scientific information that could be obtained on phase transitions and structures at pressures approaching the 10 Mb limit by such a device, and the general study of physical phenomena, like electrical resistance, at such tremendous pressures approaching 10 Mb.

KEYWORDS

- **dynamic elasticity**
- **elastic modulus**
- **glitter**
- **zero-pressure**

REFERENCES

1. Cohen, M.L. *Phys. Rev. B*, **1985,** *32*, 7988.
2. Bragg, W.L. In: *The Development of X-ray Crystallography*, Dover: NY, 1970.
3. 124. (a) Young, T. In: *A Course of Lectures on Natural Philosophy and the Mechanical Arts*, Volumes 1 and 2, P. KellandPublishers: London, UK, 1845. (b) Cantor, G. "Thomas Young's Lectures at the Royal Institution," Notes and Records of the Royal Society of London, **1970,** *25*, 87–112.
4. Dujardin, E.; Ebbesen, T.W.; Krishnan, A.; Ylanilus, P.N.; Treacy, M.M.J. *Phys. Rev. B*, **1998,** *58(20),* 14013–14019.
5. Feynman, R.P.; Leighton, R.B.; Sands, M. In: *The Feynman Lectures on Physics*, 1st edition, Addison-Wesley, Reading, MA, 1964.
6. Halliday, Resnick, Walker, In *Fundamentals of Physics*, 5th edition, Wiley, NY, 1997.
7. Liu, A.Y.; Cohen, M.L.*S cience*, **1989,** *245*, 841.
8. A note about cosα, cos β and cos β. One can easily obtain these values from elementary vector analysis by forming the ratios of the type shown in Eq. (i):

$$\cos\theta = \frac{(hkl)(xyz)^T}{\|hkl\| \cdot \|xyz\|} \qquad \text{(i)}$$

   Where, in the case shown in Eq. (i), we are calculating the projection factor for the chemical bond vector with components (x, y, z) onto the hklth crystallographic plane.
9. Herzberg, G. In: *Molecular Spectra and Molecular Structures I: Spectra of Diatomic Molecules*, 2nd edition, Van Nostrand Co. Inc.: Princeton, 1950.

10. In the cubic diamond lattice, taking x′ = r′cosα, it is obvious that the length decrement along the *a* crystallographic axis is given by 4x′. This will be clarified later on in reference 2.
11. Dimensional analysis was used early on by Bridgman in scientific studies; P.W. Bridgman, *The Phys. of High Pressure,* G. Bell and Sons, 1949. A leading contemporary reference is given by, M.L. Cohen, *Solid State Communications*, **1994,** *92*(*1*), 45.
12. For example the static bulk modulus of the diamond lattice can be approximated as equal to 1/3(k/*a*), where k is the force constant for a carbon–carbon single bond and *a* is the lattice parameter of cubic diamond. It is not clear what the physical basis of this deceptively simple relation is, nor whether it can be extended and modified to fit to other materials easily.
13. Waser, J.; Pauling, L.*J. Chem. Phys.*, **1950,** *18*(*5*), 747.
14. Although Cohen's formulation of B_0 is in terms of d, the averaged bond distance in the unit cell, Badger has shown a relation between k, the force constant of the chemical bonds, and d, that suggests that Cohen's formula can be recast in terms of k instead of d. See Ref. [1, Chapter 1] for more details.
15. (a) Badger, R.M.; *J. Chem. Phys.*, **1934,** *2*, 128. (b) Badger,R.M. *J. Chem. Phys.*, **1935,** *3*, 710.
16. Zachariasen, W. H.; Bragg, W. L. *Zeit. f. Krist.,* **1930,** *72,* 518.

PART II

Developments in Computational and Experimental Chemistry

CHAPTER 5

MANY-BODY PERTURBATION THEORY TO SECOND ORDER APPLIED ON CONFINED HELIUM-LIKE ATOMS

J. GARZA, J. CARMONA-ESPHNDOLA, I. ALCALDE-SEGUNDO, and R. VARGAS

CONTENTS

5.1 INTRODUCTION

Two-electrons systems are the most studied systems in quantum mechanics due to the fact that they are the simplest systems that contain the electron–electron interaction, which is a challenge for the solution of the Schrödinger equation [1]. In particular, helium-like atoms are used many times as a reference to apply new theoretical and computational techniques. Additionally, in recent years the study of many-electron atoms confined spatially have a particular interest since the confinement induces important changes on the electronic structure of these systems [2, 3]. The confinement imposed by rigid walls has been quite popular from the Michels proposal made 76 years ago [4], followed by Sommerfeld and Welker one year later [5]. Such a model assumes that the external potential has the expression

$$v(r)=\begin{cases}-\dfrac{Z}{r} & r<R_c \\ \infty & r\geq R_c\end{cases} \tag{1}$$

where R_c and Z represent the radius of a sphere and the atomic number, respectively, with the nucleus centered in the sphere. Naturally, the external potential of Eq. (1) imposes a cancellation of the wave function on the surface of the sphere. In a sense, this model represents an extension of the one-dimensional particle in a box.

For this kind of confinement, the solution of the non-relativistic time independent Schrödinger equation has been tackled by different techniques. For confined many-electron atoms the density functional theory [6], using the Kohn-Sham model [7], has given some estimations of the non-classical effects [8–11], through the exchange-correlation functional. An elementary review of this subject can be found in Ref. [12], where numerical techniques are discussed to solve the Kohn-Sham equations. Furthermore, in this reference, some chemical predictors are analyzed as a function of the confinement radii.

By the wave function side, the Hartree-Fock (HF) method [13] has been applied in these systems to take into account the exchange contribution to the non-classical electron–electron interaction [14–18]. Very recently, the quantum Monte Carlo technique has been applied to study these systems [19]. However, it is well known that there are several techniques based on the wave function to estimate the correlation energy (CE) involved in the non-classical electron–electron interaction, but such techniques have not been implemented to obtain the electronic structure of confined many-electron atoms because they are computationally expensive [13]. In these days, with the algorithms developed in parallel computing and the new computational architectures, is desirable the implementation of such methods for the study of atoms confined spatially.

For two-electron atoms, many approaches have been applied; a review made by Aquino reported the techniques used up to 2009 [20]. To date, the expansion of the wave function in terms of Hylleraas-type functions is the technique that gives the lowest energies for several confinement radii [21–23], which can be used as reference when other techniques are proposed for the study of these systems. However, such a technique has not been used for atoms with several electrons, for example, beryllium. In this sense, in this chapter we test the many-body perturbation theory to second order, as a technique to estimate the CE for confined many-electron atoms. In the next section, we discuss the theory behind of the HF method, and the basis set proposed for its implementation for confined atoms. In the same section, the many-body perturbation theory to second order proposed by Møller and Plesset (MP2) [24] also is discussed, and we give some details about the implementation of our code implemented in GPUs [25]. Finally, we contrast our results for helium-like atoms with more sophisticated techniques in order to know the percent of correlation energy recovered by the MP2 method.

5.2 THEORETICAL APPROACH

5.2.1 THE HARTREE-FOCK METHOD

In the HF method the wave function is represented as a Slater determinant, its corresponding energy, in atomic units (a.u.), is [13]:

$$E_{HF} = \sum_{i=1}^{N} \left\langle \chi_i \middle| \hat{h} \middle| \chi_i \right\rangle + \frac{1}{2} \sum_{i,j=1}^{N} \left\langle \chi_i \chi_j \middle\| \chi_i \chi_j \right\rangle \quad (2)$$

with

$$\hat{h} = -\frac{1}{2} \nabla^2 + v(\vec{r}) \quad (3)$$

$$\begin{aligned} \left\langle \chi_i \chi_j \middle\| \chi_i \chi_j \right\rangle &= \langle ij \| ij \rangle = \langle ij | ij \rangle - \langle ij | ji \rangle, \\ &= \iint d\vec{x}_1 d\vec{x}_2 \frac{\chi_i^*(\vec{x}_1)\chi_j^*(\vec{x}_2)\chi_i(\vec{x}_1)\chi_j(\vec{x}_2)}{|\vec{r}_2 - \vec{r}_1|} - \iint d\vec{x}_1 d\vec{x}_2 \frac{\chi_i^*(\vec{x}_1)\chi_j^*(\vec{x}_2)\chi_j(\vec{x}_1)\chi_i(\vec{x}_2)}{|\vec{r}_2 - \vec{r}_1|} \end{aligned} \quad (4)$$

In Eq. (2), *N* represents the number of electrons (occupied orbitals) in the system and each orbital has the form $\chi(\vec{x}) = \psi(\vec{r})\sigma(\omega)$ with $\sigma = \alpha, \beta$. The orbitals that give the lowest energy must satisfy the HF equations,

$$\left(-\frac{1}{2} \nabla^2 + v(\vec{r}) + \hat{J}_i - \hat{K}_i \right) \chi_i = \varepsilon_i \chi_i, \qquad i = 1,..,\infty \quad (5)$$

where $\hat{J}_i$ and $\hat{K}_i$ represent the Coulomb and exchange operators, respectively.

One way to solve Eq. (5), satisfying the constrain imposed by Eq. (1), was proposed by Ludeña by writing each orbital as a linear combination of functions as,

$$\psi_i(\vec{r}) = \sum_{\mu=1}^{K} C_\mu^i \varphi_\mu(\vec{r}) \quad (6)$$

In this case, *K* represents the size of the basis set functions $\{\varphi_\mu\}$. In his proposal, Ludeña uses modified Slater Type Orbitals (MSTO), which are Slater Type Orbitals times a cutoff function, in this way the MSTOs have the form

$$\varphi_\mu(\vec{r}) = Y_{l_\mu, m_\mu}(\theta, \phi) f_\mu(r) \quad (7)$$

with $Y_{l,m}$ as the spherical harmonics and the radial part is written as

$$f_\mu(r) = N_\mu r^{n_\mu - 1} \exp(-\xi_\mu r)\left(1 - \frac{r}{R_c}\right) \quad (8)$$

By using Eq. (6) in Eq. (5) the HF equations become as an algebraic problem:

$$FC = SC\Lambda \quad (9)$$

where Λ is a diagonal matrix with ε_i as its entries, **S** is the overlap matrix with elements,

$$S_{\mu\nu} = \int d\vec{r}\varphi_\mu^*(\vec{r})\varphi_\nu(\vec{r}) \quad (10)$$

and F is the Fock matrix.

The form of the integral 10 is obtained by using Eqs. (7) and (8), such that

$$S_{\mu\nu} = \delta_{l_\mu l_\nu}\delta_{m_\mu m_\nu} N_\mu N_\nu \int_0^{R_c} dr\, r^{n_\mu + n_\nu - 2}\left(1 - \frac{r}{R_c}\right)^2 \exp\left[\left(\xi_\mu + \xi_\nu\right)r\right] \quad (11)$$

When this expression is expanded, the resulting integrals have the form

$$\gamma(t, \theta, R_c) = \int_0^{R_c} drr^t \exp(-\theta r) \quad (12)$$

which is related directly with the incomplete gamma function [26],

$$\gamma(t, \theta, R_c) = \frac{t!}{\omega^{t+1}}\left[1 - \exp(-\theta R_c)\sum_{n=0}^{t}\frac{(\theta R_c)^n}{n!}\right] \quad (13)$$

There are integrals as Eq. (12) also for the Fock matrix and for the two-electron contribution there are integrals of the form

$$radial(l, \mu, \nu, \lambda, \sigma, R_c) = I_1(2 + l, \mu, \nu, R_c)I_1(1 - l, \lambda, \sigma, R_c) + I_2(\mu, \nu, \lambda, \sigma, 1 - l, 2 + l, R_c) - I_2(\mu, \nu, \lambda, \sigma, 2 + l, 1 - l, R_c) \quad (14)$$

with

$$I_1(k, \mu, \nu, R_c) = N_\mu N_\nu \Big[\gamma(k + n_\mu + n_\nu - 2, \xi_\mu + \xi_\nu, R_c) - \frac{2}{R_c}\gamma(k + n_\mu + n_\nu - 1, \xi_\mu + \xi_\nu, R_c) + \frac{1}{R_c^2}\gamma(k + n_\mu + n_\nu, \xi_\mu + \xi_\nu, R_c)\Big] \quad (15)$$

$$I_2(\mu,\nu,\lambda,\sigma,a,b,R_c)=N_\mu N_\nu\left[I_3(\mu,\nu,\lambda,\sigma,n_\mu+n_\nu-2+a,b,R_c)-\right.$$
$$\left.-\frac{2}{R_c}I_3(\mu,\nu,\lambda,\sigma,n_\mu+n_\nu-1+a,b,R_c)+\frac{1}{R_c^2}I_3(\mu,\nu,\lambda,\sigma,n_\mu+n_\nu+a,b,R_c)\right] \quad (16)$$

$$I_3(\mu,\nu,\lambda,\sigma,a,b,R_c)=N_\lambda N_\sigma\left[I_4(\mu,\nu,\lambda,\sigma,a,b+n_\lambda+n_\sigma-2,R_c)-\right.$$
$$\left.-\frac{2}{R_c}I_4(\mu,\nu,\lambda,\sigma,a,b+n_\lambda+n_\sigma-1,R_c)+\frac{1}{R_c^2}I_4(\mu,\nu,\lambda,\sigma,a,b+n_\lambda+n_\sigma,R_c)\right] \quad (17)$$

$$I_4(\mu,\nu,\lambda,\sigma,k,m,R_c)=\frac{m!}{(\xi_\lambda+\xi_\sigma)^{m+1}}\left[\gamma(k,\xi_\mu+\xi_\nu,R_c)-\right.$$
$$\left.\sum_{n=0}^{m}\frac{(\xi_\lambda+\xi_\sigma)^n}{n!}\gamma(k+n,\xi_\mu+\xi_\nu+\xi_\lambda+\xi_\sigma,R_c)\right] \quad (18)$$

Evidently, the evaluation of two-electron integrals involves a large part of the computational effort in the HF equations solution.

5.2.2 MANY-BODY PERTURBATION THEORY TO SECOND ORDER

By using the time-independent perturbation theory and the HF method as starting point, Møller and Plesset obtained, to second order (MP2), an estimation to the correlation energy, CE_{MP2}, as

$$CE_{MP2}=\frac{1}{2}\sum_{abrs}\frac{\left|\langle ab|rs\rangle\right|^2-\langle ab|rs\rangle\langle rs|ba\rangle}{\varepsilon_a+\varepsilon_b-\varepsilon_r-\varepsilon_s} \quad (19)$$

where a and b represent occupied orbitals and r and s unoccupied ones. This expression is written in different ways depending on the multiplicity of the system.

For open-shell systems, the Eq. (19) adopts the form

$$CE_{MP2}=\frac{1}{2}\sum_{a_\alpha b_\alpha r_\alpha s_\alpha}\frac{\left|(a_\alpha r_\alpha|b_\alpha s_\alpha)\right|^2}{\varepsilon_a^\alpha+\varepsilon_b^\alpha-\varepsilon_r^\alpha-\varepsilon_s^\alpha}+\frac{1}{2}\sum_{a_\beta b_\alpha r_\beta s_\alpha}\frac{\left|(a_\beta r_\beta|b_\alpha s_\alpha)\right|^2}{\varepsilon_a^\beta+\varepsilon_b^\alpha-\varepsilon_r^\beta-\varepsilon_s^\alpha}$$

$$+\frac{1}{2}\sum_{a_\alpha b_\beta r_\alpha s_\beta}\frac{\left|\left(a_\alpha r_\alpha \middle| b_\beta s_\beta\right)\right|^2}{\varepsilon_a^\alpha+\varepsilon_b^\beta-\varepsilon_r^\alpha-\varepsilon_s^\beta}+\frac{1}{2}\sum_{a_\beta b_\beta r_\beta s_\beta}\frac{\left|\left(a_\beta r_\beta \middle| b_\beta s_\beta\right)\right|^2}{\varepsilon_a^\beta+\varepsilon_b^\beta-\varepsilon_r^\beta-\varepsilon_s^\beta}$$

$$-\frac{1}{2}\sum_{a_\alpha b_\alpha r_\alpha s_\alpha}\frac{\left(a_\alpha r_\alpha \middle| b_\alpha s_\alpha\right)\left(r_\alpha b_\alpha \middle| s_\alpha a_\alpha\right)}{\varepsilon_a^\alpha+\varepsilon_b^\alpha-\varepsilon_r^\alpha-\varepsilon_s^\alpha}-\frac{1}{2}\sum_{a_\beta b_\beta r_\beta s_\beta}\frac{\left(a_\beta r_\beta \middle| b_\beta s_\beta\right)\left(r_\beta b_\beta \middle| s_\beta a_\beta\right)}{\varepsilon_a^\beta+\varepsilon_b^\beta-\varepsilon_r^\beta-\varepsilon_s^\beta} \quad (20)$$

In this expression the notation for the integrals is

$$\left(ij \middle| kl\right)=\iint d\vec{r}_1 d\vec{r}_2\,\frac{\varphi_i^*(\vec{r}_1)\varphi_j(\vec{r}_1)\varphi_k^*(\vec{r}_2)\varphi_l^*(\vec{r}_2)}{\left|\vec{r}_2-\vec{r}_1\right|} \quad (21)$$

For closed-shell systems, the expression for Eq. (21) is drastically reduced to

$$CE_{MP2}=2\sum_{a,b=1}^{N/2}\sum_{r,s=N/2+1}^{K}\frac{\left|\left(ar \middle| bs\right)\right|^2}{\varepsilon_a+\varepsilon_b-\varepsilon_r-\varepsilon_s}-\sum_{a,b=1}^{N/2}\sum_{r,s=N/2+1}^{K}\frac{\left(ar \middle| bs\right)\left(rb \middle| sa\right)}{\varepsilon_a+\varepsilon_b-\varepsilon_r-\varepsilon_s} \quad (22)$$

5.2.3 IMPLEMENTATION ON GPUS

The evaluation of the integrals involved in the solution of the HF equations has been implemented in the MEXICA-C code [14], which is designed with C programming techniques [27] based on MPI libraries [28]. Naturally, the MEXICA-C code was designed to solve the HF equations for confined and free atoms with internal basis set functions based on those reported by Clementi-Roetti [29], Bunge [30] or Thakkar [31], although the user has the possibility to define his own basis set functions. The main reason to design a code in C language is that it can be translated relatively easy to CUDA [25]. In this way, we have implemented the evaluation of the two-electron integrals on GPUs by using CUDA-C. For this purpose, we have built a kernel in CUDA to transform four indexes expressions, like the two-electron integrals, to two indexes. In this way, the threads used in a GPU are used as a mesh in two dimensions. Additionally, we have computed the CE_{MP2} also on GPUs by using almost the same idea. The development and test of this new code, with the name MEXICA-CUDA, was performed in a NVIDIA Quadro 5000 GPU, which is not the most robust GPU at these days, but it is enough to accelerate the computation of the electronic structure of confined atoms.

In this chapter, we show the MP2 performance to estimate the CE in confined helium-like atoms, like H^-, He and Li^+. The main reason to use these systems stems in the fact that sophisticated methods have been employed to estimate the CE and therefore, we are able to know how much of the CE is taken into account in the MP2 method.

5.3 MP2 METHOD APPLIED ON CONFINED HELIUM-LIKE ATOMS

5.3.1 DESIGN OF THE BASIS SET FUNCTIONS FOR NON-CONFINED ATOMS

The definition of the basis set functions is a significant step to perform calculations with the MEXICA-CUDA code. According to Eqs. (20) and (21) the number of unoccupied orbitals is important in the CE estimation, for that reason we have used for H^-, He and Li^+ atoms a basis set functions composed by 20 MSTOs, these are: 3(1s) 2s3s4s5s6s2p3p4p5p6p-3d4d5d6d4f5f6f. Thus, this is a triple zeta basis set. Evidently, each atom has its own exponents set, which are reported in Table 1.

TABLE 1 Exponents obtained for H^-, He and Li^+ atoms.

Basis set functions	Atom		
	H–	He	Li+
1s	0.5639	11.90	11.12
1s	1.4500	2.32	4.50
1s	1.5550	1.41	2.41
2s	9.5150	8.03	4.85
3s	15.380	3.25	5.90
4s	21.1500	10.90	11.40

TABLE 1 *(Continued)*

Basis set functions	Atom		
	H–	He	Li+
5s	0.6260	10.00	12.00
6s	1.3000	9.70	13.00
2p	0.9950	2.33	3.48
3p	1.1590	5.82	16.35
4p	1.2490	12.76	7.99
5p	9.5500	9.09	12.21
6p	11.8500	10.77	13.35
3d	1.2920	2.97	4.91
4d	2.9100	5.82	12.89
5d	4.8200	14.33	10.50
6d	6.9300	9.97	13.50
4f	1.8950	4.31	7.50
5f	1.6500	7.90	10.41
6f	5.0800	11.89	12.20

These exponents were obtained for non-confined atoms and clearly they exhibit different values depending on the considered atom. In fact, for the H^- atom the exponents are sensitive to the four digits, whereas that for He and Li^+ two digits are enough. The corresponding total energies (TE) for each atom are reported in Table 2 and they are compared with results obtained with the NWChem suite code v6.1 [32, 33] by using the aug-cc-pVTZ and aug-cc-pVQZ basis sets [34, 35]. The number of functions used with the aug-cc-pVTZ and aug-cc-pVQZ basis sets is reported in parenthesis in the same table. Additionally to TE, the HF energy and CE_{MP2} are also reported for the three basis set functions.

From Table 2, it is clear that the MSTOs give deeper energies than aug-cc-pVTZ and aug-cc-pVQZ basis set functions. In particular, for the Li^+ atom the CE estimated by the augmented correlated-consistent basis set functions is far away from the estimation by the MSTOs, which predict the biggest CE for this atom and the lowest one for the H^- atom. Evidently, the MSTOs exhibit a better performance than the aug-cc-pVQZ basis set functions, which is expected because the STOs are suitable for atoms. For a fair comparison, it is necessary for a bigger number of gaussian functions, unfortunately the aug-cc-pV5Z is not reported for the Li atom. Furthermore, the augmented correlated-consistent basis set functions were optimized for neutral atoms and the MSTOs were optimized for the ions.

TABLE 2 Hartree-Fock (HF) energy, correlation energy (CE) and total energy (TE) for non-confined H^-, He and Li^+ atoms estimated by the modified Slater type orbitals (MSTO) andw the correlated consistent basis set functions aug-cc-pVTZ and aug-cc-pVQZ. All energies are reported in Hartrees.

Energy	Atom		
	H^-	He	Li^+
MSTO			
HF	–0.487925	–2.861680	–7.236415
CE	–0.029531	–0.036428	–0.039271
TE	–0.517456	–2.898108	–7.275685
	aug-cc-pVTZ		
HF	–0.487646	–2.861223	–7.236383
CE	–0.028322	–0.033815	–0.019128
TE	–0.515968 (25)	–2.895037 (25)	–7.255511 (55)
	aug-cc-pVQZ		
HF	–0.487827	–2.861539	–7.236386
CE	–0.029326	–0.035940	–0.027082
TE	–0.517153 (55)	–2.897480 (55)	–7.263469 (105)

5.3.2 CONFINED HELIUM-LIKE ATOMS

The MSTOs basis set defined for H^-, He and Li^+ free atoms is applied for the case when these atoms are confined spatially imposing the external potential defined in Eq. (1). The HF energies, CE and TE for the He atom are reported in Table 3. In the same table are reported some TE reported previously obtained by Hylleraas-type functions.

The exponents optimization mentioned in the last section was performed for the free atom and such a basis set was used for the confined one. From Table 3, the difference obtained between the TE obtained by the MSTOs and those with highly accurate techniques [22. 23] is preserved for large R_c values and important discrepancies are observed for other confinements. However, for the biggest discrepancy, $R_c = 3.5$ a.u., the difference represents just the 0.3%. Evidently, this difference could be reduced if the exponents optimization is carried out on each confinement radius.

TABLE 3 Hartree-Fock (HF) energy, correlation energy (CE) and total energy (TE), for the confined He atom estimated by using modified Slater type orbitals. All energies are in Hartrees and the confinement radius, R_c, in atomic units.

Rc	HF	HF[a]	CE	CE[b]	TE	TE[c]	TE[d]
2.0	−2.5619	−2.5626	−0.0366	−0.0401	−2.5985	−2.6040	−2.6040
2.5	−2.7647		−0.0361		−2.8008	−2.8078	
3.0	−2.8279	−2.8311	−0.0360	−0.0402	−2.8639	−2.8725	−2.8725
3.5	−2.8484	−2.8519	−0.0361		−2.8845	−2.8936	
4.0	−2.8557	−2.8586	−0.0362	−0.0407	−2.8919	−2.9004	−2.9005
5.0	−2.8599	−2.8614	−0.0363	−0.0408	−2.8962	−2.9034	−2.9034
6.0	−2.8609	−2.8617	−0.0364	−0.0409	−2.8973	−2.9037	−2.9037
8.0	−2.8614	−2.8617	−0.0364	−0.0409	−2.8978	−2.9037	−2.9037
10.0	−2.8616	−2.8617	−0.0364		−2.8980	−2.9037	−2.9037
∞	−2.8617		−0.0364		−2.8981	−2.9037	−2.9037

[a]Ref. [15]. [b]Ref. [36]. [c]Ref. [22]. [d]Ref. [23

In the exponents optimization process, the total MP2 energy was considered and consequently the HF energies are not necessarily the lowest ones. In previous report Garza and Vargas have optimized the MSTOs exponents for the HF method [15]. In Table 3 these results are included, where clearly it is shown that an exponents optimization for the MP2 method does not include an exponents optimization for the HF method. In fact, when the exponents are optimized by using the HF method the energy changes are not observed for $R_c \geq 6$ a.u. However, when the MP2 method is used, the HF energy present important energy changes for large confinement radii.

One observation made by Ludeña for the confined He atom is that the CE, obtained with the configuration interaction method, is almost constant when the helium atom is confined. From Table 3, it is clear that with the MP2 method this result is preserved. Comparing the CE_{MP2} with the CE obtained by Ludeña, it is observed that the MP2 method recover at least the 90% of the correlation energy for any confinement radii.

The corresponding results for the confined H^- atom are reported in Table 4.

TABLE 4 Hartree-Fock (HF) energy, correlation energy (CE) and total energy (TE), for the confined H^- atom estimated by using modified Slater type orbitals. All energies are in Hartrees and the confinement radius, R_c, in atomic units.

R_c	HF	CE	TE	TE[a]	TE[b]	ΔE[c]	ΔE[d]
2.0	0.7665	–0.0368	0.7297	0.7240	0.7231	0.8	0.9
2.5	0.1800	–0.0340	0.1460	0.1394	0.1388	4.7	5.2
3.0	–0.1041	–0.0323	–0.1364	–0.1431	–0.1435	4.7	4.9
3.5	–0.2556	–0.0311	–0.2867	–0.2934	–0.2938	2.3	2.4
4.0	–0.3421	–0.0301	–0.3722	–0.3790	–0.3794	1.8	1.9
5.0	–0.4258	–0.0290	–0.4548	–0.4620	–0.4623	1.6	1.6
6.0	–0.4595	–0.0286	–0.4881	–0.4958	–0.4958	1.6	1.6
8.0	–0.4811	–0.0287	–0.5098	–0.5186	–0.5180	1.7	1.6
10.0	–0.4831	–0.0287	–0.5118	–0.5247	–0.5239	2.5	2.3
∞	–0.4879	–0.0295	–0.5175	–0.5277	–0.5278	1.9	2.0

[a]Ref. [22]. [b]Ref. [37]. [c]Relative error between TE and TE[a]. [d]Relative error between TE and TE[b].

For this atom the CE presents important differences between the confinements imposed by R_c; clearly, this atom exhibits a challenge for any method of electronic structure since the correlation energy represents around the 5% of the total energy. Consequently, the MP2 results differ appreciably from other sophisticated techniques. The relative error, in percent, between the MP2 results and the accurate methods, are reported in Table 4, where it is evident the large difference between the MP2 results and other techniques. Such difference may be associated to the exponents optimization or to the incompletes of the basis set functions, in particular for large R_c values. It is interesting that the smallest difference is obtained for $R_c = 2.0$ a.u., suggesting that for high confinements the MP2 method can recover an important part of the CE.

The HF energy, CE and TE for the confined Li^+ atom are reported in Table 5. For this confined atom the CE obtained for any confinement radius is almost constant and bigger than those results obtained for He and H^- atoms. In a sense the nuclear charge gives a contribution to the spatial confinement since the variation of the TE is not observed from R_c values smaller than those observed for He or H^-. Consequently, the MP2 total energy differs slightly from the method based on the generalized Hylleraas functions, as it is corroborated with the relative error, in percent, of the MP2 method with respect to the generalized function method reported in Table 5.

TABLE 5 Hartree-Fock (HF) energy, correlation energy (CE) and total energy(TE), for the confined Li^+ atom estimated by using modified Slater type orbitals. All energies are in Hartrees and the confinement radius, R_c, in atomic units.

R_c	HF	CE	TE	TE[a]	ΔE[b]
2.0	–7.1849	–0.0390	–7.2239	–7.2383	0.2
2.5	–7.2245	–0.0391	–7.2636	–7.2740	0.1
3.0	–7.2327	–0.0392	–7.2719	–7.2791	0.1
3.5	–7.2348	–0.0392	–7.2740	–7.2798	0.1
4.0	–7.2356	–0.0392	–7.2748	–7.2799	0.1
5.0	–7.2361	–0.0393	–7.2754	–7.2799	0.1
6.0	–7.2363	–0.0393	–7.2756	–7.2799	0.1
8.0	–7.2364	–0.0393	–7.2757	–7.2799	0.1
10.0	–7.2364	–0.0393	–7.2757	–7.2799	0.1
∞	–7.2364	–0.0393	–7.2757	–7.2799	0.1

[a]Ref. [22]. [b]Relative error between TE and TE[a].

5.4 PERSPECTIVES

From the results discussed in the previous section it is clear that for atoms confined by rigid walls an exponents optimization, for each confinement radius, is necessary when the MP2 is used. In particular, for anions where the external electrons exhibit low binding energies. In this exponents optimization process, the basis set functions for the HF method must be optimized first, and then the corresponding basis set for the MP2 method must be optimized. Evidently, for atoms with several electrons the CE will show a rich structure since the orbitals crossing is involved for strong confinements [9, 12, 18], and consequently the computational cost will be important since for each R_c the exponents optimization is required.

KEYWORDS

- **Hartree-Fock**
- **Hylleraas-type function**
- **Kohn-Sham**
- **triple zeta**

REFERENCES

1. Bethe, H. A.; Salpeter, E. E. In *Quantum Mechanics of One- and Two-Electron Atoms*; Dover Publications, Inc.: Mineola, N. Y., 2008.
2. Jaskolski, W. *Phys. Rep.,* **1996**, *271*, 1–66.
3. Sabin, J. R.; Brandas, E. In *Advances in Quantum Chemistry*; Elsevier, Inc., 2009; Vol. 57.
4. Michels, A.; De Boer, J.; Bijl, A. *Physica,* **1937**, *4*, 14.
5. Sommerfeld, A.; Welker, H. *Ann. Der Physik.,* **1938**, *32*, 56–65.
6. Parr, R. G.; Yang, W. In *Density-Functional Theory of Atoms and Molecules*; Oxford University Press, Inc.: New York, 1989.
7. Kohn, W.; Sham, L. J. *Phys. Rev.,* **1965**, *140*, 1133–1138.
8. Garza, J.; Vargas, R.; Aquino, N.; Sen, K. D. *J. Chem. Sci.,* **2005**, *117*, 379–386.
9. Garza, J.; Vargas, R.; Vela, A. *Phys. Rev., E* **1998**, *58*, 3949–3954.
10. Garza, J.; Vargas, R.; Vela, A.; Sen, K. D. *J. Mol. Str.-Theochem.,* **2000**, *501*, 183–188.
11. Sen, K. D.; Garza, J.; Vargas, R.; Vela, A. *Chem. Phys. Lett.,* **2000**, *325*, 29–32.

12. Garza, J.; Vargas, R.; Sen, K. D. In *Chemical Reactivity Theory: A Density Functional View*; Chattaraj, P. K., Ed.; Taylor and Francis Group, LLC: Boca Raton, Florida, 2009.
13. Szabo, A.; Ostlund, N. S. *Modern Quantum Chemistry: Introduction to Advanced Electronic Structure Theory*; McGraw-Hill, Inc.: New York, 1989.
14. Garza, J.; Hernandez-Perez, J. M.; Ramirez, J. Z.; Vargas, R. *J. Phys. B-At. Mol. Opt. Phys.*, **2012**, *45*, 6.
15. Garza, J.; Vargas, R. In *Advances in Quantum Chemistry, Vol. 57*; Elsevier Academic Press Inc: San Diego, 2009; 241–254.
16. Ludeña, E. V. *J. Chem. Phys.*, **1977**, *66*, 468–470.
17. Ludeña, E. V. *J. Chem. Phys.*, **1978**, *69*, 1770–1775.
18. Connerade, J. P.; Dolmatov, V. K.; Lakshmi, P. A. *J. Phys. B-At. Mol. Opt. Phys.*, **2000**, *33*, 251–264.
19. Sarsa, A.; Le Sech, C. *J. Chem. Theor. Comp.*, **2011**, *7*, 2786–2794.
20. Aquino, N. In *Advances in Quantum Chemistry, Vol. 57*; Sabin, J. R., Brandas, E., Cruz, S. A., Eds.; Elsevier Academic Press Inc: San Diego, 2009; 123–171.
21. Aquino, N.; Garza, J.; Flores-Riveros, A.; Rivas-Silva, J. F.; Sen, K. D. *J. Chem. Phys.*, **2006**, *124*, 8.
22. Flores-Riveros, A.; Rodriguez-Contreras, A. *Physics Lett. A.*, **2008**, *372*, 6175–6182.
23. Laughlin, C.; Chu, S. I. *J. Phys. A-Math. Theor.*, **2009**, *42*, 265004(1)–265004(11).
24. Moller, C.; Plesset, M. S. *Phys. Rev.*, **1934**, *46*, 618–622.
25. Sanders, J.; Kandrot, E. In *CUDA by Example*; Addison-Wesley: Boston, 2011.
26. Arfken, G. In *Mathematical Methods for Physicists. Third Edition*; Academic Press, Inc.: San Diego, 1985.
27. Oualline, S. In *Practical C Programming, Third Edition*; O'Reilly Media: California, 1997.
28. MPI Forum. *MPI: A Message-Passing Interface Standard*, University of Tennessee, 1994.
29. Clementi, E.; Roetti, C. *Atomic Data and Nuclear Data Tables* **1974**, *14*, 301.
30. Bunge, C. F.; Barrientos, J. A.; Bunge, A. V.; Cogordan, J. A. *Phys. Rev. A.*, **1992**, *46*, 3691–3696.
31. Koga, T.; Tatewaki, H.; Thakkar, A. J. *Theor. Chimi. Acta.*, **1994**, *88*, 273–283.
32. Valiev, M.; Bylaska, E. J.; Govind, N.; Kowalski, K.; Straatsma, T. P.; Van Dam, H. J. J.; Wang, D.; Nieplocha, J.; Apra, E.; Windus, T. L.; de Jong, W. *Computer Physics Communications*, *181*, 1477–1489.
33. van Dam, H. J. J.; de Jong, W. A.; Bylaska, E.; Govind, N.; Kowalski, K.; Straatsma, T. P.; Valiev, M. *Wiley Interdisciplinary Reviews-Computational Molecular Science*, *1*, 888–894.
34. Dunning, T. H. *J. Chem. Phys.*, **1989**, *90*, 1007–1023.
35. Woon, D. E.; Dunning, T. H. *J. Chem. Phys.*, **1994**, *100*, 2975–2988.
36. Ludeña, E. V.; Gregori, M. *J. Chem. Phys.*, **1979**, *71*, 2235–2240.
37. Joslin, C.; Goldman, S. *J. Phys. B-At. Mol. Opt.*, **1992**, *25*, 1965–1975.

CHAPTER 6

DISTANCE DEPENDENCE OF MAGNETIC FIELD EFFECT INSIDE THE CONFINED HETEROGENEOUS ENVIRONMENT: A CASE STUDY WITH ACRIDINE AND N, N-DIMETHYL ANILINE INSIDE AOT REVERSE MICELLES

M. K. SARANGI and S. BASU

CONTENTS

6.1 INTRODUCTION

In this article, we emphasize on the distance dependence of magnetic field effect (MFE) on donor-acceptor (D-A) pair inside confined environment of AOT/H_2O/n-heptane reverse micellar (RMs) system. For this study N, N-dimethyl aniline (DMA) is used as an electron donor while the protonated form of Acr is treated as an electron acceptor. We report the occurrence of an associated excited state proton transfer with the photoinduced electron transfer between Acr and DMA forming corresponding radical pair (RP) and radical ion pairs (RIP). The fate of these reaction products has been tested in the presence of an external magnetic field (~0.08T) by varying the size of the RMs. The MFE between Acr and DMA has been compared with the results of the interactions between Acr and TEA. We accentuate the importance of the localization of D and A inside the RMs, and the intervening distance between the pair to be the critical component for observing substantial MFE.

Magnetic effect on photo-induced electron transfer (PET) has been studied by many scientists around the globe over more than three decades and is still going on. The importance of MFE by radical pair mechanism lies on the fact that this method gives plenty of information regarding the mechanistic pathway of a reaction that proceeds through radical formation. Magnetic fields can alter the rates and yields of chemical reactions that proceed via spin-correlated radical pair intermediates and so provide information on the structures, dynamics and kinetics, reactivity of free radicals, orientation and dielectrics of the environment [1–3]. Radical pair mechanism requires the following sequence of events: first, creation of a pair of radicals, with correlated electron spins, in a pure singlet (S) or pure triplet (T) state; second, coherent evolution of the radical pair between the near-degenerate S and T spin states; and third, reaction of the S and T radical pairs to form different products (or the same product at different rates). It is most common that the electron transfer, which involves formation of geminate spin-correlated radical ion pairs (RIPs) as primary transient intermediates containing free electrons, is susceptible to a magnetic field (MF), either internal or external. It is during the second of these steps that the MF acts, via the electron Zeeman interaction, altering the extent and frequency of S↔T inter-conversion and hence the relative yields of reaction products and/or the lifetime of the radical pair. This process can be influenced by an applied MF via the electron Zeeman interaction. MFE is basically an interplay between spin dynamics and diffusion dynamics, rather than thermodynamic, and may be detected for magnetic fields whose Zeeman energies are much smaller than the average thermal energy per molecule, k_BT. By diffusion, the RIPs can separate to an optimum distance where the exchange interaction $J \sim 0$. In this situation, the electron-nuclear hyperfine coupling induces efficient mixing between the triplet ($T_{\pm}$, T_0) and the S states. The application of an external MF removes the degeneracy of the triplet states and reduces intersystem crossing thus resulting in an increase in the population of the initial spin state. So the MFE is very sensitive to the distance between the participating radical ions because the hyperfine induced spin flipping depends on J, which in turn has exponential distance dependence. When the RIP is in contact, the S↔T splitting caused by J is much stronger than the hyperfine coupling energies so that spin evolution cannot occur by this mechanism. On the other hand, at a distance where J is sufficiently small and S↔T conversion can occur, the separation of the two radicals may be already too far for geminate recombination to occur. Therefore the requirement of an optimum separation such that both spin flipping and recombination are feasible becomes a very crucial factor in controlling the MFE. This distance dependence of MFE has been demonstrated earlier in a detailed and quantitative manner by several workers using covalently linked acceptor donor systems [4, 5]. Enormous works have been done to study MFE in different organized assemblies in the last two decades to replace the linked D-A systems. Organized assemblies like micelles, reverse micelles and vesicles are of particular of interest. For proper utilization of MFE, the photo-generated ions should be prevented from subsequent rapid recombination, a prevalent event in homogeneous media. Organized assemblies such as micelles, RMs, and vesicles have the potential to prolong the lifetime of charge-transfer states and thus increase the efficiency of charge separation by partitioning of the reactants and/or products. These micro-heterogeneous systems provide a fundamental understanding of how PET dynamics is influenced by restricted system geometry. Moreover, understanding PET in simple organized assemblies can lead to a better understanding of similar processes in biological systems. In this chapter, we emphasize the role of distance on the MFE inside the confined environment of AOT/H_2O/*n*-heptane RMs system in acridine-DMA pair. Choosing the confined environment of RMs has many advantages. RMs are nano-shaped self-assemblies, formed by mixture of water, surfactant and non-polar solvents combined at the appropriate concentration. The unique microenvironment of RMs has been used as model systems for studying interactions in confinement as they mimic the immediate local environment of various bioaggregates. RMs has been widely used in chemical catalysis, enzymatic reactions, drug delivery, nano-cluster synthesis, as they create large interface between hydrophilic and hydrophobic region allowing both polar and non-polar reactants to be brought together. The perturbing effects of confinement, interfaces, counter ions, solutes etc. make the interior of the RMs extremely inhomogeneous [6, 7]. Both experimental and theoretical studies on RMs suggest that the water at the interface largely differs from those of the core and has been assigned to the differential H-bonding behavior of water in two different regions of RMs. The advantage of RMs for MFE studies is that their size can be varied just by changing w_0, ($w_0 = [H_2O]/[AOT]$, the diameter is $d = 0.29w_0 + 1.1$ nm). Therefore, RMs can serve as a very convenient medium to highlight the importance of the optimum separation distance of the RPs/RIPs of interacting molecules to maximize MFE. By different steady-state and time-resolved spectroscopic technique we have also established the role of localization and partitioning of the reacting molecules inside the hydrophilic and hydrophobic environment of the heterogeneous RMs systems.

6.2 EXPERIMENTAL SECTION

6.2.1 REAGENTS AND SAMPLE PREPARATION

Acr, AOT and *n*-heptane are purchased from Sigma–Aldrich and used without further purification (Fig. 1). DMA is obtained from Cisco Research Laboratory and used after proper distillation. Water is triply distilled before use. Solutions are made by dissolving measured amount of solid AOT in *n*-heptane by volumetric method. Subsequently triple distilled water is added to the solution to obtain appropriate value of w_0, where w_0 denotes the ratio of molar concentration of water to AOT. The w_0 values are kept as

2.5, 5, 10, 15, 20, 30 and 40. The concentration of AOT in all RMs is maintained at 0.1 M. The mixtures are then thoroughly shaken and sonicated for 5 min to obtain a transparent, homogeneous and thermodynamically stable solution. The concentration of Acr and DMA/TEA are varied between 10^{-5} and 10^{-4} M and 10^{-4} and 10^{-3} M, respectively. All our measurements are performed at room temperature. All the experiments are repeated multiple times and contain an experimental error of ±10%.

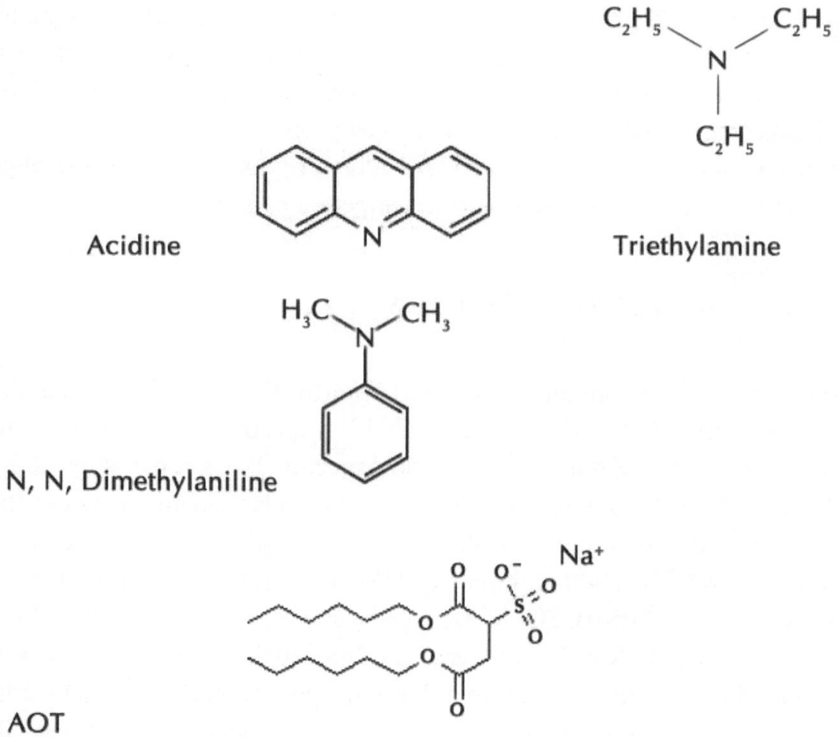

FIGURE 1 Compounds used for this study.

6.2.2 *ABSORPTION AND FLUORESCENCE MEASUREMENTS*

Steady-state absorption and fluorescence spectra of MTC in different media are recorded by using JASCO V-650 absorption spectrophotometer and Spex Fluoromax-3 Spectro-fluorimeter, respectively, using a 1.0 cm path length quartz cuvette. For a given sample, the wavelength at which absorbance is maximum (λ_{max}) is used as excitation wavelength for corresponding emission scan. Time-resolved emission spectra of MTC are performed using a picosecond pulsed diode laser based TCSPC fluorescence spectrometer with $\lambda_{ex} \sim 340$ nm and MCP-PMT as a detector. The emission from the samples is collected at a right angle to the direction of the excitation beam maintaining magic angle polarization (54.7°) with a bandpass of 2 nm. The full width at half maximum (FWHM) of the instrument response function was 270 ps and the resolution is 28 ps per channel. The data are fitted to multi-exponential functions after deconvolution of the instrument response function by an iterative reconvolution technique using IBH DAS 6.2 data analysis software in which reduced χ^2 and weighted residuals serve as parameters for goodness of fit. All the steady-state and time-resolved measurements are performed at room temperature (298°K).

6.2.3 LASER FLASH PHOTOLYSIS

Nanosecond flash photolysis set-up (Applied Photophysics) containing Nd:YAG laser (Lab series, Model Lab 150, Spectra Physics) is used for the measurement of transient absorption spectra. The sample is excited at 355 nm (FWHM = 8 ns) laser light using a 1 cm path length quartz cuvette. Transients are monitored through absorption of light from a pulsed xenon lamp (250 W). The photomultiplier (1P28) output is fed into a Tektronix oscilloscope (TDS 3054B, 500 MHz, 5 Gs/s) and the data are transferred to a computer using TEKVISA software. The MFE (0.08T) on transient spectra is studied by passing direct current through a pair of electro-magnetic coils placed inside the sample chamber. The software ORIGIN 7.5 is used for curve fitting. The solid curves are obtained by connecting the points using B-spline option. The samples are de-aerated by passing pure Argon gas for 20 min prior to each experiment. No degradation of the samples is observed during the experiment.

6.3 RESULTS AND DISCUSSIONS

Acr and its derivatives are the members of the polynuclear N-heteroaromatic family, and have a number of significant pharmaceutical uses [8]. Spectroscopic measurements of excited state *pKa* of Acr indicate that, the singlet and triplet states of Acr are stronger bases than the ground state due to large increase in the electron density over the N-atom ($pKa = 5.4, pKa^* = 9.2$) [9, 10]. Hence at suitable pH, it abstracts proton from the surrounding solvent and gets protonated ($AcrH^+$). Both Acr and $AcrH^+$ remain in dynamic protonation equilibrium (1) with each other, and show distinct spectroscopic behavior,

$$Acr^* + H_3O^+ \leftrightarrow AcrH^{+*} + H_2O \quad (1)$$

However, the photophysics of the tautomeric forms and the protonation equilibrium (1) alters with changes in the local properties like change in pH, dielectric, temperature, etc. The sensitivity of Acr to such micro environmental conditions makes it an effective probe for studying immediate local properties of its surrounding. The fluorescence of different Acr derivatives is strongly dependent on the medium, as it is extremely weak in low dielectric solvent like hydrocarbon and become moderately strong in protic solvents. In protic medium, fluorescence quantum yield of $AcrH^{+*}$ is comparatively higher than the deprotonated forms, as the radiative decay pathway in the former is predominant over the non-radiative decay and the reverse is true for the rest. Moreover intersystem crossing quantum yield for Acr* (0.39 in alkaline water) and Acr*···H–O (0.42 in ethanol) is very high, but negligible for $AcrH^{+*}$ [11, 12]. This is also reflected in their singlet excited state lifetimes, with the protonated form having a very high lifetimes (~30 ns) compared to its deprotonated counterpart (few ns for Acr*, even less than 1 ns for Acr···H–O). The photo-physical behavior of Acr in wide variety of homogeneous solvents is well documented in the literature, however in complex heterogeneous self-assembled structure is very rare. Herein we attempt to unravel some of the important photo-induced interactions with amines inside the inhomogeneous microenvironment and their susceptibility to an external magnetic field.

Figure 2 depicts the absorption spectra (inset) and optical density corrected normalized fluorescence spectra of Acr in AOT/H_2O/*n*-heptane RMs with $w_0 = 5.0$. The typical absorption spectrum of Acr shows maximum absorption around 356 nm for π–π* transition. The transition at 375–430 nm depends on the polarity of solvent. Although absorption maxima of all the different isomeric forms of Acr appear at 356 nm, the spectrum of $AcrH^+$ is characterized by a broad shoulder within 370–430 nm range. Previous studies on Acr suggest that the broad shoulder is the characteristic of the protonated form but not for that of aggregated Acr [9]. Similarly the fluorescence maximum of Acr inside the RMs corresponds to the protonated species of Acr. Buchviser et al. by steady-state and time-resolved fluorescence reported the formation of $AcrH^+$ inside heterogeneous environment of both SDS micelles and reverse micelles [13]. Moreover the protonation of Acr is facilitated with augment in the size nano-sphere. With increase in the waterpool inside the RMs, Acr from the nonpolar region is partitioned to the aqueous region and is protonated at the Stern layer of the RMs shifting the protonation equilibrium towards right. We perform triplet–triplet (T–T) transient absorption study of Acr in presence of two different amines inside such complex heterogeneous environment with simultaneous applications of an external MF of 0.08 T.

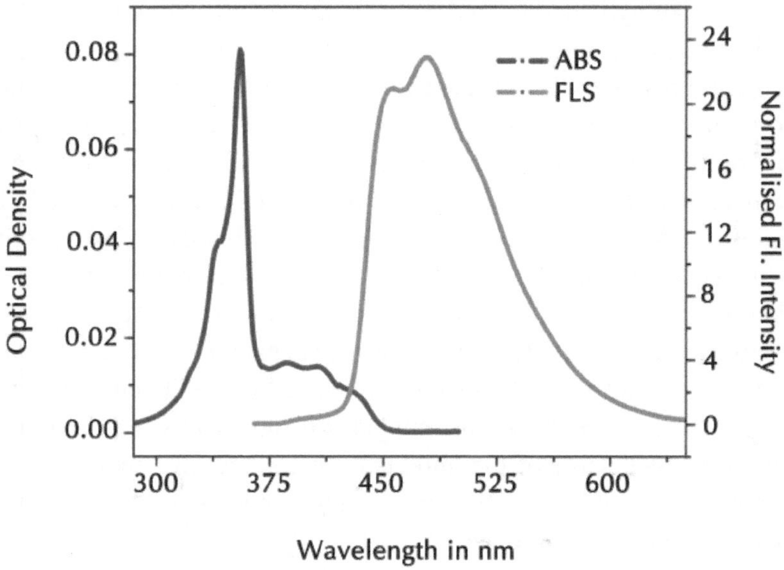

FIGURE 2 Steady-state absorption and normalized fluorescence spectra ($\lambda_{ex} = 356$ nm) of Acr (10 μM) in 0.1M AOT/H_2O/n-heptane RMs with w_0 =7.5.

6.3.1 T–T TRANSIENT ABSORPTION OF ACR INSIDE RMS

Figure 3 shows a typical T–T transient absorption spectra of Acr in confined AOT/water/*n*-heptane RMs of w_0 = 8 and the inset shows the corresponding normalized decay profiles at 480 nm (black line) and 440 nm (red line), respectively. The spectrum depicts two distinct absorption peaks, one at 440 nm and the other around 480 nm. The peaks at 440 nm and 480 nm correspond to the T–T absorption transients of ^{3}Acr and ^{3}AcrH$^+$ as reported in the literature [14, 15]. Similar to the results obtained in the steady state, the nature of T–T absorption spectra and lifetimes of ^{3}Acr and ^{3}AcrH$^+$ at 440 and 480 nm varied drastically with changes in immediate vicinity of its local environment. With corresponding increase in the sizes of the RMs, the peak at 480 nm increases with simultaneous decrease in the peak at 440 nm, similarly with decrease in the size of the RMs nanopool 440 nm peaks becomes prominent with substantial drop in the 480 nm. These results are in good agreement with the steady-state and time-resolved fluorescence data of Acr, which infer the protonation of Acr at the interfaces of such micro-heterogeneous nanoassemblies. We intend to study the behavior of these two differently localized species of Acr in presence of DMA and TEA, which are preferentially found in the hydrophobic and hydrophilic environment of the RMs.

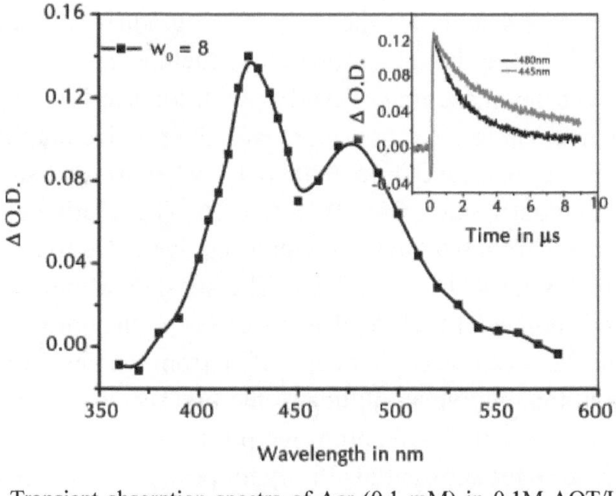

FIGURE 3 Transient absorption spectra of Acr (0.1 mM) in 0.1M AOT/H_2O/n-heptane RMs with w_0 = 7.5 at 1 μs after the laser flash at 355 nm. The inset shows the lifetime traces of ^{3}Acr and ^{3}AcrH$^+$ at 440 and 480 nm, respectively.

6.3.2 PHOTO-PHYSICS OF ACR IN PRESENCE OF AMINES

Due to their differential solubility in water, the two amines are separately partitioned inside the RMs. Being soluble in water, TEA is found in the core of the RMs, and while DMA being sparingly soluble in water is more likely to be in the Stern layer of the RMs interface. However with variation in the sizes of the RMs and with excess amount of the quenchers, the localization of the two amines is expected to be increased in both the extreme regions of the RMs. Due to their different localization and partitioning, the amines will interact with the two prototropic isomers of Acr differently. Being at the interfaces of the micellar head groups, $AcrH^+$ is likely to interact with both DMA and TEA, while Acr at the core will have a better access to interact with TEA. However such photo-induced interactions between Acr will be severely affected by the variations in the sizes of the nanopool.

Figure 4a and 4b represent the time-resolved fluorescence decay profiles of Acr excited at 377 nm in absence and presence of varied concentrations of DMA and TEA, respectively, in RMs with $w_0 = 7.5$. The decay curves fit to multi exponential functions, with more than 95% contribution coming from a long lifetime species, which infers that the probe molecule, Acr, produces number of transients in the excited state in such inhomogeneous RMs media. The longest lifetime is 27 ns due to $AcrH^{+*}$ and the shorter lifetime of 4 ns is for the Acr*. With gradual addition of DMA from 0 to 3.75 mM, although an overall quenching is observed for both the species, yet the quenching of $AcrH^{+*}$ with longest lifetime (27 ns) is faster in comparison to the other. Moreover, the relative amplitude of the protonated species decreases from 0.98 to 0.7, whereas Acr* shows a gradual increase in contribution from 0.18 to 0.26. The relative amplitude of $AcrH^{+*}$ decreases from 0.98 to 0.78 with an equivalent amount of increase in that of Acr* with addition of TEA. The substantial quenching in the fluorescence lifetimes of both Acr* and AcrH+* with addition of quenchers DMA and TEA can be explained by PET from the bases to Acr, while the changes in the relative amplitude of the two species are explained by an excited state proton transfer from the protonated species to the bases, as both the bases have substantial affinity for protons. It is to be mentioned here that the concentration of TEA is quite low compared to that of DMA for observation of a considerable amount of proton transfer. This is due to the fact that being sparingly soluble in water, the partitioning DMA inside the nanopool of water is very low, and a large concentration of DMA is required for a substantial proton transfer, however being soluble in water, the effective concentration of TEA in the nanopool is very high and a very small concentrations of TEA is good enough for a substantial proton transfer. Moreover, as DMA is more likely to be near the Stern layer of the micellar interface, collisional probability between the $AcrH^+$ and DMA increases in comparison to the expected collision between Acr at the core and DMA at the interface. Moreover previous reports suggest that the distribution of DMA between oil / RM interface is 13.8 M^{-1} and between interface / water pool is 0.012 M^{-1} [16]. Similarly for TEA, the collision with Acr* at the core is more likely to be observed than with $AcrH^+$. It is to be mentioned here that, as proton transfer between donor and acceptor requires the presence of intervening water molecule, ESPT is expected to be at the interfacial region of such heterogeneous self-assemblies. For an extensive analysis of the above-mentioned PET and ESPT, we perform laser-flash photolysis to trace out the transient radicals produced during such photo-induced processes (Table 1).

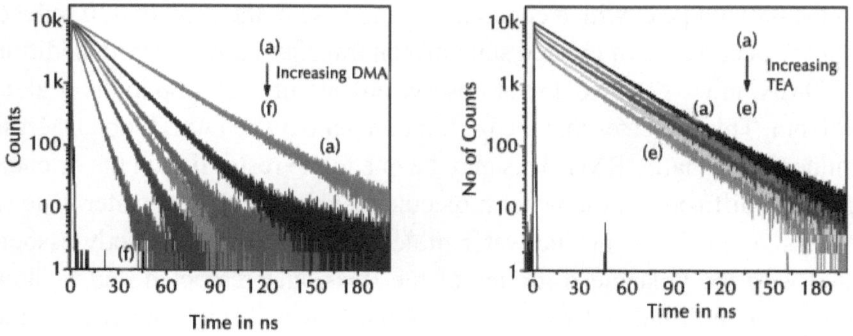

FIGURE 4 (a) Time-resolved fluorescence decay spectra of Acr (0.1 mM) with increasing concentrations of DMA (a) 0 mM, (b) 0.75 mM, (c) 1.50 mM, (d) 2.25 mM, (e) 3.00 mM, (f) 3.75 mM in RMs with $w_0 = 7.5$ ($\lambda_{ex} = 377$ nm and $\lambda_{em} = 478$ nm). (b) Time-resolved fluorescence decay spectra of Acr (0.1 mM) with increasing concentrations of TEA (a) 0 mM, (b) 0.04 mM, (c) 0.08 mM, (d) 0.13 mM, (e) 0.18 mM in RMs with $w_0 = 7.5$. ($\lambda_{ex} = 377$ nm and $\lambda_{em} = 478$ nm).

TABLE 1 Lifetime data of Acr (1×10^{-4}M) with quenchers DMA and TEA (λ_{ex} = 370 nm and λ_{em} = 478 nm); a ± 5%.

[DMA]	τ_1^a	B1	$_2^a$	B2	[TEA]	τ_1^a	B1	τ_2^a	B2
0	27	0.98	4.8	0.02	0	27	0.98	4.8	0.02
0.75	16.8	0.96	3.7	0.04	0.04	27	0.90	3.5	0.1
1.50	12.4	0.91	3.2	0.09	0.08	26	0.81	3.0	0.19
2.25	9.8	0.87	2.8	0.13	0.13	26	0.88	2.1	0.12
3.00	8.04	0.82	2.7	0.18	0.18	25	0.75	1.8	0.25
3.75	5.2	0.74	2.4	0.26	0.23	25	0.66	1.4	0.34

Figure 5a–d represents the transient absorption spectra of Acr (black line) in RMs with w_0 = 2.5, 5, 10 and 20, respectively, in presence of TEA (red line) and DMA (blue line). The two peaks at 440 and 480 nm are due to ^{3}Acr and its protonated form ^{3}AcrH$^+$, which vary with the variations of RMs size. With increase in the size of the RMs, the peak at 480 nm increases with a substantial drop in the 430 nm peak. This reflects of a water-assisted protonation of Acr with enhancement in the nanopool of water. In all the RMs size from w_0 = 2.5–20, in presence of TEA we observe a drop in the 480 nm peak with a concomitant increase in the peak of 430 reflecting the occurrence of excited state proton transfer. However with addition of DMA, in w_0 =2.5, we do not observe any significant drop in the peak at 480 nm. This indicates that the proton transfer between AcrH$^+$ and DMA is hindered in smaller RMs. This may be due to the restriction in the orientation and diffusion of the water molecule inside nanopool of water. Due to excessive confinement, the water molecules are more compactly associated with the polar head groups of the RMs, and restricted due to slow solvation and hindered rotational motion. However with increase in the nanopool environment, ESPT between AcrH$^+$ and DMA gets facilitated. Though the extent of proton transfer is not equivalent to the ESPT between AcrH$^+$ and TEA, still we observe a comparable ESPT at a higher RMs size. This discrepancy in the extent of the ESPT can be explained by the localization and partitioning of the donors and acceptors inside the inhomogeneous and complex interior of the RMs. As explained in the time-resolved fluorescence study, we observe an enhanced ESPT between AcrH$^+$ and

TEA, as both of them are found near the interface and the intervening water molecules acts as intermediate for the proton to move from $AcrH^+$ to TEA. In case of DMA, as $AcrH^+$ and DMA are found in the two opposite end of the Stern layer, the proton transfer between the two is hindered due to sluggish movement of intervening water molecules. However, with increase in the sizes of the RMs and in enhancement in the concentrations of DMA, more and more DMA are partitioned into the interfaces of the RMs, and hence an increase in the proton movement is observed. Moreover with enhancement of the RMs nanopool water, it acquires the bulk like water properties with free rotational and translational motion, which enriches the proton motion between the two.

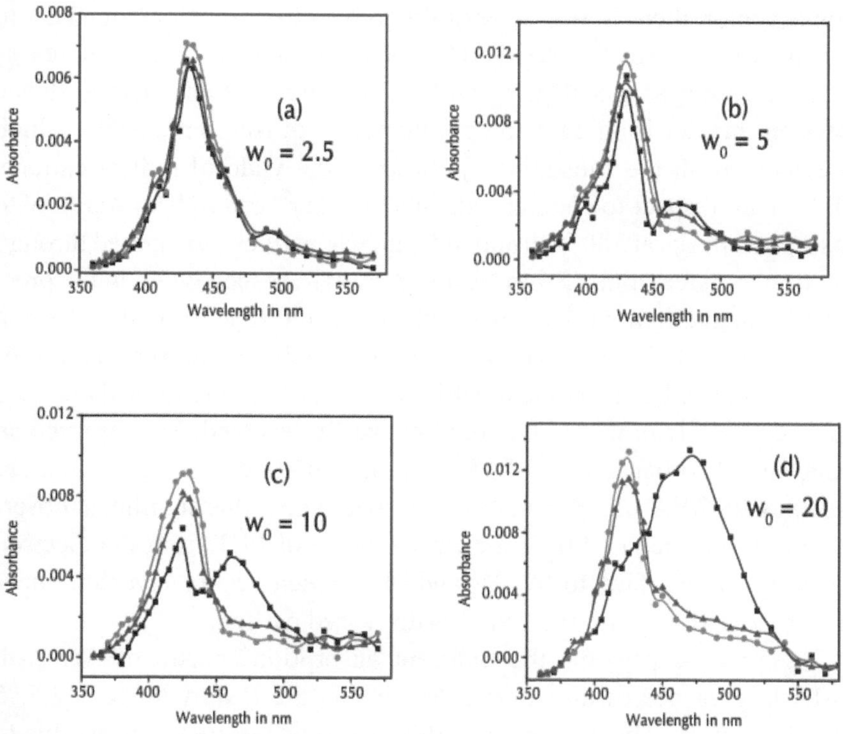

FIGURE 5 T–T absorption spectra of Acr (0.1 mM) (■), Acr (0.1 mM) with TEA (0.2 mM) (●) and Acr (0.1 mM) with DMA (0.2 mM) (▲) in RMs with w_0 2.5 (a), 5 (b), 10 (c) and 20 (d) at 1 μs after the laser flash at 355 nm.

Figure 6a–c presents the transient absorption spectra of Acr with TEA in presence of an external MF inside the RMs with w_0 = 2.5, 5 and 10, respectively. The red line shows the enhanced peak of Acr after proton transfer with TEA. The blue line reflects the different peaks of the radical intermediates observed in the presence of an external magnetic field. The results in absence of magnetic field are quite straight forward, but become complicated in the presence of field. The magnetic field enhances the peak at 380 nm along with an increase in absorbance around 440 nm and 480 nm, which specifies the generation of spin-correlated RPs/RIPs with transient absorptions in these regions. However, the field effect varies with w_0. A prominent MFE is observed in case of w_0 = 2.5, which decreases with gradual increase in the size of RMs and vanishes beyond $w_0 \geq 10$. Since TEA is an electron donor and Acr can act as electron acceptor, there is also a possibility of electron transfer from TEA to ^{3}Acr as well as from TEA to $^3AcrH^+$. The partitioning of $^3AcrH^+$ increases with increase in size of RMs as evident from the transient spectra. Hence electron transfer from TEA to both the forms of Acr is plausible, which is evident from the substantial increase in the yield of radical cations, $TEA^{\bullet+}$ with respect to counter radical anion $Acr^{\bullet-}$ and radical $AcrH^{\bullet}$ with absorption peaks at 380, 430 and 480 nm, respectively, generated through electron transfer from TEA to ^{3}Acr and $^3AcrH^+$, respectively, in the presence of magnetic field. However with slight increase in w_0 the electron transfer from TEA to $^3AcrH^+$ will be more facile as the protonation of Acr is enhanced. A prominent MFE is observed for the radicals in w_0 = 2.5 and 5, while in $w_0 \geq 10$ it could be hardly resolved. We observed an enhanced absorbance around 380 nm, 430 nm and 480 nm, which are assigned for $TEA^{\bullet+}$, $Acr^{\bullet-}$ and $AcrH^{\bullet}$, respectively. This significant observation in presence of MF is a clear evidence of PET from the electron donating amines TEA to $^3AcrH^+$ and ^{3}Acr which occurs as a simultaneous process along with the ESPT as discussed earlier.

Figure 7a–c presents the transient absorption spectra of Acr with DMA in presence of an external MF inside the RMs with w_0 = 2.5, 5 and 10, respectively. As can be readily observed we are not been able to observe any significant MFE even in any of the RMs. A diminished MFE is observed only with w_0 = 2.5 which vanishes with increase in sizes of the nanopool. A small enhancement in the peaks at 460 and 480 nm is observed for w_0 = 2.5. The increment at 460 nm indicates the presence of $DMA^{\bullet+}$, an obvious electron transfer product between Acr and DMA. The field effect dies out very quickly with even a slightest of change in the sizes of the RMs. As discussed in the introduction, MFE is product of the both spin and diffusion dynamics. Photo-induced electron transfer between $^3AcrH^+$ and DMA results in the formation of water mediated spin-correlated-solvent-separated ion pair (SSIP), which is stabilized by the solvent cage. The stability of SSIP increases inside smaller RMs due to excess confinement and restricted mobility of bound water. However with changes in the sizes of RMs, the water regains its bulk property with faster dynamics which distorts the stability of SSIP, hence the spin-correlation between the solvent caged RIPs becomes nonspecific. Moreover, as DMA is localized at the Stern layer of the RMs, the electron transfer between ^3Acr and DMA is less plausible, and hence cannot contribute to the net MFE. However, with increase in w_0 the generation of $AcrH^+$ from Acr is enhanced and hence the yield of SSIPs formed through electron transfer from DMA to ^3Acr is even more reduced. In small sized RMs the SSIPs formed from electron transfer from DMA to $^3AcrH^+$, which are relatively stable, can undergo diffusion and S–T spin conversion through hyperfine interactions. The latter is maximized when the partners of RIPs are separated by a distance where exchange interaction becomes negligible. In the presence of an external magnetic field, the S–T spin conversion will be perturbed by Zeeman splitting in triplet sublevels causing an increase in the SSIPs. It is to be remembered that the separation distance between the radicals/radical ions in the SSIPs is very important and has to be maintained in such a way so that the exchange interactions between geminate electrons should be minimized, which is an important factor to observe MFE. The significance of distance between the RPs/RIPs in MFE has been explained by several workers using linked donor–acceptor systems. In larger RMs water shows bulk like property and a floppy encounter complex is formed due to free diffusion of water, which brings destabilization and loss in spin-correlation in the geminate SSIPs. Thus the radicals formed during the photo-induced interactions diffuse freely inside the core of the RMs, hence loose geminate inter-radical spin correlation. This results in insignificant MFE with DMA in both small as well as large sized RMs.

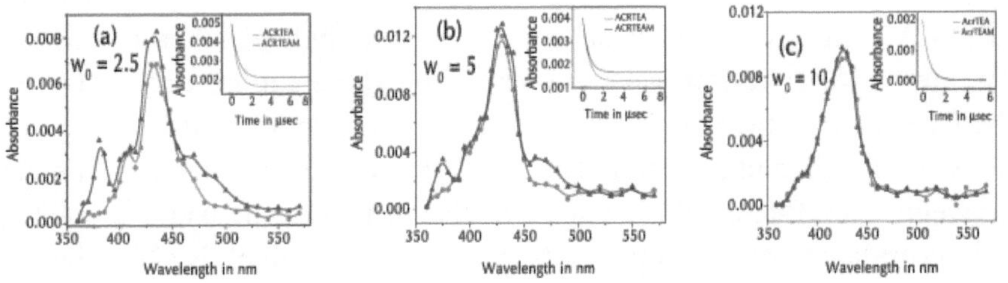

FIGURE 6 T–T absorption spectra of Acr (0.1 mM) with TEA (0.2 mM) (•) and Acr (0.1 mM) with TEA (0.2 mM) in presence of magnetic field (▲) in RMs with w_0 2.5 (a), 5 (b) and 10 (c) at 1 μs after the laser flash at 355 nm. The inset of the corresponding figure shows the decay profile at 480 nm in absence (red) and presence (blue) of magnetic field.

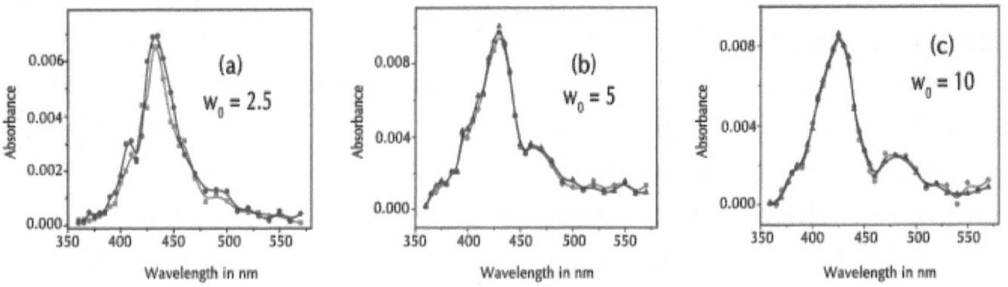

FIGURE 7 T–T absorption spectra of Acr (0.1 mM) with DMA (0.2 mM) (•) and Acr (0.1 mM) with DMA (0.2 mM) in presence of magnetic field (▲) in RMs with w_0 2.5 (a), 5 (b) and 10 (c) at 1 μs after the laser flash at 355 nm.

6.4 CONCLUSION

In this chapter, we have investigated the role of localization of the donors and acceptors and the dependence of distance between the two as an important parameter for tuning the MFE. The distance dependence and partitioning between the radicals are monitored by varying the sizes of the RMs. With increase in the size of the RMs, the distance between the reactant becomes large enough to destroy the spin correlation between the radical ion pair and hence the MFE. Proton transfer from $AcrH^+$ for the both the quenchers are observed in all the RMs, however, electron transfer along with the MFE is observed prominently for TEA inside small sized RMs. The insignificant MFE with DMA and in large sized RMs are assigned to the loss of spin correlation between the generated electron transfer radical ion pairs due to inaccessibility of the donors and acceptors and the random diffusion of RIP inside the large nanopool of water. Too far a separation between the RIP spoils the MFE, even if the donors and acceptors are well partitioned inside such heterogeneous RMs.

KEYWORDS

- **S–T spin conversion**
- **Tektronix oscilloscope**
- **triplet–triplet (T–T) transient absorption**
- **Zeeman interaction**

REFERENCES

1. Steiner, U. E.; Ulrich, T. *Chem. Rev.,* **1989**, *89*, 51–147.
2. Brocklehurst, B. *Chem. Soc. Rev.*, **2002**, *31*, 301–311.
3. Turro, N. J. *Proc. Natl. Acad. Sci. USA*, **1983**, *80*, 609–621.
4. Sengupta, T.; Choudhury, S. D.; Basu, S. *J. Am. Chem. Soc.*, **2004**, *126*, 10589–10593.
5. Sarangi, M. K.; Basu, S.; *Chem. Phys. Lettrs*., **2011**, *506*, 205–210
6. De, T. K.; Maitra, A. *Adv. Colloid Interface Sci.,* **1995**, *59*, 95–193.
7. Levinger, N. E. *Science,* **2002**, *298*, 1722–1723.
8. Denny, W. A. *Curr. Med. Chem.,* **2002**, *9*, 1655–1665
9. Encaranacion, I. N.; Arce, R.; Jimenez, M. *J. Phys. Chem. A,* **2005**, *109*, 787–797
10. Ryan, E. T.; Xiang, T.; Johnston, K. P; Fox, M. A. *J. Phys. Chem. A,* **1997**, *101*, 1827.
11. Tokumura, K.; Kikuchi, K.; Koizumi, M. *Bull. Chem. Soc. Jpn*., **1973**, *46*, 1309–1315.
12. Kellmann, A. *J. Phys. Chem*., **1977**, *81*, 1195–1198.
13. Buchviser, S. F.; Gehlen, M. H. *J. Chem. Soc., Faraday Trans*., **1997**, *93*, 1133–1139.
14. Oelkrug, D.; Uhl, S.; Wilkinson, F.; Willsher, C. J. *J. Phys. Chem.,* **1989**, *93*, 4551–4556.
15. Sarangi, M. K.; Bhattacharyya, D.; Basu, S. *Chem. Phys. Chem*., **2012**, *13*, 525–534.
16. Falcon, R. D.; Correa, N. M.; Biasutti, A.; Silber, J. J. *J. Colloid. Int. Sci.,* **2006**, *296*, 356–364.

CHAPTER 7

APPLICATION OF NON-THERMAL PLASMA FOR SURFACE MODIFICATION OF POLYESTER TEXTILES

H. DAVE, L. LEDWANI, and S. K. NEMA

CONTENTS

7.1 INTRODUCTION

Global production of textiles products has increased more than 72.5 million metric tonnes of which synthetic fibers form an important part [1]. In fact, the production of natural fibers surpasses by synthetic fibers with more than 55% of products in the textile industry [2]. Polyester is the most used synthetic fiber first commercially produced by the du Pont Company. Polyesters are made from chemical substances found mainly in petroleum and are manufactured in three basic forms—fibers, films and plastics of which polyester fibers used to make fabrics is the largest segment. Polyethylene terephthalate (PET) (Fig. 1) is the most common polyester used for textile manufacturing. PET is synthesized from ethylene glycol with either terephthalic acid (Fig. 2) or its methyl ester in the presence of an antimony catalyst under high temperature and vacuum to achieve high molecular weights needed to form useful fibers. Production of polyester accounted of 30.3 million metric tonnes in 2008 representing about 72% of the entire man-made fiber [1]. The 2009 global aggregate polyethylene terephthalate output summed up 49.2 m tonnes, of which PET fiber accounted for about two-thirds, whereas PET for packages and films 34% [1]. In 2010, the world textile industry has experienced the most potent growth in 25 years. Manufacturing volumes of natural and manmade fibers rocketed upwards by 8.6%, or 6.4 million tonnes, at 80.8 million tonnes of polyester industrial filament yarn jumped up by spectacular 37%. Viscose fibers produced a record-breaking growth of 17% [2]. PET fibers are expected to account for 50% of the world fibers production in 2012 [1].

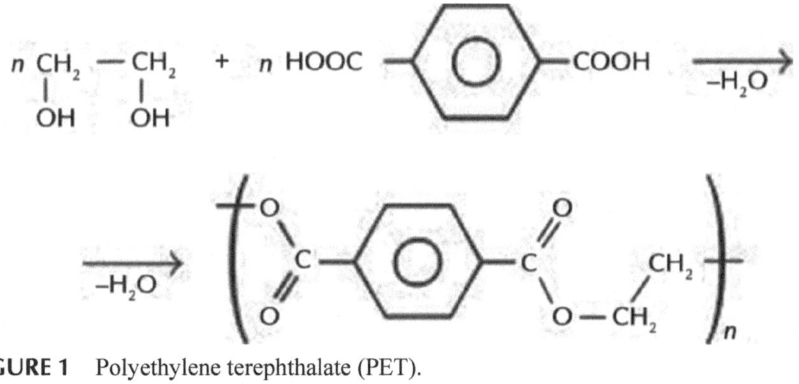

FIGURE 1 Polyethylene terephthalate (PET).

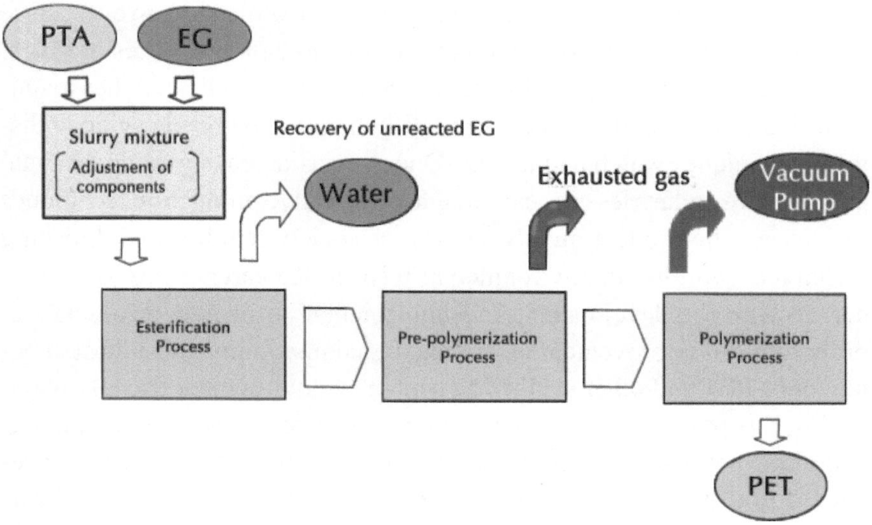

FIGURE 2 Production of PET.

Polyester textiles have major advantages of high modulus and strength, excellent chemical, physical and mechanical properties, stiffness, stretch, wrinkle and abrasion resistances, relatively low cost, tailorable performance and easy recycling but a great disadvantage of polyester is hydrophobicity and inert nature of the constituent polymer which cause variety of difficulties in finishing specifically in dyeing and at consumer use due to poor moisture regains, less wearing comfort, build-up of electrostatic charge, the tendency to pilling, and insufficient washability associated with their hydrophobic nature [3, 4]. Surface properties of polyester materials have strong effects on most of their practical applications. Indeed many properties, such as adhesion, gloss, wettability, permeability, dye ability, printability, surface cleanliness, bonding of different components, biocompatibility, or mechanical performance and antistatic behavior are related more to the surface than to the bulk of the material and therefore, to correct this, the strategy of surface modification can be implemented. Polyester surfaces can thereby be changed to achieve a variety of goals, including increasing adhesion, improving wettability, reducing friction, reducing susceptibility to harsh chemicals or environmental agents, and increasing dye absorption. Conventionally, polyester textile are modified by various thermal, mechanical, and chemical treatments, however, these methods often damage excellent mechanical and bulk properties of fibers; also have a lot of energy and water consumption as well as higher manufacturing costs; and there is more waste generation requiring suitable disposal techniques which add to cost. Due to ever-increasing environmental concern now industries are searching for innovative production techniques to improve the product quality, as well as society requires new finishing techniques working in environmental respect. Researchers are now concerned with the development and implementation of new techniques in order to fulfill improvement in wetting, biocompatibility and adhesion behavior of PET and other synthetic fibers in an environment friendly way.

For this purpose, many great studies have been carried out including: chemical methods (enzymatic modification, treatment with different reagents, grafting of different monomers, application of supercritical carbon dioxide and micro-encapsulation techniques); blending of polymers with different organic and inorganic compounds in fiber production and physical methods (microwave, ozone-gas, gamma, UV, laser and plasma functionalization) [4]. Nonetheless, most of the treatments still are not satisfactory in terms of performance and can result in increased waste production, unpleasant working conditions and higher energy consumption. In this sense, plasma technology is claimed to replace some wet chemical applications as an environmental friendly process [4]. In the past decade, the use of non-thermal plasmas for selective surface modification has been a rapidly growing research field. Non-thermal plasmas are particularly suited to apply to textile processing because most textile materials are heat sensitive polymers. The reactive species of plasma, resulting from ionization, fragmentation, and excitation processes, are high enough to dissociate a wide variety of chemical bonds, resulting in a significant number of simultaneous recombination mechanisms. Thus plasma opens up new possibilities for polymer industrial applications where the specific advantages of plasma processing provide effective surface modification of polymers in environment friendly manner over convention treatment. This technology has other advantages of textile surface modification and leaving the bulk characteristics unaffected. For instance, plasma technology provides an attractive means of on-line treatment with wide applicability, low quantities of input needed in terms gases and electricity, making it non-energy intensive and environment. Plasma technology being dry technique and does not require water; it involves extremely low quantities of starting materials; the process is realized in gas phase, process duration is low; it provides energy saving and its low temperature avoids sample destruction [4].

The pre-treatment and finishing of textiles by non-thermal plasma technologies becomes more and more popular as a versatile surface modification technique where a large variety of chemically active functional groups can be incorporated into the textile surface. Activation of textile surface results in improved wettability, adhesion of coatings, printability, induced hydro- and/or oleophobic properties, changing physical and/or electrical properties, cleaning or disinfection of fiber surfaces etc. Moreover, non-thermal plasma surface modifications can be achieved over large textile areas. In literature, some excellent reviews on plasma treatment of textile are available which gives overview of different all types of non-thermal plasma treatments of textiles in general specifically for effect of plasma on textile substrate without focusing on application aspect [5–7]. In the last decade, considerable efforts have been devoted on surface modification of PET textile with plasma technology to enhance wettability, wick ability and dyeing and printability etc. As non-thermal plasma technology is widely studied for PET textile modification with enormous amount of potential uses of non-thermal plasma for the modification of textile products, categorizing the applications is difficult, and therefore, a review is given on applications of plasma treatment on the PET textile. This chapter on application of non-thermal plasma treatment of polyester textile gives overviews different application strategies of plasma treatment on PET textile and ways to obtain these effects reported during the last two decades.

7.2 NON-THERMAL PLASMA

Broadly the plasma state can be considered to be a gaseous mixture of oppositely-charged particles with a roughly zero net electrical charge (Fig. 3) Ionization processes can occur when for instance molecules of a gas are subjected to high-energy radiation, electric fields or high caloric energy.

During these processes the energy levels of particles composing the gas increase significantly and as a result electrons are released and charged heavy particles are produced. Sir William Crooks suggested the concept of the 'fourth state of matter,' in 1879, for electrically discharged matter and Irving Langmuir first used the term 'plasma' to denote the state of gases in discharge tubes.

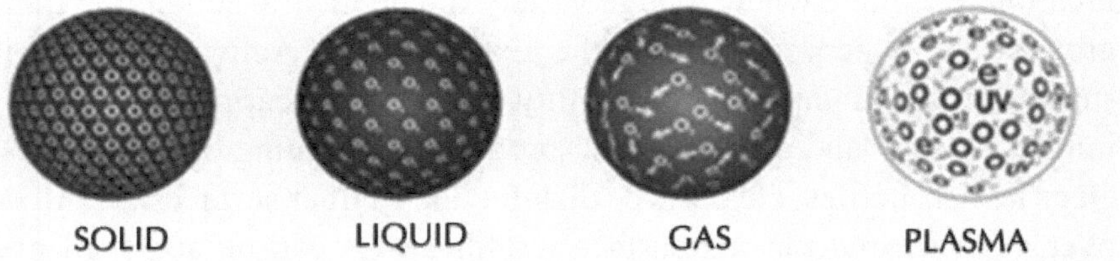

FIGURE 3 Plasma the forth state of matter.

Plasma states can be divided in two main categories: Hot Plasmas (near-equilibrium plasmas) and Cold Plasmas (non-equilibrium plasmas) (Fig. 4). Hot plasmas are characterized by very high temperatures of electrons and heavy particles, both charged and neutral, and they are close to maximal degrees of ionization (100%). Cold plasmas are composed of low temperature particles (charged and neutral molecular and atomic species) and relatively high temperature electrons and they are associated with low degrees of ionization (10^{-4}–10%). Hot plasmas include electrical arcs, plasma jets of rocket engines, thermonuclear reaction generated plasmas, etc., while cold plasmas include low-pressure direct current (DC) and radio frequency (RF) discharges (silent discharges), discharges from fluorescent (neon) illuminating tubes, corona discharge and dielectric barrier discharge (DBD) (Fig. 4). It is obvious that the high temperatures used in thermal plasmas are destructive for polymers, so most applications for textile surface modification make use of non-thermal or cold plasmas. Most of the prior research related to discharge-mediated surface-modification reactions involved low-pressure cold-plasma environments. However, in recent last years, atmospheric pressure non-equilibrium plasma installations have been designed, developed and tested with great success. The atmospheric pressure plasma process has attracted great interest due to its low cost, high processing speed and simple system of operation without vacuum equipment's for industrial applications. A review on applications of atmospheric pressure plasma for textile processing is given in literature [5].

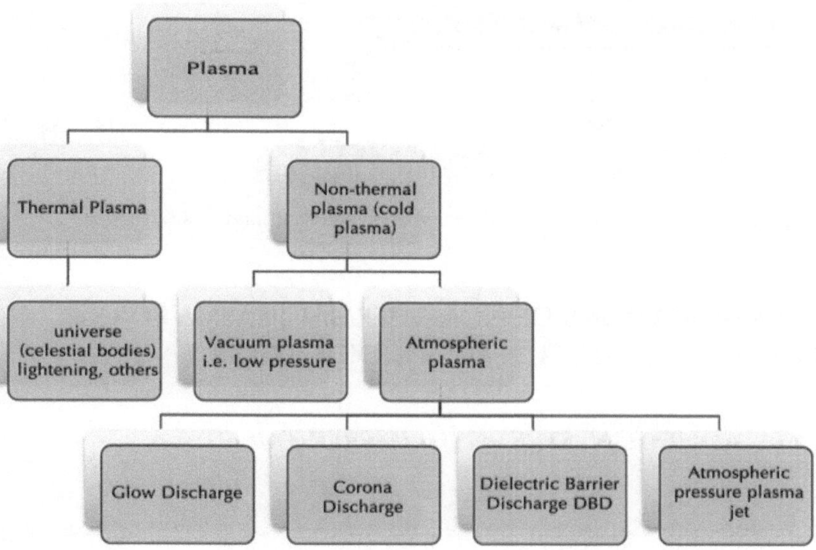

FIGURE 4 Classification of plasma.

Plasma treatment in general generates different plasma constituents like electrons, ions, free radicals, meta-stables and photons (Fig. 5) [8]. These either directly or indirectly participate in plasma-chemical reactions which introduce reactive groups and free radicals onto the surface, thus improving the adhesion of chemicals and polymers mostly by improved physical interaction. These interactions are normally seen in the form of hydrogen bonds, Van der Waals forces or dipolar interactions. It can produce different radicals and reactive groups on the surface by the use of different gas mixtures. Thus it achieves surface modification whilst maintaining the bulk properties. The low treatment temperature avoids deterioration of delicate organic samples. The process is water free and avoids the need for costly effluent treatment.

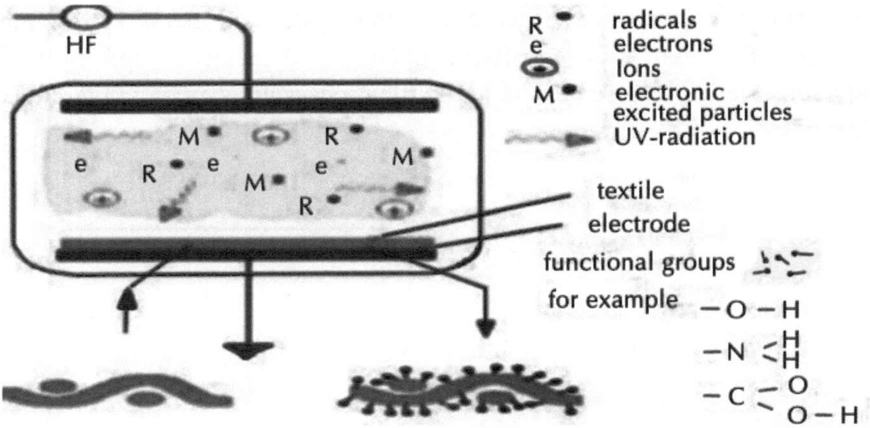

FIGURE 5 Interaction of plasma with textile substrates (*Source*: Ref. [8]).

7.3 APPLICATIONS OF NON-THERMAL PLASMA FOR POLYESTER TEXTILE SURFACE MODIFICATION

7.3.1 ADHESION STRENGTH IMPROVEMENT

Polyesters show strong hydrophobic behavior and low surface energy. Increasing the surface energy of the fiber could increase the polymer–fiber interaction and improve the binding strength. The polar groups creation at the polymer surface and the surface roughness modification could increase coating adhesion. Plasma modification of a textile especially at atmospheric pressure to increase surface energy to improve adhesion is favorable over other methods for several reasons. Garg et al., have reported improvement of the ability of the polyester fabrics pretreated with atmospheric pressure glow discharge (APGD) to bond with anthraquinone-2-sulfonic acid doped conducting polypyr role coating using a range of APGD gas mixtures and treatment times. Conducting polymer-coated textiles can be made to possess a combination of desirable properties, suggest potential applications in the fields of sensors and production of electronic textiles, key impediments for commercial application of conductive polymer coated textiles have been degradation and poor adhesion of the coating to the fiber surface. Investigation showed that APGD treated fabrics exhibited better hydrophilicity and increased surface energy. Surface treatment by an APGD gas mixture of 95% helium/5% nitrogen yielded the best results with respect to coating uniformity, abrasion resistance and conductivity [9]. Varieties of electromagnetic shielding conductive fabrics are often produced by electroless plating of polyester and polypropylene nonwoven fabrics. In the standard electroless plating procedure, a catalytically active substance, usually containing palladium, is first deposited on the fiber surface from an aqueous solution. The fabric is then metallized by immersing in a chemical metallizing solution. Pre-treatment with plasma renders the fiber surface hydrophilic facilitates absorption to reach necessary levels of uniformity and adhesive strength of the metal layer. Šimor et al., in their study found atmospheric-pressure nitrogen plasma generated in the surface barrier discharge has been to be very effective in the treatment of PET nonwoven fabric for subsequent electroless plating by nickel. The results obtained show that an extremely short 1-s plasma treatment is sufficient to hydrophilize the fabric fibers surface and to obtain a clear improvement in the nickel-plating uniformity and adhesion. Because of atmospheric-pressure operation, very short treatment times, and its robustness the method has the potential to be used in line with standard metal plating lines [10]. Similar work is carried out using low-pressure plasma treatment with oxygen and argon gases. Overall, the plasma treatment can successfully improve the electroless nickel-plating performance. In addition, oxygen plasma treatment is comparatively more efficient than argon plasma treatment [11]. When compared to untreated, the electroless nickel plating with oxygen plasma treatment was significant to improve the performance of nickel-plated polyester fabrics as reflected by the scanning electron microscopy, tensile strength, ultraviolet protection as well as, fabric weight. On the contrary, it was also enhanced the fabric thickness and color fastness to crocking. Moreover, the application of plasma treatment slightly reduced the performance of contact angle and wrinkle recovery property [12].

Leroux et al., showed that in the case of polyester (PET), the degree of increase in the surface energy depends on the textile fabric structure. Adhesion between silicon or fluorine compounds used to produce hydrophobic textile is very low on polyester fibers yielding a poor washing fastness. They reported that adhesion of the fluoropolymer to the woven PET was greatly enhanced by the air plasma treatment resulting in increased the wet-picking of a padding process with fluoropolymer solution and therefore, the treatment washing fastness [13]. They further reported, atmospheric air plasma treatment based DBD technology can oxidize various polyester material such as film, nonwoven and woven fabric, which studied by wettability in order to follow the polyester polar and dispersive surface energy components. The polyester surface chemical composition investigated by X-ray Photoelectron Spectroscopy (XPS) showed polar groups created at the polyester surface by atmospheric plasma treatment, which enhances adhesion with silicon resin containing adhesive primer. For each polyesters material, a critical treatment power (TPc) necessary to obtain the best results either for oxidation or silicon adhesion reported and was correlated with the structure of the material [13]. Parvinzadeh and Ebrahimi studied influence of air APGD treatment on performance of nano-emulsion silicone softener on polyethylene terephthalate fibers and reported that the plasma pre-treatment modifies the surface of fibers and increases the reactivity of substrate toward nano-emulsion silicone. Moisture regain and microscopic tests showed that the combination of plasma and silicone treatments on polyethylene terephthalate can decrease moisture absorption due to uniform coating of silicone emulsion on surface of fibers [4].

7.3.2 WETTABILITY IMPROVEMENT

Troubles connected with minimizing of the textile hydrophobicity are usually being solved by the textile fibers' surface chemical modification, but from ecological point of view modification of fabric with low temperature plasma is superior to classical chemical wet processes. Wróbel et al., firstly reported improvement in wettability of polyester textile. In their study, PET fabric was treated by low-pressure glow plasma initiated in various gases: nitrogen, oxygen, air, carbon dioxide and ammonia. Plasma-treated fabric showed a considerable change in surface structure and wettability which closely dependent on the gas type and treatment conditions and a good correlation exits between surface structure and wettability. The wetting time of plasma treated fabric considerably drops in comparison to untreated fabric and the best results were obtained by treatment in nitrogen, oxygen and air plasma [14]. Surface chemical and physical modifications of PET fibers induced by low pressure RF air plasma exhibited a remarkable increase in hydrophilicity, accompanied by extensive etching and by the implantation of both oxygen- and nitrogen-containing polar groups [15]. RF plasma of nitrogen, air and oxygen gas also reported to improve hydrophilicity of polyester fabric [16]. To compare the effectiveness of different plasma sources Vatuňa et al., performed a series of experiment both in RF and microwave (MW) plasmas. For working gas nitrogen, oxygen and their mixtures were employed. Results showed that obtain comparable PET fabric modification effect either in oxygen and nitrogen plasma, the nitrogen plasma treatment must be longer or higher discharge input power is necessary. The costs/modification efficiency ratio seems to be the most economical in case of air–oxygen mixtures while RF plasma modification seems to be more effective than that of the MW plasma [17]. Wei et al., reported PET fibers were treated in low pressure oxygen plasma introduce the polar groups on fiber surfaces and so reduce the advancing and receding contact angles of the PET fibers. The contact angle hysteresis of plasma treated PET fibers is found to be altered by roughening of the fiber surface [18]. Wardman and Abdrabbo studied effect of low pressure oxygen plasma treatment on two polyester fabric types, polylactic acid and standard polyester (PET), and the influence on their respective wetting characteristics is investigated by a novel analytical system, based on image analysis, was developed for measuring the rate of spreading and dynamic movement of liquid over the fabrics. The analyses showed that the oxygen plasma treatment abraded the surface of the PLA fabric, but did not alter their chemical nature, whilst the surfaces of the PET fabric were less abraded, but had enhanced polarity due to an increase in carbonyl groups. The increased surface abrasion made little difference to the wetting or wicking rates of water on PLA fabric, but the increased polarity made a large difference to the rates on PET fabric [19]. Costa et al., treated PET fabrics in low pressure and temperature plasma in order to study the effect of gas composition on the hydrophilicities. Different plasma atmosphere were used in this work, leaving the gas type (oxygen, nitrogen, methane and hydrogen) as plasma variables, other parameters like pressure, exposure time, voltage and current were not variables in the process. The polyester fabrics treated with different plasma gases exhibited different morphological changes. The capillary method was applied to evaluate the improvement in water uptake of polyester fabrics, which indicated a good wettability to all atmospheres used in this work, except for a mixture containing methane, hydrogen and nitrogen [20].

Permanent hydrophilic properties can be achieved by plasma polymerization of acetylene mixed with ammonia (C_2H_2/NH_3) by RF low-pressure discharge where both deposition and etching processes took place yielding a nano-porous, cross-linked network with accessible functional groups. Using plasma co-sputtering of a silver target, Ag nano particles can be in situ embedded within the growing plasma polymer yielding a well-defined size and distribution of nano particles at the coating surface and thus an anti-microbial activity achieved. This can also be useful for substrate-independent dyeing. When plasma treated polyester dyed with acid dyes, increasing color intensity with film thickness proved that accessible amine groups were deposited within a nano-porous hydrocarbon matrix. Multifunctional textile surfaces can thus be obtained by adjusting combined properties such as wettability, functional group density as well as antibacterial and bio-responsive surfaces [21]. The effects of low pressure O_2 and NH_3 plasma on the morphology and topometry of fabrics on four different length scales, as well as the influence of the topographical changes of textile structures on the resulting water spreading and absorption rates were investigated. The results of the topographic characterization using two non-contact optical methods and wettability measurements indicate that the modification of filament nano-topography cannot satisfactorily explain the drastic changes observed in wettability. Dimensional changes (relaxation and shrinkage) as well as changes in warp morphology and inter-yarn spaces are more decisive for inducing hydrophilicity in polyester woven plain fabrics than an increase in the surface nano-roughness of their filaments [22].

Xu and Liu reported that polyester fabric could be modified to be more hydrophilic by corona discharge irradiation. Good hydrophilic property was observed around 10 kV when the fabric was treated at speed of 5 cm/min. After the treatment, dye-uptake ratio and dyeing speed was also improved, this can shorten the dyeing time and reduce the dyeing cost. Also, affinity of the fabric with starch can be greatly improved, this shows the polyester yarn can be sized with green sizing agent like, modified starch effectively after plasma treatment, thus it can replace polyvinyl alcohol (PVA) which is widely used in the polyester sizing [23]. Samanta et al., reported continuous treatment with APGD plasma in argon, helium, air and oxygen can significantly improve wettability along with increase in oil absorbency after plasma treatment. Water absorption explained on the basis of concurrent increase in surface energy due to the chemical changes occurred on the surface, while this increase in oil absorbency for all the treated samples was due to the enhanced capillary action on the fabric surface caused by the formation of horizontal and vertical nano-sized channels after plasma treatment [24]. De Geyter et al., reported that a polyester non-woven treatment with air DBD at medium pressure (5.0 kPa) results in significant wettability improvement, and wettability increase with discharge powers [25]. Píchal and Klenko in their the applicability study of the atmospheric pressure air DBD for synchronous treatment of several sheets of fabric reported hydrophilicity of individual sheets of fabric has distinctly increased after plasma treatment [26].

Wang et al., reported that the atmospheric pressure plasma jet (APPJ) treatment was effective in improving the wettability of polyester fabrics due to surface etching and chemical composition change. One of the biggest difference between APPJ and other plasma surface treatment is that only one side (top) of substrate is contacted with plasma jet at atmospheric pressure while in other plasmas two sides (top and bottom) of substrate are both contacted with plasma. The modification of the bottom side of woven fabric treated by APPJ is largely dependent on the penetration of active species in plasma jet, which is accordingly affected by plasma parameters and the structure of materials. They found in their study that penetration of surface modification of multiple layers of fabrics exposed to helium/oxygen APPJ was possible. To study effect of pore size on penetration depth, eight-layer stack of woven polyester fabrics of two different pore size (200 μm and 100 μm) were exposed to a helium/oxygen APPJ. The pore size of the fabric affected plasma treatment effect on each fabric layer and its penetration depth into the textile structure. After APPJ treatment, the topside of the polyester fabric became more hydrophilic which gradually decrease, as fabric layer got deeper. The penetration of plasma surface modification into the fabric layers was deeper for fabrics with larger average pore sizes. It was found that helium/oxygen APPJ was able to penetrate eight layers of polyester fabrics with pore sizes of 200 μm while it was only able to diffuse through six layers of the fabrics with an average pore size of 100 μm [27]. In order to determine the relationship between the treatment duration of APPJ and the penetration depth of the surface modification into textile structures, a four-layer stack of polyester woven fabrics (200 μm poresize) was exposed to helium/oxygen APPJ for different treatment durations. The water-absorption time of the top and the bottom sides of each fabric layer was reduced from 200 s to almost 0 s indicating no difference in wettability improvement between the two sides of each fabric layer after APPJ treatment, which demonstrated that interaction of chemically active species generated in plasma jet with the substrate surface was not only on the surface directly in contact with the plasma jet but also observed on surfaces which were not facing the plasma jet. As the layer number increased, the effectiveness of the APPJ treatment gradually reduced. There was a linear relationship between the capillary flow height and the treatment time and the rate of the capillary flow height increasing with the treatment time decreased linearly as the layer number increased. An empirical model constructed using the experimental data predicted that the maximum penetration depth was six layers of fabrics with reasonable treatment duration as adopted in the study [28]. In order to investigate the influence of pore size on penetration of surface modification into woven fabric treated with APPJ, four kinds of polyester woven fabrics with different pore sizes were used as the model porous medium. Two groups of parallel polyester fibers are respectively and tightly pasted on the top and bottom side of each fabric. Penetration of plasma effects through the pores was detected by changes in contact angle on the bottom side before and after APPJ treatment. The degree of penetration of APPJ surface modification was increased with the increasing pore size. Complete penetration was realized in fabric with pore size larger than 200 μm and nearly no penetration was found in fabric with the pore size smaller than 10 μm [29].

7.3.3 DESIZING IMPROVEMENT

Plasma treatment can serve as an alternative clean desizing technology. The work of Bae et al., indicate that vacuum O_2 plasma treatment can be applied to reduction of water pollution from conventional desizing process. Desizing of polyester fabric is revealed by weight loss and Scanning electron microscopy (SEM) pictures. Treatment effectively removed sizing agents such as PVA and polyacrylic acid (PAA) and their mixtures. This treatment also increased hydrophilic groups, which gave water solubility enhancement. After O_2 plasma treatment, the PET fabric was subjected to conventional desizing process at different temperatures. Except for the PET fabric sized with PVA, plasma treated fabrics showed more efficient desizing results when compared with untreated fabrics. Furthermore, the desizing effluent from the treated fabric gave lower TOC, COD and BOD values [30]. However, use of vacuum limits its industrial applications, atmospheric pressure plasma developed in recent years can be one of the suitable alternatives. The influence of He/O_2 APPJ treatment on subsequent wet desizing of PAA on PET fabrics was studied by Li et al. Plasma treatment could directly remove some of PAA size on PET fabrics because of plasma etching. SEM Image showed that the fiber surfaces were as clean as unsized fibers after 35 s treatment followed by $NaHCO_3$ desizing. The percept desizing ratio (PDR) results showed that more than 99% PDR was achieved after 65 s plasma treatment followed by a 5 min $NaHCO_3$ desizing indicating that compared to conventional wet desizing, plasma treatment could significantly reduce desizing time [31].

7.3.4 HYDROPHOBIC TEXTILE PREPARATION

Hydrophobic properties of textiles are usually obtained by padding or coating processes using silicon or fluorine compounds. These traditional methods are not environmentally friendly since a huge amount of water and chemical products are used, and lot of energy is required to evaporate the water. Leroux et al., in their study deposited a fluorocarbon coating PET woven fabric using pulse discharge plasma treatment by injecting a fluoropolymer directly into the plasma dielectric barrier discharge which modifies the textile surface with a small quantity of chemical product and water [32]. Sulphur hexafluoride (SF_6) plasma treatments and hexamethyl disiloxane (HMDSO) plasma polymerization were performed on PET meshes and the resulting wettability against liquids having very different surface tensions were investigated at the light of a possible use of the materials in the fuel/water separation technology, moreover, an increase in hydrophobic performances was achieved with HMDSO plasma polymerization followed by SF_6 plasma treatment [33].

Ji et al., in their work report the formation of water-repellent superhydrophobic surface on (PET) fiber via plasma polymerization at atmospheric pressure. Polyester fiber made water-repellent by treating with atmospheric pressure middle frequency (MF) and RF plasma system using Argon and HMDSO [34]. The PET sample with the RF plasma treatment passes of 20 times showed the water repellency of 90 rating from the AATCC standard spray method. From the Fourier transform infrared spectroscopy (FT-IR) and SEM results, it is identified that the improvement of water-repellency was attributed by Si–O–Si, Si–$(CH_3)_2$, and Si–C chemical functional groups produced from the HMDSO fragmentation and the plasma chemistry reactions between PET and plasma. The plasma system in this work does not require any vacuum instruments and is operated not in a batch mode but an in-line mode. So it can be easily scaled up for application to large substrate surfaces or continuous processing [35].

7.3.5 DYEING IMPROVEMENT

PET is widely used in the field of the fabrics industry and has excellent mechanical characteristics, chemical inertness and heat resistance however, polyester exists several serious problems in the producing of textile products due to nonpolar groups on the fiber surface. Firstly, polyester only can be dyed by disperse dye under high temperature (around 130°C) and high pressure; this needs lot of energy and special equipment with high funds [23]. Secondly, PET reflects a large quantity of light from the surface due to its intrinsic large refractive index (n = 1.725) and it is difficult to obtain full shaded color especially in black or other dark colors [36]. Initially, studies have been performed to obtain full color intensity on dyed PET fabric surfaces by low-pressure plasma polymerization, sputter etching and low refractive index chemical resin coating. Plasma polymerization has been performed to increase color intensity of dyed PET fabrics using a low refractive index (n = 1.3–1.4) monomer. The sputter-etching process on dyed PET fabrics increases color intensity due to the physical formation of irregularity on the fabric surface, which induces light scattering and decreases reflectance of light on the surface. Deposition of low refractive index resins on fabrics also improved color intensity. In their work, the effect of the plasma polymerization process on the change of color intensity of dyed PET fabrics discussed by Lee et al. Hexamethyldisilane (HMDS) and tris(trimethylsilyloxy)vinylsilane (TTMSVS) organic sources were used for the effective polymerization on PET fabrics. In addition, sputter-etched PET fabrics that were treated using He and O_2 gases also be discussed in terms of the change of color intensity comparing with those of the polymerized PET samples. HMDS and TTMSVS polymerized films on PET are reasonably good for the deep coloring of PET fabric and the He-O_2 sputtering effect is also pretty clear for increasing the color intensity. Anti-reflection layer is a major role for the increase of color intensity in the polymerized PET surface. Minimum reflectance has been observed at the polymerized film thickness of 1500–2000 Å using HMDS and TTMSVS organic sources [36]. Corona plasma treatment of polyester fabric is reported to improve dye-uptake ratio and dyeing speed also with disperse dyes, thus plasma treatment can shorten the dyeing time and reduce the dyeing cost [23]. As the treating voltage increased, the fabric dye-uptake ratio was also improved. This means the dyeing time can be shortened and the dyeing temperature can be decreased, which is generally important for the polyester fabric dyeing. Polyester fabric is usually dyed at high temperature under high pressure, and the dyeing time is also long, this is partly because that the fabric hydrophilic property is poor and the dye cannot be quickly absorbed onto the fiber surface, after the corona discharge treatment, the dye can be absorbed by the fiber surface quickly and can transfer into the inner part of the fiber, thus it can reduce the dyeing energy [23].

Low pressure RF air plasma surface treatment was used as an effective tool to increase surface hydrophilicity and roughness of PET fibers. Dyeing of weaved fabrics was proceeding in much better way after plasma treatment. A new recipe for dyeing bath without any use of sodium sulphate was found with the same result of reflectance of final dyed fabrics. Thus, due to the effect of plasma treatment we could eliminate sodium sulphate in dyeing bath to obtain the same intensity of color after dyeing what is very important from both ecological and economical point of view [37]. Low-pressure plasma was used to deposit onto polyester fiber multi-functional thin film from ammonia/ethylene or acetylene mixture. The coated polyester showed fiber was acid dyeable and showed high color strength values per film thickness. Moreover, the plasma-deposited and dyed polyester fabrics showed a good rubbing and washing fastness demonstrating the coating-functional permanency. The excellent abrasion resistance confirmed that the coating was permanently adhered to the substrate [21, 38]. An increase in color depth upon dyeing was obtained after treating PET fabrics with low-pressure air and Ar RF plasma employed by Raffaele-Addamo et al. This may be easily related to optical effects connected to the plasma-induced increase of surface roughness, which contributes to the increase of K/S values of dyed PET specimens by decreasing the fraction of light reflected from treated surfaces respect to more smooth surfaces. Other effects, such as the increased surface area and the modifications of the partition equilibrium of the dye between the dyeing bath and the macromolecular surface in contact with it, can also play a role. Indeed, the introduction of hydrophilic groups, induced by both reactive and chemically inert plasmas, may increase the water swelling capability and the affinity of PET fibers for dyes containing polar groups [39]. The formation of negatively charged groups on poly (ethylene terephthalate) fabrics, PET, surfaces by using low-pressure oxygen plasma and subsequently, the positively charged polyelectrolyte poly(diallyldimethylammonium chloride), poly-DADMAC, was applied. This leads to stable surface modification with accessible quaternary ammonium groups that provide a high substantivity to acid dyes via complementary electrostatic attractions [40]. In this work, two different strategies to endow polyester fabrics (PET) with accessible primary amino groups are compared. (a) NH_2 groups were produced directly using low-pressure ammonia plasma. (b) Negatively charged groups were introduced by low pressure oxygen plasma to hydrophilize the fabric surfaces and used as anchor groups for the immobilization of waterborne polyelectrolyte copolymers poly(vinyl aminecovinyl amide) (PVAm).These carboxylic groups can act as an anchor for the adsorption of cationic polyelectrolytes from aqueous solution to PET fabric surface such as PVAm. It was demonstrated that the anchoring PVAm layer onto PET fabric surface modified with low-pressure oxygen plasma is an efficient approach for improving the interactions of PET fabrics with different anionic dyes. PET fabric modified with this synthetic route present the highest concentration of primary amino groups as determined by XPS measurements. This provides an efficient way to improve the surface interactions with anionic dyes as determined by K/S measurements and good stability [41].

Improvement in dyeing with suitable natural dyes is also demonstrated after plasma treatment, which is remarkable, in-terms that hazardous synthetic dyes can be replaced by biodegradable eco-friendly natural dyes. Recently, the attention for natural dyes, which are friendlier to the environment as compared to synthetic dyes, increased. Natural dyes can exhibit better biodegradability and generally have a higher compatibility with the environment. The potential use of natural dyes in textile coloration as anti-UV and anti-microbial has been investigated, however, natural dyestend to be hydrophilic and have limited substantivity for hydrophobic substrates, like polyester which is crystalline linear aromatic polymer with a small number of amorphous regions available for sorption and diffusion. In their work Shahidi and Ghoranneviss coated the samples with a layer of Platinum using a magnetron sputter device with a DC sputter source. The sputtered metal particles were deposited on the surface of the fabrics, which were placed on the platinum anode. The argon gas at pressure of 2×10^{-2} torr is used as discharge producing gas. Then the dyeability of this fabric to basic, acidic, disperse, madder and henna was investigated and compared with an untreated sample. The results show that, by coating samples with platinum, more basic dye can be absorbed by the fabric. However, platinum coating on surface of PET fabric has no effect on dyeability of PET by disperse dyes. By plasma sputtering, the relative color strength and fastness properties of dyed samples improve significantly. The improvement for natural dyes is more significant than for synthetic dye. According to antibacterial activity tests, platinum itself has no antibacterial effect against *Staphylococcus aureus*. However, the natural dyed Pt-coated samples show moderate antibacterial effect [42].

Plasma treatment with dielectric barrier discharge (DBD) in air was carried out on polyester substrate, to confer durable wettability and increase dye uptake with natural dye eco-alizarin. DBD at atmospheric pressure has attracted great interest in textile industry due to its low cost, high processing speed, and simple operating system that does not require costly vacuum equipment. SEM images (Fig. 6) of plasma treated fabric indicated surface of untreated fiber were smooth with only trace of impurities, while after plasma treatment some lamellar structures appeared on fiber surface. The observed variations in features may be due to differences in physical properties and crystallinity. Possible effects that explain the observed morphology was the occurrence of some amorphization of crystalline domains of PET provoked by the plasma polyester interaction. Partially crystalline polymers, as polyester, are built up as a two-phase structure, being composed of layer-like crystallites, which are separated by disordered regions. The impinging plasma species can transfer enough energy to the polymer matrix to enable the destruction of crystalline domains, which might relax in a disordered amorphous form, causing volume differences at surface [43].

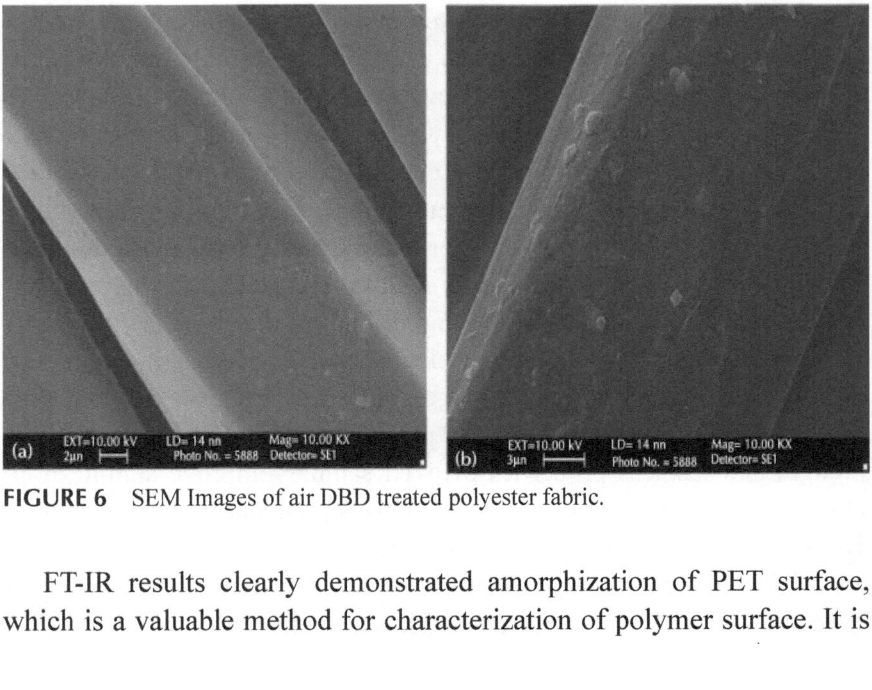

FIGURE 6 SEM Images of air DBD treated polyester fabric.

FT-IR results clearly demonstrated amorphization of PET surface, which is a valuable method for characterization of polymer surface. It is clear from the FT-IR spectra (Fig. 7) that there was no change in overall structure of the polymer, but change in intensity has been observed. The main characteristic absorption peaks of polyesters recorded are the C=O bond at 1710 cm^{-1}, vibration of aromatic ring at 1406, 1015, and 872 cm^{-1}, bending vibration of $-CH_2$ groups at 1340 and 1177 cm^{-1}, stretching vibration of C–O bonds at 1234 and 969 cm^{-1}, stretching vibration of C–O bonds due to amorphous and crystalline structure, respectively, at 1126 and 1082 cm^{-1}. The chain scission by plasma treatment reflected in the FTIR spectra by decrease in absorption bond corresponding to the carbonyl ester group at 1710 cm^{-1} and increase in intensity of terminal C–H bond at 1363 cm^{-1}. Increase in absorption intensity at 1556 cm^{-1} and 1577 cm^{-1} are assigned to asymmetrical stretching vibration of COO^-, and coupling interaction between the two equivalent carbon bonds of carboxylate anion, while increased absorbance signal at 2500–3600 cm^{-1} assigned to stretching vibration of O–H. The appearance of increase in intensity of these bands suggests oxygen containing polar groups like COO^-, OH introduce onto the PET macromolecular backbone. Amorphization of the crystalline fraction of the polymer and scission of the main chain at para position of disubstituted benzene rings were assigned to increase intensity of bands at 1472 and 1454 cm^{-1}, respectively. The conformational composition can be obtained from bands at 1340 cm^{-1} (*trans*-crystalline) and 1370 cm^{-1} (*gauche*-amporphous). The Intensity of *gauche* band increased relative to *trans* band after plasma treatment. The fraction of *trans* conformer (T) of fabric can be calculated, taking in account the intensities of the two bands using following equation. The fraction of *trans* conformer T (%) decrease from 21.77 for untreated to 19.74 after 15 min plasma treatment indication amorphization of fabric due to plasma treatment. Surface amorphization provides the sites for diffusion of dyes into polyester fibers, and thus increase dye uptake [43].

$$T = \frac{A1340}{A1340 + 6.6\,'\,A1370}$$

K/S value was increased significantly from 5.29 to 9.25 for air treated polyester fabric dyed in eco-alizarin dye. Wash and light fastness improved from fair to good. This has clearly indicated an increase in color intensity on plasma exposed polyester fabric as well as an increase in the dye uptake as compared with the untreated fabric. Thus, increase in hydrophylicity and incorporation of polar groups by oxidation of polymer surface and amorphization by plasma treatment increase dye uptake of polyester fabric with natural dye like Eco alizarin even at low temperature and atmospheric pressure dyeing condition [43].

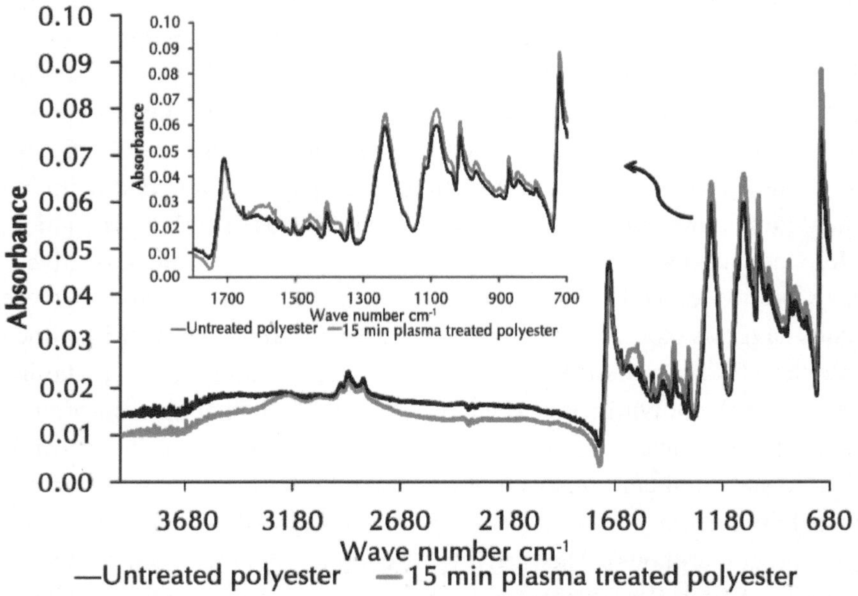

FIGURE 7 FT-IR spectra of untreated and plasma treated polyester fabric.

7.3.6 PRINTABILITY IMPROVEMENT

Due to its special characteristics such as superior strength and resilience, polyester fabric is often used as inkjet printing substrate, which is becoming increasingly important and popular for the printing of textiles. Nevertheless, patterns that are directly printed on polyester fabrics with pigment inks have poor color yields and bleed easily. Therefore, pre-treatment of fabric must be done before printing to obtain better inkjet printing effects. Traditional methods cost long time and engender huge consumption of energy and water. At the same time, the generation of toxic substance and wastewater may cause serious environmental problems. Plasma surface treatment of polyester fabrics carried out in composite atmosphere with air and 10%. Argon under different experimental conditions and printing with pigment inks results in to enhanced color yields and excellent pattern sharpness, much better antibleeding performance and fresher color, on plasma surface-modified polyester fabrics compared with untreated samples. The optimum treatment conditions obtained were exposure time of 150 s with a distance of 3 mm between the two discharging boards and a working power around 300 W. exhibited. SEM and XPS analyses indicated that this improved color performance was mainly contributed by not only the etching effect but also oxygen containing polar groups induced onto fiber surfaces through plasma treatment. Thereby the surface modification of polyester fabrics using air/Argon plasma offers a potential way to fabric pre-treatment for pigment inkjet printing with the advantages of environmental friendly and energy saving over traditional pre-treatment methods [44]. XPS analysis revealed that air + 50%Ar plasma introduced more oxygen-containing groups onto the fabric surface than air plasma. Although atomic force microscopy (AFM) images indicated that etching effects generated by air plasma treatments were more evident, the air/Ar plasma treated sample has higher K/S value and better color performance. These studies have also shown that the chemical modification of plasma appears to be relatively more significant for improving the effect of inkjet printing [45].

7.3.7 ANTI-STATIC PROPERTY IMPROVEMENT

The static problem is commonly found in synthetic fiber, especially polyester fabric and is significant in dry and low humidity condition. As for polyester, it has lower degree of amorphous region and higher degree of crystalline region, resulting in lower moisture regains and therefore, polyester has the most significant static problem. Generation of static charges is accomplished by motion such as rubbing fabric, walking on carpet and sitting down etc. During this movement, the fabric surfaces contact with each other, resulting in the generation of positive charges on one surface and negative charges on the other surface. The generated and accumulated charges will stay on the fabric and make the wearer uncomfortable. Polyester fibers widely employed for making filter fabric because of its excellent strength, anti-crease, good abrasion resistance and stable performance in acid or alkali environment. When the fabric is used as filter to separate solid particles, the flowing movement of finely powdered substances or mechanical agitation can build-up static electricity, potential endanger exists with accumulated static charge. Discharge of static electricity can create severe hazards in those industries dealing with flammable substances, where a small electrical spark may ignite explosive mixtures. Many methods have been employed to improve the anti-static properties of the polyester fabric like fiber modification; fabric anti-static treatment and including evenly spaced metal filaments in fabric are typical examples which are costly and ecologically harmful [46, 47].

Kan and Yuen applied low temperature plasma treatment to improve the anti-static property of polyester fabric treating with low-pressure RF plasma generated in oxygen under different conditions. Changes in surface properties of fabric provided more capacity for polyester to capture moisture and hence increase the dissipation of static charges. The relationship between moisture content and half-life decay time for static charges was studied and the results showed that the increment of moisture content would result in shortening the time for the dissipation of static charges. Moreover, there was a great improvement in the anti-static property of the low temperature plasma-treated polyester fabric after comparing with that of the polyester fabric treated with commercial anti-static finishing agent [47]. Chongqi et al., investigated atmospheric pressure plasma technique to improve its anti-static behaviors as an environmentally friendly process because no chemicals are involved and released in the treatment. The action of plasma is mainly interactive (such as oxidation and etching) on the surface of textile materials, hence the surface properties together with those associated with the surface characteristics modified and thus short time exposure to plasma could greatly reduce the charge density of the fabric. Further improvement of the fabric anti-static property was noticed after both plasma and acrylic acid treatments. Influences of some parameters during the treatments on the fabric charge density were evaluated. The research concluded that plasma discharge power of 50 V, electrode gap of 4 mm, acrylic acid concentration of 60%, acid immersion duration of 30 s and acid temperature of 70°C were recommended as the optimum selections for a better anti-static property of the fabric [47].

7.3.8 ANTIMICROBIAL FINISHING

Due to large surface area and ability to retain moisture, textile materials are likely to be affected by microorganism's growth and thus potential carriers of microorganisms, which can cause several problems like strength deterioration, staining of textiles, odor generation and health concerns caused by microorganisms. Textiles used as apparels, medial, and as protective clothing of workers or air filters in food processing industries are susceptible for cross-contaminations of pathogenic bacteria. During the last several years, antimicrobial finishing of textiles received a lot of attention, especially for use in long-lasting contact with a skin like Sportswear, medical, and protective textiles, which are particularly important in prevention and healing of sensitive skin prone to bacterial and fungal infections. Various agents based on quaternary ammonium salts, metal salts solutions, and antibiotics were used for imparting antimicrobial properties to textile materials. However, poor efficiency or high toxicity made them unsuitable for the long-term use. On the contrary, silver in different forms exhibit outstanding antimicrobial activity showing low toxic impact to mammalian cells. The manufacturing of sportswear, medical, and protective textiles based on PET fibers continuously increased but these fibers exhibit hydrophobic behavior and low surface energy which results in poor incorporation of various antimicrobial finishes as well as its durability.

Higher accessibility of hydrophobic fibers to various chemical species can be obtained by plasma functionalization and/or plasma etching. Yang et al., in their study activated nonwoven PET by argon gas plasma generated by applying RF frequency and subsequently grafted with water-soluble monomers (acrylamide and itaconic acid) onto the PET by UV induced surface graft polymerization to modify it into a hydrophilic surface. Subsequently, the nonwoven PET can be further incorporated with biocides such as $AgNO_3$ solution complexes, Vinyl quaternary ammonium salt (VQAS) and chitosan. For Ag^+ ion impregnation and VQAS grafting, the amount of these biocidal materials on the surface increases when the monomer concentration increases. However, for the chitosan immobilization, the chitosan amounts seemingly decrease with an increase in the chitosan concentration. For biocidal properties, the Ag^+ ions produce best result. However, the VQAS grafting is most durable against the flushing water [48]. The chemical deposition of silver-clusters onto polyamide–polyester based fabric surfaces previously activated by low pressure RF plasma reported to provide excellent antibacterial activity by Yuranova et al., [49]. The low-pressure RF plasma used in the above mentioned study limits its industrial applications, as it is not suitable form in line processes. The limitation of low-pressure plasma can be overcome by using atmospheric pressure plasma treatment based on DBD, one of the most apposite technologies for textile modification. In their work, Onsuratoom et al., treated the woven PET surface by dielectric barrier discharge (DBD) under various operating conditions (electrode gap distance, plasma treatment time, input voltage, and input frequency) and various gaseous environments (air, O_2, N_2, and Ar) in order to improve its hydrophilicity. It was experimentally found that a decrease in electrode gap distance and an increase in input voltage increased the electric field strength, leading to higher hydrophilicity of the PET surface characterized by wickability and contact angle measurements. In comparisons among the studied environmental gases, air gave the highest hydrophilicity, being comparable to O_2, while Ar and N_2 gave lower hydrophilicity of the woven PET surface. For air DBD discharge, the optimum conditions for a maximum hydrophilicity of the PET surface were an electrode gap distance of 4 mm, a plasma treatment time of 10 s, an output voltage of 15 kV, and a frequency of 350 Hz. After the plasma treatment under the obtained optimum conditions, the woven PET was loaded with Ag particles using a $AgNO_3$ aqueous solution, which exhibited good antimicrobial activity against both *E. coli* (gram-negative bacteria) and *S. aureus* (gram-positive bacteria) [50]. In another work, a woven PET with an antimicrobial activity was prepared by depositing chitosan on its hydrophilic surface achieved by a plasma treatment using DBD in air. The plasma-treated PET specimen deposited with chitosan by immersing in a chitosan acetate aqueous solution and the effects of temperature, chitosan concentration, and number of rinses on the amount of deposited chitosan on the PET surface was investigated. The chitosan deposited plasma-treated woven PET possessed an exceptionally high antimicrobial activity against both *E. coli* (gram-negative bacteria) and *S. aureus* (gram-positive bacteria). In XPS Study, it was clearly observed polar functional groups formed by plasma treatment involve in interacting with the chitosan [51].

Plasma treatment can also open new possibility for the introduction of nanotechnologies in textile processing, using Ag nanoparticles (NPs) (supported or not to be) used as antimicrobial/bacterial agent. The recent progress in synthesis of Ag (NPs) without stabilizers opened up new possibilities for advanced textile finishing which is favorable in comparison with other silver forms since the significant portion of Ag atoms on the surface of the NPs is exposed toward surrounding medium, providing a desired antimicrobial effect, large area-to-volume ratio and high reactivity being of practical use when supported on a variety of surfaces. Improved loading of Ag NPs from colloids and enhanced interaction between hydrophilic Ag NPs and hydrophobic polyester fabrics can be achieved by plasma treatment [52]. In their study Ilić et al., researched on possibility of using the corona treatment for fiber surface activation (introduction of new reactive groups, oxidation, free radicals, increase of solid surface tension, etc.), which can facilitate the loading of Ag NPs from colloids onto the PET fabrics and thus, improve their antifungal activity against *Candida albicans* which is responsible for a wide range of itching skin infections with yeasts particularly in the skin folds. Corona treatment positively influenced the binding of Ag NPs to the surface of PET fabrics, leading to an increase in the amount of silver. The content of Ag on corona pre-treated PET fibers the content is 60% higher than on untreated PET fabric. The increased number of carboxyl groups confirmed by XPS analysis along with existence of benzene rings in polymer structure indicates the possibility of strong interaction between PET fibers and Ag NPs. Corona pre-treated polyester fabrics loaded with Ag nanoparticles showed better antifungal properties compared to untreated fabrics. The advantage of corona treated fabrics became even more prominent after washing test, particularly for polyester fabrics. Antifungal efficiency of polyester fabrics loaded with Ag NPs were almost unaffected by dyeing process with disperse dyes. Thus, atmospheric plasma treatments such as corona discharge might be more suitable for textile processing, avoiding more complex handling of textile materials through the vacuum systems in case of low pressure plasma

[52]. The possibility of using DBD air plasma treatment for fiber surface activation to facilitate deposition of nano silver onto polyester fabric was investigated. Nano silver in powder form (1 g/100 g fiber) were distributed carefully either on the polyester samples or on the electrode disk before subjecting to the plasma treatment. In other experiment, some samples were padded in a solution containing 1% (owf) of nano-silver, squeezed to pick up 100%, air dried and then subjected to air low temperature plasma treatments at discharge powers of 1.3 and 2 watt for 1 and 2 min. Treatment with plasma-nano Ag was evaluated by the measurements of antibacterial activity which was found to be improved. The second treatment technique (padding) is found to be more effective on enhancing the antibacterial activity. It is noticed that the maximum inhibition zone could be attained upon using the second technique at discharge power of 1.3 watt for 2 min, while the minimum value was observed for sample treated by the first technique [53].

Another alternative plasma processing is the sputtering in a vacuum chamber used to deposit very thin Ag films (few nanometers to several micrometers width) on substrates. It is performed by applying a high voltage across a low-pressure gas (usually argon) to create "plasma," which consists of electrons and gas ions in a high-energy state. The deposition of silver by Sputtering allows the deposition of Ag-films showing high bactericide activity and adhesion. Depositing Ag by magnetron sputtering has been recently used as outperforming physical vapor deposition (PVD) having has such advantages like having stronger binding between the film and its substrate and no potentially polluting waste products unlike in case of conventional treatment. Magnetron sputtering has developed over the last few years due to the increasing demand of high quality functional films showing improved wear resistance, corrosion resistance and defined optical properties. Chadeau et al., in their study have deposited thin layers of silver particles (10–100 Å) by DC magnetron sputtering on textiles mainly composed with cotton or polyester in order to obtain antimicrobial properties. These textiles presented a strong antimicrobial activity against *L. innocua*. It could be interesting to reduce the quantity of silver deposited on textile because both of the possible toxicity of silver and the resulting colorization of textile. In the present study, it was shown that the thinner thicknesses of silver layers were as effective as larger silver layers.

This methodology has potential to develop antibacterial textiles for an application in food-processing industry for dressing or air filtration [54].

Baghriche and their co-workers have reported that the sputtering deposition of Ag leads to uniform nano-films showing a high bactericide activity and adequate film adhesion on polyester substrate avoiding features found Ag-films deposited some years ago in their laboratory by wet techniques leading to non-uniform Ag-films, with a low antibacterial activity and adhesion. In the study focuses on DC-magnetron sputtering (DCMS) and pulsed magnetron sputtering (DCPMS) as effective methods to functionalizing polyester fibers containing silver nanoparticles for inactivation of *E. coli*, they compared Ag-samples sputtered in the same chamber and with the same geometry varying the sputtering energy and pulse wave either in the DCMS or DCPMS. The deposition of Ag on the polyester fiber was observed to be the function of the type of sputtering used and this has direct implications on the *E. coli* inactivation kinetics. The shortest *E. coli* inactivation was observed within two hours when sputtering with DCPMS Ag-layers 160 nm thick ~ at 300 mA within 2.7 min. This is equivalent to a growing coating rate of 74 nm/min. By transmission electron microscopy (TEM) Ag-polyester fibers it was found that Ag sputtered by DCPMS penetrated more deeply into the polyester full fiber compared to DC-sputtering leading to a more favorable *E. coli* inactivation kinetics. The *E. coli* inactivation times were found to be a function of the applied DCMS and DCPMS energies and higher sputtering energies reduced the *E. coli* inactivation times [55]. They have further compared DCMS Ag-films on polyester with high power impulse magnetron sputtering (HIPIMS), which is an important development in the sputtering field and is gaining acceptance in many applications for surface treatments of metallic surfaces due to the HIPIMS higher sputtering energy of several kW/cm^2 and higher electronic densities of $1018/m^3$ compared to $1014/m^3$ by DCMS. This enhances the film uniformity, the production of highly ionized metal-ions, the film adherence and compactness. HIPIMS technology produces high power homogeneous plasma glow at high currents up to 2000 V and 10 A. The amounts of Ag needed to inactivate *E. coli* by HIPIMS sputtering were an order of magnitude lower than with DCMS indicating a significant saving of noble metal and concomitantly a faster *E. coli* inactivation was observed compared to samples sputtered with DCMS. Higher current densities applied with DCMS led to shorter *E. coli* inactivation times and this trend was observed also for HIPIMS sputtered samples. By DCMS the thicker layers needed to inactivate *E. coli* comprised slightly larger Ag-aggregates compared to the thinner Ag-layers sputtered by HIPIMS to inactivate *E. coli* within short times. Longer sputtering times by DCMS and HIPIMS lead to optically darker Ag-deposits reaching the absorption edge of silver absorption of ~1000 nm. Mass spectroscopic analyses indicated that HIPIMS produced a much higher amount of Ag^{1+} and Ag^{2+} compared to DCMS due to the higher peak discharge current employed in the former case [56].

7.3.8 FLAME RETARDANCY

Carosio et al., studied plasma surface activation by low-pressure oxygen plasma at different process parameters (namely, power and etching time) in order to increase nanoparticle adsorption (i.e., a natural montmorillonite) to create a functional coating for thermal stability and flame retardant properties. Plasma activated fabrics have shown a remarkable improvement in terms of time to ignition (up to 104%) and a slight reduction of the heat release rate (ca. 10%) as compared to neat PET [57]. The possibility of using atmospheric pressure air DBD plasma to facilitate the deposition of Al_2O_3, onto polyester fabric was investigated. Aluminum oxide powder (1 g/100 g fiber) was distributed carefully either on the polyester samples or on the electrode disk before subjecting to the plasma treatment. Treatment of PET with plasma-Al_2O_3 indicated that whiteness index and surface roughness were dependent on the time of exposure. The TGA analysis and flame retardancy tests at 1.3 watt for 2 min indicate that this treatment caused an improvement in thermal stability and flame retardancy which was found to be more effective upon using the technique of spreading the powder on the sample rather than on the electrode disk. Also dyeing and printing properties improved slightly due to roughness increase by DBD treatment [53].

7.3.9 UV PROTECTION

The possibility of using atmospheric plasma DBD to facilitate the deposition of nano TiO_2 onto polyester fabric was investigated. Nano titanium dioxide powder (1 g/100 g fiber) was distributed carefully either on the polyester samples or on the electrode disk before subjecting to the plasma treatment. It was supposed that TiO_2 NPs formed complexes with oxygen atom built on fiber surface after plasma treatment. TiO_2 NPs could be adsorbed on surface with –COOH groups by forming hydrogen bond or by bounding with two oxygen atoms Plasma-nano TiO_2 treatment resulted in a noticeable enhancement in Ultraviolet Protection Factor of polyester fabric [53].

7.4 CONCLUSION

In this chapter, various approaches of plasma surface modification of polyester textile have been described. This chapter demonstrated that plasma treatments of polyester look very promising which can be used both in substitutions of conventional processes and for the production of innovative multifunctional polyester with properties that cannot be achieved via conventional processing. Plasma treatments have potential to solve problems with synthetic polyester textile to expand their usefulness for examples problems such as low absorbency of water, flammability, and pilling, low dyeability, and static problems during production and usage can be solved. Both the improved and the newly developed finishes without significant alteration of bulk properties can be produced by plasma treatment with advantages like fast and extremely gentle process, as well as environmentally friendly, being dry processes characterized by low consumption of chemicals and energy. If used as pre-treatments, plasma treatment can reduce markedly the amount of chemicals required by the process and the concentration of pollutants in the effluents. Polyester textile industry being largest in synthetic textile will undergo revolution in next ten years with applications of plasma treatment for textile processing due to great advances of the last two decades in the field of the plasma polyester processing. Also, environmental aspects are going to play a more and more important role. With these perspectives, plasma processing is certainly going to supersede many conventional processes [58–60].

KEYWORDS

- **fourth state of matter**
- **padding**
- **plasma**
- **polyester**

REFERENCES

1. Polyethylene Terephthalate Global Production and Market, Eurasian Chemical MarketInternational Magazine [online] 2011; *8(56)*, article/2422/. http://www.chemmarket.info/en/home/article/2422/ (accessed 5-12-2012).
2. Oerlikon Textile GmbH & Co. KG (Germany), The Fibre Year 2011; http://www.oerlikontextile.com/desktopdefault.aspx/tabid-1763/ (accessed 5/12/2012).
3. Donelli, I.; Freddi, G.; Nierstrasz, V. A.; Taddei, P. *Polym. Degrad. Stabil.*, **2010**, *95*, 1542–1550.
4. Parvinzadeh, M.; Ebrahimi, I. *Appl. Surf. Sci.*, **2011**, *257*, 4062–4068.
5. Kale, K. H.; Desai, A. N. *Indian J. Fibre Text.*, **2011**, *36*, 289–299.
6. Morent, R.; De Geyter, N.; Verschuren, J.; De Clerck, K.; Kiekens, P.; Ley, C. *Surf. Coat. Tech.*, **2008**, *202*, 3427–3449.
7. Tomasino, C.; Cuomo, J. J.; Smith, C. B.; Oehrlein, G. *J. Ind. Textil.*, **1995**, *25*, 115–127.
8. Morshed, A. M. A., An Overview of the Techniques of Plasma Applications in Textile Processing. *Bangladesh textile today* [online] 2011; Mar-Apr issue, http://www.textiletoday.com.bd/magazine/printable.php?id=175 (accessed 5-12-2012).
9. Garg, S.; Hurren, C.; Kaynak, A. *Synt. Met.*, **2007**, *157*, 41–47.
10. Šimor, M.; Ráhel', J.; Černák, M.; Imahori, Y.; Štefečka, M.; Kando, M., *Surf. Coat. Tech.*, **2003**, *172*, 1–6.
11. Yuen, C. W. M.; Jiang, S. Q.; Kan, C. W.; Tung, W. S. *Appl. Surf. Sci.*, **2007**, *253*, 5250–5257.
12. Kan, C. W.; Yuen, C. W. M. *Nucl. Instr. and Meth. in Phys. Res. B*, **2007**, *265*, 501–509.
13. Leroux, F.; Campagne, C.; Perwuelz, A.; Gengembre, L. *Surf. Coat. Tech.*, **2009**, *203*, 3178–3183.

14. Wróbel, A. M.; Kryszewski, M.; Rakowski, W.; Okoniewski, M.; Kubacki, Z.; *Polymer,* **1978**, *19*, 908–912.
15. Riccardi, C.; Barni, R.; Selli, E.; Mazzone, G.; Massafra, M. R.; Marcandalli, B.; Poletti, G. *Appl. Surf. Sci.*, **2003**, *211*, 386–397.
16. Ferrero, F. *Polym. Test.*, **2003**, *22*, 571–578.
17. Vatuňa, T.; Špatenka, P.; Píchal, J.; Koller, J.; Aubrecht, L.; Wiener J. *Czech. J. Phys.* **2004**, *54 (Suppl. C),* C1–C8.
18. Wei, Q.; Liu, Y.; Hou, D.; Huang, F. *J. Mater. Process. Technol.*, **2007**, *194*, 89–92.
19. Wardman, R. H.; Abdrabbo, A. *AUTEX Res. J.*, **2010**, *10(1)*, 1–7.
20. Costa, T. H. C.; Feitor, M. C.; Alves Jr., C.; Freire, P. B.; de Bezerra, C. M. *J. Mater. Process. Technol.*, **2006**, *173*, 40–43.
21. Hegemann, D.; Hossain, M. M.; Balazs, D. *J. Prog. Org. Coat.*, **2007**, *58*, 237–240.
22. Calvimontes, A.; Saha, R.; Dutschk, V. *AUTEX Res. J.*, **2011**, *11*, 24–30.
23. Xu, W.; Liu, X. *Eur. Polym. J.*, **2003**, *39*, 199–202.
24. Samanta, K. K.; Jassal, M.; Agrawal, A. K. *Surf. Coat. Tech.*, **2009**, *203*, 1336–1342.
25. De Geyter, N.; Morent, R.; Leys, C. *Surf. Coat. Tech.*, **2006**, *201*, 2460–2466.
26. Píchal, J.; Klenko, Y. *Eur. Phys. J. D,* **2009**, *54*, 271–279.
27. Wang, C. X.; Ren, Y.; Qiu, Y. P. *Surf. Coat. Tech.*, **2007**, *202*, 77–83.
28. Wang, C. X.; Liu, Y.; Xu, H. L.; Ren, Y.; Qiu, Y. P. *Appl. Surf. Sci.*, **2008**, *254*, 2499–2505.
29. Wang, C. X.; Du, M.; Qiu, Y. P. *Surf. Coat. Tech.*, **2010**, *205*, 909–914.
30. Bae, P. H.; Hwang, Y. J.; Jo, H. J.; Kim, H. J.; Lee, Y.; Park, Y. K.; Kim, J. G.; Jung, J. *Chemosphere* **2006**, *63*, 1041–1047.
31. Li, X.; Lin, J.; Qiu, Y. *Appl. Surf. Sci.*, **2012**, *258*, 2332– 2338.
32. Leroux, F.; Campagne, C.; Perwuelz, A.; Gengembre, L. *Appl. Surf. Sci.*, **2008**, *254*, 3902–3908.
33. Zanini, S.; Massini, P.; Mietta, M.; Grimoldi, E.; Riccardi, C. *J. Colloid Interface Sci.*, **2008**, *322*, 566–571.
34. Ji, Y. Y.; Hong, Y. C.; Lee, S. -H.; Kim, S.-D.; Kim, S.-S. *Surf. Coat. Tech.*, **2008**, *202*, 5663–5667.
35. Ji, Y. Y.; Chang, H. K.; Hong, Y. C.; Lee, S.-H. *Curr. Appl. Phys.*, **2009**, *9*, 253–256.
36. Lee, H.-R.; Kim, D.; Keun-Ho Lee, K.-H. *Surf. Coat. Tech.*, **2001**, *142–144*, 468–473.
37. Lehocký, M.; Mráček, A. *Czech. J. Phys.*, **2006**, *56(Suppl. B)*, 1277–1282.
38. Hossain, M. M.; Herrmann, A. S.; Hegemann, D. *Plasma Processes Polym.*, **2007**, *4*, 135–144.
39. Raffaele-Addamo, A.; Selli, E.; Barni, R.; Riccardi, C.; Orsini, F.; Poletti, G.; Meda, L.; Massafra, M. R.; Marcandalli, B. *Appl. Surf. Sci.*, **2006**, *252*, 2265–2275.
40. Salem, T.; Uhlmann, S.; Nitschke, M.; Calvimontes, A.; Hund, R.-D.; Simon, F. *Prog. Org. Coat.*, **2011**, *72*, 168–174.
41. Salem, T.; Pleul, D.; Nitschke, M.; Müller, M.; Simon, F. *Appl. Surf. Sci.*, **2012**, http://dx.doi.org/10.1016/j.apsusc.2012.10.014.
42. Shahidi, S.; Ghoranneviss, M. *Prog. Org. Coat.*, **2011**, *70*, 300–303.
43. Dave, H.; Ledwani, L.; Chandwani, N.; Kikani, P.; Desai, B.; Chowdhuri, M. B.; Nema, S. K. *Compos. Interfaces,* **2012**, *19*, 219–229.
44. Zhang, C.; Fang, K. *Surf. Coat. Tech.*, **2009**, *203*, 2058–2063.
45. Fang, K.; Zhang, C. *Appl. Surf. Sci.*, **2009**, *255*, 7561–7567.

46. Kan, C. W.; Yuen, C. W. M. *Nucl. Instr. and Meth. in Phys. Res. B,* **2008**, *266*, 127–132.
47. Chongqi, M.; Shulin, Z.; Gu, H. J. *Electrostat.* **2010**, *68*, 111–115.
48. Yang, M.-R.: Chen, K.-S.; Tsai, J.-C.; Tseng, C.-C.; Lin, S. F. *Mater. Sci. Eng. C,* **2002**, *20*, 167–173.
49. Yuranova, T.; Rincon, A. G.; Bozzi, A.; Parra, S.; Pulgarin, C.; Albers, P.; Kiwi, J. J. *Photochem. Photobiol. A,* **2003**, *161*, 27–34.
50. Onsuratoom, S.; Rujiravanit, R.; Sreethawong, T.; Tokura, S.; Chavadej, S. *Plasma Chem. Plasma Process.,* **2010**, *30*, 191–206.
51. Sophonvachiraporn, P.; Rujiravanit, R.; Sreethawong, T.; Tokura, S.; Chavadej, S. *Plasma Chem. Plasma Process.,* **2011**, *31*, 233–249.
52. Ilić, V.; Šaponjić, Z.; Vodnik, V; Molina, R.; Dimitrijević, S.; Jovančić, P.; Nedeljković, J.; Radetic, M. *J. Mater. Sci.,* **2009**, *44*, 3983–3990.
53. Raslan, W. M.; Rashed, U. S.; El-Sayad, H.; El-Halwagy, A. A. *Mat. Sci. Appl.,* **2011**, *2*, 1432–1442.
54. Chadeau, E.; Oulahal, N.; Dubost, L.; Favergeon, F.; Degraeve, P. *Food Control,* **2010**, *21*, 505–512.
55. Baghriche, O.; Ruales, C.; Sanjines, R.; Pulgarin, C.; Zertal, A.; Stolitchnov, I.; Kiwi, J. *Surf. Coat. Tech.,* **2012a**, *206*, 2410–2416.
56. Baghriche, O.; Zertal, A.; Ehiasarian, A. P.; Sanjines, R., Pulgarin, C.; Kusiak-Nejman, E.; Morawski, A. W.; Kiwi, J. *Thin Solid Films,* **2012**, *520*, 3567–3573.
57. Carosio, F.; Alongi, J.; Frache, A. *Eur. Polym. J.* **2011**, *47*, 893–902.
58. BioZone Europe, Purifying Plasma, http://www.biozone-europe.eu/v2/en/technologies/cold-plasma (accessed 5-12-2012).
59. Hitachi Plant technologies, Ltd., Production process for polyethylene terephthalate (PET), http://www.hitachi-pt.com/products/ip/process/pet.html (accessed 5/12/2012).
60. Nuovaplast, PET, http://www.asaplast.com/pet-en.html, (accessed 5/12/ 2012).

CHAPTER 8

EQUALIZATION "PRINCIPLES" IN CHEMISTRY

A. CHAKRABORTY, R. DAS, and P. K. CHATTARAJ

CONTENTS

8.1 INTRODUCTION

In this chapter, we present and analyze three important equalization principles in chemistry *viz.*, electronegativity equalization principle (EEP), hardness equalization principle (HEP), and principle of electrophilicity equalization (PEE).

The basic philosophy towards the fruitful completion of a chemical reaction and hence, the formation of a stable chemical bond is "spontaneity." According to the second law of thermodynamics spontaneity originates from a potential difference, which ceases at the equilibrium attained through a spontaneous process. In thermodynamic equilibrium simultaneously thermal (equalization of temperature), mechanical (equalization of pressure) and material (both phase and reaction, equalization of chemical potential) equlibria are attained. A spontaneous chemical process at constant temperature and pressure is usually accompanied by a favorable lowering of the Gibbs free energy of the entire system leading to the formation of a strong chemical bond. A chemical reaction is devised on the concept of reorganization of the electron density around the constituent interacting atoms to create new molecules as products. A reaction pathway is therefore envisaged as the mode in which the electron density distribution around the reactant atoms changes to eventually offer a new arrangement of atoms. Such electron distribution and redistribution are commonly ascribed to the phenomena of rupture of older bonds followed by the formation of newer ones. Therefore, Chemistry, attributed to the science of bond-breaking and bond-making actually follows the logic of changes in the electron density distribution pattern in systems upon any external perturbation/excitation. Now, the manner in which the electron density in an atom or molecule changes upon an external effect neatly depends on some important electronic properties, which are conceived on a purely theoretical basis. In this regard, conceptual density functional theory, CDFT [1–5] as a unique theoretical algorithm deserves to be mentioned specially. The DFT bears its roots from the two theorems stated and proved by Hohenberg and Kohn [6] and is further nourished and flourished under the leadership of Professor Robert G. Parr [1] who is known to be a front runner in the development of the DFT school and applications of the conceptual DFT methodology in chemistry. The given mathematical algorithm attempts to elucidate an interacting system of fermions via its probability density, ρ(r) instead of the many-body wave function, $\psi(r_1, r_2, \ldots\ldots\ldots, r_N)$. For an N-electron system, the corresponding Hamiltonian is completely determined by the number of electrons (N) and the external potential υ(r), which entails that all the properties of the given system may be ascertained by appropriate variations of N and υ(r). Conceptual DFT (CDFT) boasts off several mathematical response functions also known as descriptors. These descriptors serve as important indices, which quantitatively determine the chemical reactivity of a molecule as a whole or that of a particular atomic site therein. Thus these chemical reactivity descriptors fall into two categories, viz. global reactivity descriptors (which determine the reactivity pattern of an entire species) and local reactivity descriptors (that determine the reactivity pattern of a particular atomic site in a molecule). The global reactivity descriptors nicely mimic the molecular properties as dictated by the popular qualitative chemical concepts like electronegativity (χ) [7–9], hardness (η) [10–12] and electrophilicity (ω) [13–16].

For an N – electron system with total energy E, electronegativity (χ) [7–9] and hardness (η) [10–12] are defined as the following energy derivatives, within a conceptual density functional theory framework:

$$\chi = -\mu = -\left(\frac{\partial E}{\partial N}\right)_{v(\mathrm{r})} \tag{1}$$

and

$$\eta = \left(\frac{\partial^2 E}{\partial N^2}\right)_{v(\mathrm{r})} \tag{2}$$

In the above equations μ and v (r) represent the chemical and external potentials, respectively.

Parr et al. [13] have defined an electrophilicity index (ω) as follows:

$$\omega = \frac{\mu^2}{2\eta} = \frac{\chi^2}{2\eta} \tag{3}$$

So, a variation in the corresponding electronegativity (χ) [7–9], hardness (η) [10–12] and electrophilicity (ω) [13–16] values of a given system upon chemical response affirms a solid theoretical rationale towards understanding molecular reactivity. More inroads into a deeper understanding of reactivity trends can be explained from a vivid scrutiny of the corresponding local variants like atomic charges (Q_k) [17] and Fukui functions (f_k) [18], which further play a key role in ascertaining the local site-selectivity in a molecule.

A spontaneous chemical process is therefore accompanied by favorable changes in the values of the reactivity descriptors (χ, η and ω) of the interacting system(s) which eventually leads to the formation of stable product(s). The given values for the products seem to bear a mathematical rationale with the corresponding values of the reactants. Such a relationship triggers the existence of an equalization phenomenon in between the reactants and products involved in the chemical reaction. An equalization phenomenon in fact means the changes in the value of a given global reactivity parameter that is observed during a favorable chemical process. Therefore, the change in the values of these reactivity parameters during an electron-transfer process (a reaction) is a signature of its spontaneity. This ideology sets up the scientific rationale of proposing equalization principles, which act as further guiding parameters to determine the extent to which the reactivity descriptors actually influence the reaction behavior of a chemical species. In this chapter, we discuss the three important equalization principles and their influence on molecular reactivity.

8.2 ELECTRONEGATIVITY EQUALIZATION PRINCIPLE

The introduction of the term electronegativity (χ) to chemical literature is credited to Professor Linus Pauling, a pioneer towards the development of the valence bond theory (VBT) [19, 20]. Pauling, in an attempt to explain the factor of "extra stabilization" of a heteronuclear chemical bond A – B (between two different atoms A and B) obtained from its allied homonuclear analogues (A – A and B – B) during a chemical combination process, proposed that, there actually occurs a difference in the aptitude of electron donation (or acceptance) between the interacting atoms. He defined this behavior as the ability of an atom (or a functional group) in a molecule to attract bonded electrons (or electron density) towards itself, which is named as electronegativity. Another generalized definition of the electronegativity is also proposed [21] as: "An atom, for which the E vs. N plot (E = energy, N = number of electrons) has a given slope at the origin $(-dE/dN)_{N=0}$ will take electrons away from any atom which has a smaller slope, because in the process of doing so, the energy of the system as a whole will be lowered. Since this is the behavior of a more electronegative atom toward a less electronegative atom, one may identify the term electronegativity with this slope: $\chi = (-dE/dN)_{N=0}$. Similarly, the electronegativity of an ion would be given by the slope of the curve corresponding to that ion; …". Mulliken's prescription [19, 20, 22, 23] of the electronegativity (χ) is however based on an arithmetic mean of two experimental observables viz., first ionization potential (*IP*) and the electron affinity (*EA*). Thus, the electronegativity (χ) according to Mulliken's formalism can be mathematically expressed as:

$$\chi = \frac{IP + EA}{2} \tag{4}$$

This definition of the electronegativity (χ) is independent of any arbitrary relative scale and therefore, unlike Pauling's recipe (of χ) is invariant with the chemical environment. Mulliken's electronegativity is also termed as "absolute electronegativity" by Pearson [24]. Therefore, the drift of electrons during a chemical process occurs from a less electronegative center to a more electronegative center accompanied by a gradual lowering of energy of the system. The question therefore arises as to how long will this electron shifting between the two species occur? Well, definitely till the system reaches an energy minimum. A minimum energy situation is also accompanied by the attainment of a unique electronegativity value in the resulting product. It thus seems as if that the electronegativity values of the interacting atoms adjust among themselves (increase or decrease) and eventually give rise to a common, single value in the product. This idea conceptualizes the electronegativity equalization principle (EEP) and is credited to R. T. Sanderson [25–27]. The Sanderson principle states that "when two or more atoms initially different in electronegativity combine chemically, they become adjusted to the same intermediate electronegativity within the compound." Such an equalization of the electronegativity values of the interacting chemical systems closely mimic the "principle of calorimetry" as applicable to the macromolecules. This intermediate electronegativity (χ_{GM}) is given by the geometric mean of the individual electronegativities of the isolated component atoms [28–30] and is expressed as:

$$\chi_{GM} \approx \left(\prod_{k=1}^{P} \chi_k \right)^{\frac{1}{P}} \quad (5)$$

where the molecule contains P atoms (same or different) and { χ_k , $k = 1, 2$,......., P } signify their isolated atom electronegativities.

The rationale of electronegativity equalization based on an exponential model of the variation of energy with respect to the number of electrons was provided by Parr and Bartolotti [29]. It shows an exponential decay in the atomic energies for several atoms as a function of the number of electrons which, eventually generates an almost constant decay parameter ranging between 2.15 ± 0.59. The protocol of electronegativity equalization using both local approach (exploiting the local changes in electron density), global approach (using the global electron transfer phenomenon) as well as the atoms-in-a-molecule (AIM) approach [31, 32] (particularly for diatomic molecules) is also elaborated [31]. The migration of electron density from the less electronegative (and hence more electropositive) atom to the more electronegative (and thus less electropositive) one produces a partial positive charge on the former and an equal and opposite partial negative charge on the latter. The existence of a positive charge on the electropositive atom increases its effective nuclear charge thereby enhancing its electronegativity value. The same trend happens in the opposite direction for the more electronegative atom, until the two interacting systems acquire the same electronegativity. The process of electronegativity equalization therefore occurs through the adjustment of the bond polarities. Again as the chemical potential (μ) of a system corresponds to the "escaping tendency" of the electron cloud from the same, its obvious significance in this regard should be noted. The charge transfer between the two interacting systems occurs due to the presence of a chemical potential gradient and will continue till their chemical potentials are equalized to a unique value. The Sanderson principle therefore connotes the equalization of both the electronegativity (χ) and chemical potential (μ) of two interacting species. Further validations [28, 29] also show that the equalization of the electronegativity and the global chemical potential are similar concepts.

8.3 HARDNESS EQUALIZATION PRINCIPLE

The proposal of a hardness equalization principle [33] possibly arises from a dual consequence of the electronegativity equalization principle and the fact that global molecular softness (S) can be expressed as a mean of the corresponding local softness (S_i) of the constituent atoms [34]. It is already known that the hardness (η) varies directly with the electronegativity (χ). i.e., η µ χ, for both atoms [30, 34] and molecules [34]. Therefore, Eq. (1) can be similarly expressed in terms of hardness as [33, 35, 36]:

$$\eta_{GM} \approx \left(\prod_{k=1}^{P} \eta_k \right)^{\frac{1}{P}} \tag{6}$$

where { η_k , k = 1, 2 ,……, P } correspond to the associated isolated atom hardness values. Molecular hardness of various systems is computed from their respective atomic hardnesses using the above formalism and a close correspondence between the calculated and experimental hardness values is obtained. Such an observation corroborates with the existence of a hardness equalization principle as a corollary of the widely discussed EEP. However, in spite of equality among the local and global hardness values in many cases, ambiguities regarding the universal validation of the HEP do remain due to an indistinct definition of the concept of local hardness [37–46].

8.4 PRINCIPLE OF ELECTROPHILICITY EQUALIZATION

An electrophile in the chemical sense simply means an electron-deficient species, which has got a special affinity towards attracting electrons. It may attain some positive charge or even exist as a free radical with a strong inclination towards attracting electron-rich systems (nucleophiles). The interaction between the electrophiles and nucleophiles helps us to design the mechanistic pathway of a plethora of chemical reactions. Thus, an electrophile upon reacting with a nucleophile tends to "pull off" electron density from the latter with the formation of a stable covalent bond along with a consequent lowering of energy of the system. The qualitative concept of such an energy lowering of an electrophilic system upon chemical attack (by a nucleophile) has been quantified by Parr et al. [13] with the prescription of a new reactivity descriptor called electrophilicity index (ω). Parr's postulate of electrophilicity index (ω) is based on thermodynamic considerations and is a measure of the favorable change in energy upon saturation of a system with electrons. The electrophilicity index (ω) as defined by Parr et al. [13] is:

The above expression for electrophilicity (ω) mimics the equation of electrical power (W) in classical physics: $W = V^2/R$, where V and R denote the voltage and electrical resistance, respectively. Thus ω represents the "electrophilic power" of a species. Some recent reviews on electrophilicity index (ω) [14–16] have made an in-depth study regarding its genesis and rigorous applications towards an understanding of chemical reactivity.

Now, when an electrophile (an electron-deficient species) reacts with a nucleophile (an electron-rich species), the electrophilicity of the former is decreased while for the latter it is increased respectively owing to a charge transfer process. Such a phenomenon calls for an equalization of the electrophilicity values between the interacting systems and hence like that of the chemical potential (discussed earlier) conceptualizes an electrophilicity equalization principle [47]. The recipe of the equalization of electrophilicity (ω) originates from the fact that both the chemical potential (μ) and hardness (η), of which it (ω) is qualitatively conceived [13] are also equalized during a chemical process. Hence a geometric mean principle for the equalization of ω can be conceived. To be precise, simultaneous validity of EEP and HEP implies that of a principle of electrophilicity equalization. The equalized electrophilicity (ω_{GM}) is mathematically expressed as:

$$\omega_{GM} = \frac{\chi_{GM^2}}{2\eta_{GM}} = \left(\prod_{k=1}^{P} \omega_k\right)^{\frac{1}{P}} \tag{7}$$

where $\omega_k = \frac{\chi_k^2}{\eta_k}, (k = 1, 2, ..., P)$ are the electrophilicities of the isolated atoms.

The equalized electrophilicity (like that of electronegativity/chemical potential and hardness) becomes equal to the geometric mean of the associated isolated atom values. The corresponding local variant [48–50], however, does not carry any appreciable chemical sense in this regard and remains constant and becomes equal to the global electrophilicity. Theoretical computations on some representative molecular systems clearly depict that the electrophilicity equalization principle follows a reasonable suite where the corresponding equalization algorithms of electronegativity (χ) and hardness (η) are also duly obeyed. The electrophilicity equalization principle, in conjunction with the EEP and HEP therefore serve as a useful qualitative benchmark towards rationalizing chemical behavior. However, the postulation of any qualitative concept or principle can also be attempted from other approaches [51, 52]. These two works [51, 52] clearly refute the claim by Szentpaly [53] to rule out any electrophilicity equalization principle. The only point to differ is that a new mathematical algorithm in spite of corroborating with the basic scientific sense might become applicable or rather exemplify an entirely different set of molecular systems. However, the introduction of a fudge-factor [51] and then to claim a strong correlation is not warranted. The main claim of this chapter [51] through an earlier chapter of these authors has been strongly criticized by Szentpaly [53]. Other than the basic theory, there seems to be as such no universality laid with the application of the electrophilicity equalization principle. It mostly serves as a good qualitative index towards rationalizing molecular reactivity and is duly obeyed only when the corresponding equalization algorithms of electronegativity and hardness are applicable. So to compare [48] an electrophilicity equalization principle with the stature of Heisenberg's uncertainty principle and then to advocate a tendency of "ruling out" the same [53, 54] is certainly a blatant over say. A subsequent chapter criticizing such fastidious comments [55] is also published. In the same logic EEP (Szentpaly would be the "last to disagree on its success") and many other principles should be out of the literature. As already mentioned, the principle of electrophilicity equalization (PEE) performs nicely (or poorly) at par with the electronegativity equalization principle (EEP). Moreover, as mentioned by Szentpaly [53], neither any question regarding the better validity of PEE over EEP arise nor an excellent correlation regarding the R^2 values [47] is claimed. The calculated R^2 values just attempts to reflect the qualitative accuracy of all three such "principles" over a set of molecules. Regarding the "counterintuitive" behavior of the electronegativity, hardness and electrophilicity values and the justification of "plurality" [53], it seems as if there should exist a very strong nucleophile with a high χ value which naturally conceptualizes chemistry in a new way, heuristically proposed by Szentpaly. As for the difference in the equalized electrophilicities of (i) isomers having different bond connectivities, (ii) singlet or triplet O_2, (iii) cationic electrophiles, etc., mentioned as counterexample in the reference 53 we suggest to compute the corresponding χ values in all such counter examples (cf. Table 1, Ref. 53) and then to testify the validity or otherwise of the EEP as well. The question of electrophilicity equalization amongst a set of molecular systems should usually be addressed by comparing the associated electronegativity values. And, in this regard the tautological sermon to test the usefulness of geometric averaging [53] may take several chapters including many of L. v. Szentpaly out of scientific literature. Last but not the least, a glad agreement with the maximum hardness principle (MHP) [56] clearly shows that the cited examples [53] oppose the given claim. It may be noted that there are several such "principles" including the Sanderson's electronegativity "principle." Szentpaly failed to understand that the qualitative nature of all these principles are highlighted [47, 55], numerical examples showed their qualitative nature, a small set was not chosen to show the applicability of PEE in particular and all of them and many other such prescriptions are called "principles" in the literature. He did not care to check the validity of his favorite EEP in all of his counterexamples [53] for PEE. These molecular electronic structure principles simply attempts to serve as useful guiding parameters towards gaining deeper insights into molecular behavior. The proposal of an electrophilicity equalization principle is definitely significant, so as its genesis from different mathematical approaches by holding its relation with the other equalization principles intact. Of course, whether all such propositions would be called "principles" or "prescriptions" that ought to be decided.

8.5 CONCLUSION

All of the three equalization principles, viz., electronegativity equalization principle (EEP), hardness equalization principle (HEP), and principle of electrophilicity equalization (PEE) are qualitative in nature. If first two are valid then, the third one would be automatically valid.

KEYWORDS

- **absolute electronegativity**
- **calorimetry**
- **counterintuitive**
- **Fukui functions**
- **hardness equalization principle**
- **Pauling's recipe**
- **plurality**
- **ruling out**
- **spontaneity**

REFERENCES

1. Parr, R. G.; Yang, W. In *Density Functional Theory of Atoms and Molecules;* Oxford University Press: New York, 1989.
2. Chattaraj, P. K. In *Chemical Reactivity Theory: A Density Functional View;* Taylor & Francis/CRC Press: Florida, 2009.
3. Geerlings, P.; De Proft, F.; Langenaeker, W. Conceptual Density Functional Theory. *Chem. Rev.,* **2003**, *103*, 1793–1874.
4. Chattaraj, P. K.; Giri, S. Electrophilicity Index Within a Conceptual DFT Framework. *Ann. Rep. Prog. Chem. Sect. C: Phys. Chem.,* **2009**, *105*, 13–39.
5. Chakraborty, A.; Duley, S.; Giri, S.; Chattaraj, P. K. An Understanding of the Origin of Chemical Reactivity from a Conceptual DFT approach in A Matter of Density: Exploring the Electron Density Concept in the Chemical, Biological, and Materials Sciences. Sukumar, N., Ed.; John Willey, 2012.
6. Hohenberg, P.; Kohn, W. Inhomogeneous Electron. Gas *Phys. Rev.*, **1964**, *136*, B864–B871.

7. Sen, K. D.; Jorgenson, C. K. Eds. Structure and Bonding, Vol. 66: Electronegativity; Springer: Berlin, 1987.
8. Chattaraj, P. K. Electronegativity and Hardness: A Density Functional Treatment. *J. Indian. Chem. Soc.*, **1992**, *69*, 173–184.
9. Parr, R. G.; Donnelly, R. A.; Levy, M.; Palke, W. E. Electronegativity: The Density Functional Viewpoint. *J. Chem. Phys.*, **1978**, *68*, 3801–3807.
10. Sen, K. D.; Mingos, D.M. P. Eds. Structure and Bonding, Vol. 80: Chemical Hardness; Springer: Berlin, 1993.
11. Parr, R. G.; Pearson, R. G. Absolute Hardness: Companion Parameter to Absolute Electronegativity, *J. Am. Chem. Soc.*, **1983**, *105*, 7512–7516.
12. Pearson, R. G. Chemical Hardness: Applications from Molecules to Solids; Wiley-VCH: Weinheim, 1997.
13. Parr, R. G.; Szentpaly, L. v.; Liu, S. Electrophilicity Index. *J. Am. Chem. Soc.* **1999**, *121*, 1922–1924.
14. Chattaraj, P. K.; Sarkar, U.; Roy, D. R. Electrophilicity Index. *Chem. Rev.* **2006**, *106*, 2065–2091.
15. Chattaraj, P. K.; Roy, D. R. Update 1 of: Electrophilicity Index. *Chem. Rev.* **2007**, *107*, PR46–PR74.
16. Chattaraj, P. K.; Giri, S.; Duley, S. Update 2 of: Electrophilicity Index. *Chem. Rev.*, **2011**, *111*, PR43–PR75.
17. Mulliken, R. S. I. Electronic Population Analysis on LCAO-MO Molecular Wave Functions. *I. J. Chem. Phys.* **1955**, *23*, 1833–1840.
18. Parr, R. G.; Yang, W. Density Functional Approach to the Frontier-Electron Theory of Chemical Reactivity. *J. Am. Chem. Soc.*, **1984**, *106*, 4049–4050.
19. Pauling, L. The Nature of the Chemical Bond. IV. The Energy of Single Bonds and the Relative Electronegativity of Atoms. *J. Am. Chem. Soc.*, **1932**, *54*, 3570–3582.
20. Pauling, L. The Nature of the Chemical Bond. Cornell University Press: Ithaca, NY, 1960.
21. Iczkowski, R. P.; Margrave, J. L. Electronegativity. *J. Am. Chem. Soc.*, **1961**, *83*, 3547–3551.
22. Mulliken, R. S. A New Electroaffinity Scale; Together with Data on Valence States and on Valence Ionization Potentials and Electron Affinities. *J. Chem. Phys.*, **1934**, *2*, 782–793.
23. Mulliken, R. S. Electronic Structures of Molecules XI. Electroaffinity, Molecular Orbitals and Dipole Moments. *J. Chem. Phys.*, **1935**, *3*, 573–785.
24. Pearson, R. G. Absolute Electronegativity and Absolute Hardness of Lewis Acids and Bases. *J. Am. Chem. Soc.*, **1985**, *107*, 6801–6806.
25. Sanderson, R. T. An Interpretation of Bond Lengths and a Classification of Bonds. *Science,* **1951**, *114*, 670–672.
26. Sanderson, R. T. Carbon–Carbon Bond Lengths. *Science,* **1952**, *116*, 41–42.
27. Sanderson, R. T. Partial Charges on Atoms in Organic Compounds. *Science,* **1955**, *121*, 207–208.
28. Politzer, P.; Weinstein, H. Some Relations Between Electronic Distribution and Electronegativity. *J. Chem. Phys.*, **1979**, *71,* 4218–4220.
29. Parr, R. G.; Bartolotti, L. On the Geometric Mean Principle for Electronegativity Equalization. *J. Am. Chem. Soc.*, **1982**, *104*, 3801–3803.

30. Nalewajski, R. A Study of Electronegativity Equalization. *J. Phys. Chem.*, **1985**, *89*, 2831–2837.
31. Nalewajski, R. F. Electrostatic effects in interactions between hard (soft) acids and bases. *J. Am. Chem. Soc.*, **1984**, *106*, 944–945.
32. Nalewajski, R. F.; Koninski, M. Atoms-in-a-molecule model of the chemical bond. *J. Phys. Chem.*, **1984**, *88*, 6234–6240.
33. Datta, D. Geometric Mean Principle for Hardness Eualization: A Corollary of Sanderson's Geometric Mean Principle of Electronegativity Equalization. *J. Phys. Chem.*, **1986**, *90*, 4216—4217.
34. Yang, W.; Lee, C.; Ghosh, S. K. Molecular Softness as the Average of Atomic Softnesses: Companion Principle to the Geometric Mean Principle for Electronegativity Equalization. *J. Phys. Chem.*, **1985**, *89*, 5412–5414.
35. Chattaraj, P. K.; Nandi, P. K.; Sannigrahi, A. B. Improved Hardness Parameters for Molecules. *Proc. Indian Acad Sci (Chem Sci)*, **1991**, *103*, 583–589.
36. Chattaraj, P. K.; Ayers, P. W.; Melin, J. Further Links Between the Maximum Hardness Principle and the Hard/Soft Acid/Base Principle: Insights from Hard/Soft Exchange Reactions. *Phys. Chem. Chem. Phys.*, **2007**, *9*, 3853–3856.
37. Fuentealba, P.; Chamorro, E.; Cardenas, C. Further Exploration of the Fukui Function, Hardness, and Other Reactivity Indices and its Relationships within the Kohn–Sham Scheme. *Int. J. Quant. Chem.*, **2007**, *107*, 37–45.
38. Ayers, P. W.; Parr, R. G. Local Hardness Equalization: Exploiting the Ambiguity. *J. Chem. Phys.*, **2008**, *128*, 184108(1–8).
39. Ghosh, S. K. Energy Derivatives in Density Functional Theory. *Chem. Phys. Lett.*, **1990**, *172*, 77–82.
40. Harbola, M. K.; Chattaraj, P. K.; Parr, R. G. Aspects of the Softness and Hardness Concepts of Density Functional Theory. *Isr. J. Chem.*, **1991**, *31*, 395–402.
41. Berkowitz, M; Ghosh, S. K.; Parr, R. G. On the Concept of Local Hardness in Chemistry. *J. Am. Chem. Soc.*, **1985**, *107*, 6811–6814.
42. Baekelandt,B. G.; Cedillo, A.; Parr, R. G. Reactivity Indices and Fluctuation Formulas in Density Functional Theory: Isomorphic Ensembles and a New Measure of Local Hardness. *J. Chem. Phys.*, **1995**, *103*, 8548–8556.
43. Langenaeker, W.; Deproft,F.; Geerlings, P. Development of Local Hardness-Related Reactivity Indices: Their Application in a Study of the SE at Monosubstituted Benzenes within the HSAB Context. *J. Phys. Chem.*, **1995**, *99*, 6424–6431.
44. Chattaraj, P. K.; Roy, D. R.; Geerlings,P . Torrent-Sucarrat, M. Local Hardness: A Critical Account. *Theor. Chem. Acc.*, **2007**, *118*, 923–930.
45. Torrent-Sucarrat, M.; De Proft, F.; Ayers, P. W.; Geerlings, P. On the Applicability of Local Softness and Hardness. *Phys. Chem. Chem. Phys.*, **2010**, *12*, 1072–1080.
46. Cardenas, C.; Tiznado, W.; Ayers, P. W.; Fuentealba, P. The Fukui Potential and the Capacity of Charge and the Global Hardness of Atoms. *J. Phys. Chem. A*, **2011**, *115*, 2325–2331.
47. Chattaraj, P. K.; Giri, S.; Duley, S. Electrophilicity Equalization Principle. *J. Phys. Chem. Lett.*, **2010**, *1*, 1064–1067.
48. Chattaraj, P. K.; Maiti, B.; Sarkar, U. Philicity: A Unified Treatment of Chemical Reactivity and Selectivity. *J. Phys. Chem. A*, **2003**, *107*, 4973–4975.

49. Pérez, P.; Toro-Labbe´, A.; Aizman, A.; Contreras, R. Comparison between Experimental and Theoretical Scales of Electrophilicity in Benzhydryl Cations. *J. Org. Chem.*, **2002**, *67*, 4747–4752.
50. Chamorro, E.; Chattaraj, P. K.; Fuentealba, P. Variation of the Electrophilicity Index along the Reaction Path. *J. Phys. Chem. A,* **2003**, *107*, 7068–7072.
51. Islam, N.; Ghosh, D. C. On the Electrophilic Character of Molecules Through Its Relation with Electronegativity and Chemical Hardness. *Int. J. Mol. Sci.,* **2012**, *13*, 2160–2175.
52. Fuentealba' P.; Cárdenas, C. On the Exponential Model for the Energy with Respect to the Number of Electrons. *J Mol. Model.,* doi: 10.1007/s00894-012-1708-5.
53. Szentpaly, L. v., Reply to "Comment on 'Ruling Out Any Electrophilicity Equalization Principle'." *J. Phys. Chem. A,* **2012**, *116*, 792–795.
54. Szentpaly, L. v., Ruling Out Any Electrophilicity Equalization Principle. *J. Phys. Chem. A,* **2011**, *115*, 8528–8531.
55. Chattaraj, P. K.; Giri, S.; Duley, S. Comment on "Ruling Out Any Electrophilicity Equalization Principle." *J. Phys. Chem. A,* **2012**, *116*, 790–791.
56. Parr, R. G.; Chattaraj, P. K. Principle of Maximum Hardness. *J. Am. Chem. Soc.*, **1991**, *113*, 1854–1855.

CHAPTER 9

RING OPENING METATHESIS POLYMERIZATION IS A VERSATILE TECHNIQUE FOR MAKING POLYMERIC BIOMATERIALS

N. V. RAO, S. R. MANE, and R. SHUNMUGAM

CONTENTS

9.1 INTRODUCTION

The synthesis and complete characterization of all the monomers namely norbornene derived folate (mono 1), norbornene derived doxorubicin hydrazone linker (mono 2) and Norbornene derived barbiturate (NDB) are clearly described. The monol and mono2 are processed with diblock copolymerization whereas the homopolymerization of NDB is done through ring-opening metathesis polymerization (ROMP). In the covalent approach folic acid (FA) is used as the receptor-targeted anticancer therapy. The primary purpose of using drug delivery systems with hydrogen linkages is attributed to its pH-responsive feature. So its quite evident when the drug release from micelles of FOL-DOX is significantly accelerated even at mildly acid pH of 5.5 to 6.0 compared to physiological pH of 7.4. In case of non covalent approach, the NDBH polymersomes are used as a nano carrier for anti cancer (DOX) drug delivery.

The recent development in the research domain of human health care is rampant and quite phenomenal. The most rapidly advancing area of science in which chemists and chemical engineers are contributing their major efforts is controlled drug delivery technology [1–6]. The role of polymers is very important in these delivery systems [7a, b]. Such systems often use synthetic polymers as carriers for the drugs. These delivery systems offer numerous advantages compared to conventional dosage forms including improved efficacy, reduced toxicity, and improved patient compliance and convenience [10]. It's a great challenge for researchers in the field to generate an effective system in which the release of drugs can be regulated via an appropriate stimulus such as pH or temperature with site-specific targeting properties [8a–c]. Recently, efforts have been focused on developing drugs through self-assembly or high-throughput processes to facilitate the development [11a, b]. Drugs covalently bound in polymeric micellar cores via pH-sensitive linkers such as an imine [12], acetal [13], oxime [14a–c] or orthoester [15], showed enhanced in vitro cytotoxicity compared to free drug, resulting from rapid release of drug from acidic endosomes.

Towards this application, controlled polymerization techniques are considered. Among the controlled polymerization technique, Ring Opening Metathesis Polymerization (ROMP) [16] is considered to be more promising, because it is very easy to polymerize, easy to functionalize. This is due to the nature of the Grubbs' catalyst—it is well known for tolerance of different types of functional groups such as alcohols, aldehydes, amides, esters, ketones, and carboxylic acids. Thus, it can yield polymers with extremely rich functionalities imparting water solubility [17–19] antimicrobial [20–24] and antibiofouling [25] activity, and even metal containing functional groups [26, 27].

Even though there are lot of familiar methods and procedures that can be followed to control the release rate of drugs by covalently attaching them to a hydrolytically labile bond, we are very specific about hydrazone bonds for their responsive acuteness in delivery behavior and also for its ability to the usually facile incorporation of hydrazides into delivery materials [28]. However, incorporation of hydrazone into polymer prodrugs is still synthetically challenging and not explored much except very few pioneering report on poly(aspartate hydrazone adriamycin) based polymers [29]. Typically, the release rate of drug with hydrazone linkage is governed by the pH of the surrounding media, with faster release observed in acidic (pH 5–6) environments compared with physiological media (pH 7.4).

Here, we report a simple and unique approach to design and synthesis of block copolymer (FOL-DOX) that has the potential application as drug carrier. This work investigates design and synthesis of the FOL-DOX consisting of hydrazone tethered DOX and FOLATE ligand to Sitespecific targeting capability (Fig. 1; Scheme 1). Folic acid is being used because folate receptors are over-expressed on the surface of several cancer cells compared to normal cells. In this chapter we are presenting the design and synthesis of ROMP polymers that have potential application in cancer therapy. This includes both covalent and non-covalent approach to attach the drug on to the polymeric backbone.

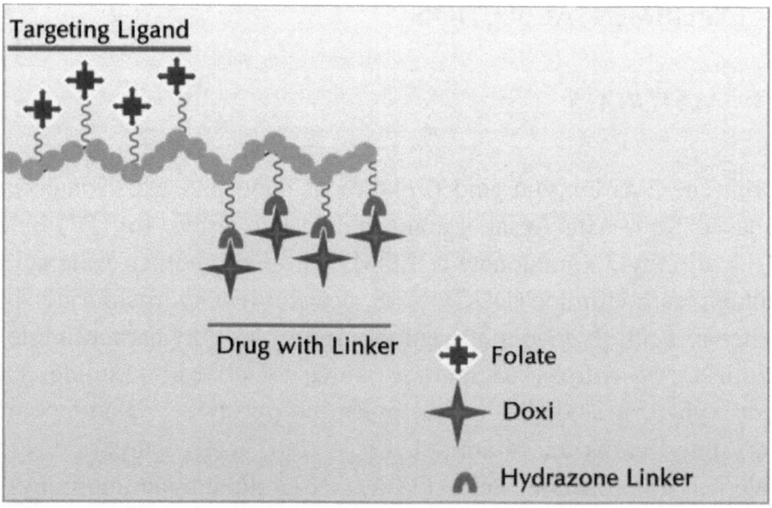

FIGURE 1 Cartoon representation of FOL-DOX.

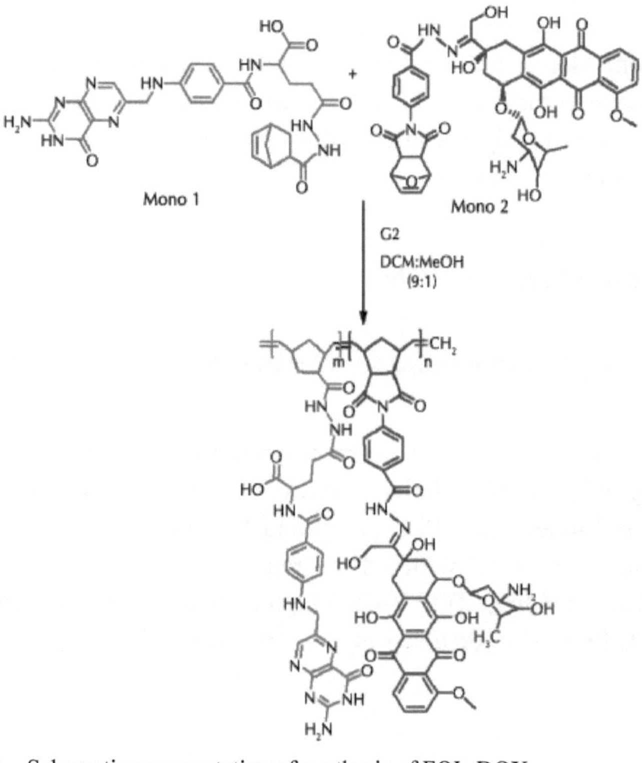

SCHEME 1 Schematic representation of synthesis of FOL-DOX.

9.2 EXPERIMENTAL SECTION

9.2.1 *MATERIALS*

5-norborene-2-carboxylic acid (mixture of endo and exo isomers), was purchased from Alfa Aesar. Furan, maleic anhydride, tin (IV) chloride ($SnCl_4$), di ethyl oxomalonate (DEOM), vinyl ethyl ether, folic acid, dicyclohexylcarbodiimide (DCC), NHS, doxorubicin hydrochloride, 4-aminobenzoic acid, N-(3-dimethylaminopropyl)-N'-ethylcarbodiimide hydrochloride, N-hydroxysuccinimide, dicyclohexyl carbodiimide (DCC), 1-hydroxybenzotriazole, 4-dimethylamino-pyridine, 2-aminoethanol, tertiary butyl carbazate, 4-aminobenzoic acid, acetic anhydride, sodium acetate, Trifluoroacetic acid (TFA), N-(3-dimethylaminopropyl)-N'-ethylcarbodiimide hydrochloride, dimethyl sulphoxide, second generation Grubbs' catalyst, $CDCl_3$ and DMSO-D_6 were purchased from Sigma Aldrich. Aniline, Urea, tetra- hydrofuran (THF), diethyl ether were purchased from Merck and used as received. Dichloromethane (DCM), acetone, toluene, pyridine were dried and used for the reactions. The sodium methoxide was freshly prepared from methanol and Na metal and used without further purification.

9.2.2 *CHARACTERIZATION*

9.2.2.1 *GEL PERMEATION CHROMATOGRAPHY (GPC)*

Molecular weights and PDIs were measured by Waters gel permeation chromatography in THF relative to PMMA standards on systems equipped with Waters Model 515 HPLC pump and Waters Model 2414 Refractive Index Detector at 35°C with a flow rate of 1 ml/min. HRMS analyses were performed with Q-TOF YA263 high resolution (Waters Corporation) instruments by +ve mode electrospray ionization.

9.2.2.2 NUCLEAR MAGNETIC RESONANCE (NMR)

The ^{1}H NMR spectroscopy was carried out on a Bruker 500 MHz spectrometer using $CDCl_3$ as a solvent. ^{1}H NMR spectra of solutions in CDC13 were calibrated to tetramethylsilane as internal standard (δH 0.00).

9.2.2.3 FOURIER TRANSFORM INFRA RED (FT-IR)

FT-IR spectra were obtained on FT-IR Perkin-Elmer spectrometer at a nominal resolution of 2 cm^{-1}.

9.2.3 COVALENT APPROACH

9.2.3.1 SYNTHESIS AND CHARACTERIZATION

9.2.3.1.1 SYNTHESIS OF FOLATE-NHS ESTER

Folic acid (5 g, 11.3 mmol) was dissolved in 100 ml of dry dimethylformamide (DMF) to which (1.55 g, 1.52 mmol) of dicyclohexylcarbodiimide (DCC) and (1.285 g, 11.3 mmol) of NHS were added. The reaction mixture was stirred for 15 h at room temperature in the dark under nitrogen atmosphere. The byproduct, dicyclohexylurea, was filtered, and 100 ml of diethyl ether was added with slow stirring at room temperature. A yellow precipitate was filtered. Then the precipitate was washed with diethyl ether for three times, the yellow color solid was used for the next step.

9.2.3.1.2 SYNTHESIS OF BOC-FOLATE

Folate-NHS ester (6.0 g, 11 mmol) was dissolved completely in 100 ml of dry pyridine, and tertiary butyl carbazate (2.25 g, 11 mmol) was slowly added for about 60 min. The mixture was stirred at room temperature in the dark for 12 h. Pyridine was concentrated under vacuum to yield a yellow color solid. The product was used for next step. ^{1}H NMR (DMSO-$D_{6,}$ 500 MHZ): δ 11.2–11.5 (s, 1H), 8.624 (s, 1H), 8.6 (s, 2H), 8.0–8.1 (d, 2 H), 7.36–7.7.38 (d, 2 H), 6.55–6.70 (d, 2H), 6.50 (s, 2H), 4.50–4.54 (m, 2H), 4.43–4.4 (m, 2H), 2.30–2.33 (t, 2H), 2.0–2.07 (m, 2H), 1.86–1.94 (m, 2H), 1.30 (s, 9H). IR (KBr, cm^{-1}): 3392, 2927, 1696, 1507, 1405, 1308, 1195, 997.

9.2.3.1.3 DEPROTECTION OF BOC-FOLATE

The resulting product (BOC-Folate) was dissolved in 20 ml of trifluoroaceti cacid (TFA) to deprotect the BOC group. Deprotection was carried out at room temperature for 2 h. TFA was removed under vacuum. Then 10 ml of diethyl ether was added to the pasty mass to get yellow color solid. ^{1}H NMR (DMSO-$D_{6,}$ 500 MHZ): 11.2–11.5 (s, 1H), 8.624 (s, 1H), 8.6 (s, 2H), 8.0–8.1 (d, 2 H), 7.36–7.7.38 (d, 2 H), 6.55–6.70 (d, 2H), 6.50 (s, 2H), 4.50–4.54 (m, 2H), 4.43–4.4 (m, 2H), 2.30–2.33 (t, 2H), 2.0–2.07 (m, 2H), 1.86–1.94 (m, 2H). IR (KBr, cm^{-1}): 3337, 2935, 1699, 1605, 1509, 1335, 1198, 1137, 947.

9.2.3.1.4 SYNTHESIS OF MONO 1

A 100 mg (0.728 mmol) of 5-Norbornene-2-exo-carboxylic acid was added into 10 ml of dimethyl formamide. A 300 mg (1.456 moml) of N-(3-dimethylaminopropyl)-N'-ethylcarbodiimide hydrochloride (EDC) and 1-hydroxybenzotriazole (HOBT) 196 mg (1.456 mmol) in to the reaction mixture. Reaction mixture was allowed to stir for 15 h at room temperature. Then it was allowed to cool at 0–5°C. De-BOC-Folate (418 mg, 0728) was dissolved in 10 ml of dimethyl formamide and this solution was added to the reaction mixture at 0–5°C. Then, the reaction mixture was stirred for another 30 min at 0–5°C. To the reaction mixture ethyl acetate was added followed by water to the reaction mixture. Organic layer was washed with 2 × 10 ml of water followed by sodium bicarbonate wash. Finally, organic layer was washed with brine solution. Organic layer concentrated under vacuum to yield a white color solid. ^{1}H NMR (DMSO-D_6, 500 MHZ):

11.2–11.5 (s, 1H), 9.6–9.8 (bs, 1H), 8.624 (s, 1H), 8.6 (s, 2H), 8.0–8.1 (d, 2 H), 7.36–7.7.38 (d, 2 H), 6.55–6.70 (d, 2H), 6.50 (s, 2H), 6.13–6.16 (s, 2H), 4.50–4.54 (m, 2H), 4.43–4.4 (m, 2H), 2.30–2.33 (t, 2H), 2.0–2.07 (m, 2H), 1.86–1.94 (m, 2H). IR (KBr, cm^{-1}): 3272, 2933, 1686, 1606, 1512, 1330, 1300, 1190, 840.

9.2.3.1.5 SYNTHESIS OF NADIC ACID

Exo-oxabicylo-[2.2.1] hept-5-ene-2, 3-dicarboxylic anhydride, [1] (1.914 g, 11.5 mmol) was added to a four-neck reaction flask. A 35 ml of acetone was added and heated until it became clear solution. To this solution, para-amino benzoic acid (1.605 g, 11.5 mmol) was added with stirring. After heating for fifteen min the reaction mixture was allowed to stir for about 30 min at room temperature. The solid was filtered and dried at 55°C under vacuum. After the drying process, the compound was then dissolved in 30 ml of dimethyl formamide and heated to 50°C. Acetic anhydride (15 ml, 158.97 mmol) and sodium acetate (0.635 g, 7.743 mmol) were added under stirring. The reaction mixture was allowed to stir for 3 h at 55°C. After stirring for 3 h, the reaction mixture was poured into 500 ml of water then acidified by addition of 5 ml concentrated HCl. White color solid was precipitated immediately. It was then filtered and dried at 90°C, under vacuum (80 % yield). ^{1}H NMR (DMSO-$D_{6,}$ 400 MHZ): δ 13.1 (bs, 1H), 8.0–8.2 (m, 2H), 7.4–7.5 (m, 2H), 6.6 (s, 2H), 3.1 (s, 2H). ^{13}C NMR (DMSO-D_6, 400 MHz): δ 176.43, 166.59, 136.65, 135.78, 130.0, 126.79, 80.86, 47.58. IR (KBr, cm^{-1}): 3236, 2635, 2073, 1954, 1826, 1780, 1729, 1698, 1607, 1515, 1418, 1218, 1144, 1125, 1020, 975, 950, 912, 883, 878, 804, 726, 672, 633, 598, 541, 521. MS (ESI) calculated for $C_8H_{10}O_2Na$ [M + H]$^+$: 285.05; observed 284.95.

9.2.3.1.6 SYNTHESIS OF BOC-NADIC

A 1 g (6.92 mmol) of nadic acid was added into 10 ml of dimethyl formamide. A 0.85g (4.46 moml) of N-(3-dimethylaminopropyl)-N'-ethylcarbodiimide hydrochloride (EDC) and 1-hydroxybenzotriazole

(HOBT) (0.62 g, 4.46 mmol) in to the reaction mixture. Reaction mixture was allowed to stir for 15 h at room temperature. Then it was allowed to cool at 0–5°C. Tertiary butyl carbazate was dissolved in dimethyl formamide and this solution was added to the reaction mixture at 0–5°C. Reaction mixture was stirred for another 30 min at 0–5°C. To the reaction mixture ethyl acetate was added followed by water to the reaction mixture. Organic layer was washed with 2 × 10 ml of water followed by sodium bicarbonate wash. Finally organic layer was washed with brine solution. Organic layer was concentrated under vacuum to yield a white color solid (700 mg, 70% yield). ^{1}H NMR (DMSO-$D_{6,}$ 400 MHZ): δ 10.30 (bs, 1H), 8.96 (s, 1H), 7.8–7.9 (m, 2H), 7.2–7.26 (m, 2H), 6.60 (s, 2H), 5.34 (s, 2H), 3.11 (s, 2H), 1.26 (s, 9H). ^{13}C NMR ($CDCl_3$, 400 MHz): δ 175.48, 165.39, 155.42, 136.64, 134.80, 128.03, 126.67, 80.85, 79.27, 47.58, 28.08. IR (KBr, cm^{-1}): 3456, 3367, 3277, 2999, 1786, 1747, 1729, 1660, 1630, 1541, 1516, 1390, 1261, 1014, 946, 912, 875, 867, 781, 722. MS (ESI) calculated for $C_8H_{10}O_2Na$ $[M + H]^+$: 399.40 ; observed 399.14.

9.2.3.1.7 DEPROTECTION OF BOC-NADIC

A 500 mg (1.754 mmol) of BOC-NADIC was dissolved in 5 ml of dichloromethane at room temperature. TFA 6 ml was added in to the reaction mixture. Reaction mixture was allowed to stirred for 1 h at room temperature. Reaction mixture was concentrated to pasty mass, then diethyl ether was added to the pasty mass. Resultant white product was collected by suction filtration, washed with 10 ml of diethyl ether and dried at 40°C under vacuum (420 mg, 84% yield). ^{1}H NMR (DMSO-$D_{6,}$ 400 MHZ): δ 11.43 (bs, 1H), 7.8–7.9 (m, 2H), 7.33–7.40 (m, 2H), 6.6 (s, 2H), 5.25 (s, 2H), 3.11 (s, 2H), 1.8–1.9 (m, 3H); IR (KBr, cm^{-1}): 3481, 2975, 1785, 1717, 1512, 1393, 1304, 1207, 1172, 1070, 880, 726. ^{13}C NMR (CDC13, 400 MHz): δ 175.48, 165.39, 155.42, 136.64, 134.80, 128.03, 126.67, 80.85, 79.27, 47.58. MS (ESI) calculated for $C_8H_{10}O_2Na$ $[M + H]^+$: 299.28 observed 299.98.

9.2.3.1.8 SYNTHESIS OF NORBORNENE DERIVED DOXORUBICIN HYDRAZONE LINKER (MONO 2)

Doxorubicin hydrochloride 29 mg (0.05 mmol), and compound nadic hydrazine 72.6 mg (0.18 mmol) were dissolved in 10 ml of anhydrous methanol. TFA (3 μl) was added to the reaction mixture. The reaction mixture was stirred at room temperature for 24 h while being protected from light. The reaction mixture was concentrated to a volume of 1 ml and added to acetonitrile (20 ml) drop wise with stirring. The resulting solution was allowed to stand at 4°C. This product was isolated by centrifugation, washed with fresh methanol/acetonitrile (1:10) and dried under vacuum to yield the hydrazone linker of doxorubicin 15 mg, 51% yield). ^{1}H NMR (DMSO-$D_{6,}$ 400 MHZ): δ 11.8 (s, 1H), 7.69–7.95 (m, 7H), 7.79–7.80 (m, 2H), 6.62 (s, 2H), 5.43 (m, 2H), 5.26 (m, 2H), 5.30 (s, 2H), 4.95–4.97 (m, 1H), 4.85–4.88 (t, 1H), 4.56–4.57 (m, 1H), 4.17–4.18 (m, 1H), 4.21 (m, 1H), 4.02 (s, 3H), 3.55–3.56 (m, 1H), 3.11 (s, 2H), 2.95–3.02 (m, 2H), 2.13–2.15 (m, 2H), 1.88–1.89 (m, 1H), 1.6–1.69 (m, 1H), 1.16–1.19 (m, 3H). ^{13}C NMR (CD_3OD, 400 MHz): δ 215.0, 187.62, 188.82, 177.20, 162.77, 158.39, 156.46, 137.83, 129.04, 127.98, 122.16, 120.49, 112.12, 100.98, 82.81, 76.86, 71.78, 67.88, 61.0, 59.0, 39, 32.0, 33.81, 29.42, 16.98. IR (KBr, cm^{-1}): δ 3474, 2992, 2419, 1718, 1799, 1674, 1604, 1567, 1525, 1416, 1299, 114, 1258, 1154, 1026, 1006, 960, 900, 891,788, 733. MS (ESI) calculated for $C_8H_{10}O_2Na$ $[M + H]^+$: 824.25; observed 824.9.

9.2.3.2 RESULTS AND DISCUSSION

9.2.3.2.1 MONOMER SYNTHESIS

Exo-5-norbornene-5-carboxylic acid was separated from the commercially available mixture of endo and exo 5-norbornene-2-carboxylic acid by the idolactization methods of Ver Nooy and Rondestvedt. Carboxylic acid peak was observed at δ 12 ppm (br, 1H) and δ 183 ppm (Figs. 2, and 3). The signals at δ 6.02–6.06 ppm (m, 2H) were corresponding to norbornene olefenic protons while the signals at δ 1.97–2.05 ppm (dt, J = 12.7Hz, 1H) and δ 1.66–1.71 (m, 1H) were belonging to norbornene bridged hydrogens, which was further confirmed by IR (Fig. 4) where the stretching mode due to carboxylic acid of exo norbonene carboxylic acid was observed at 1700 cm^{-1}.

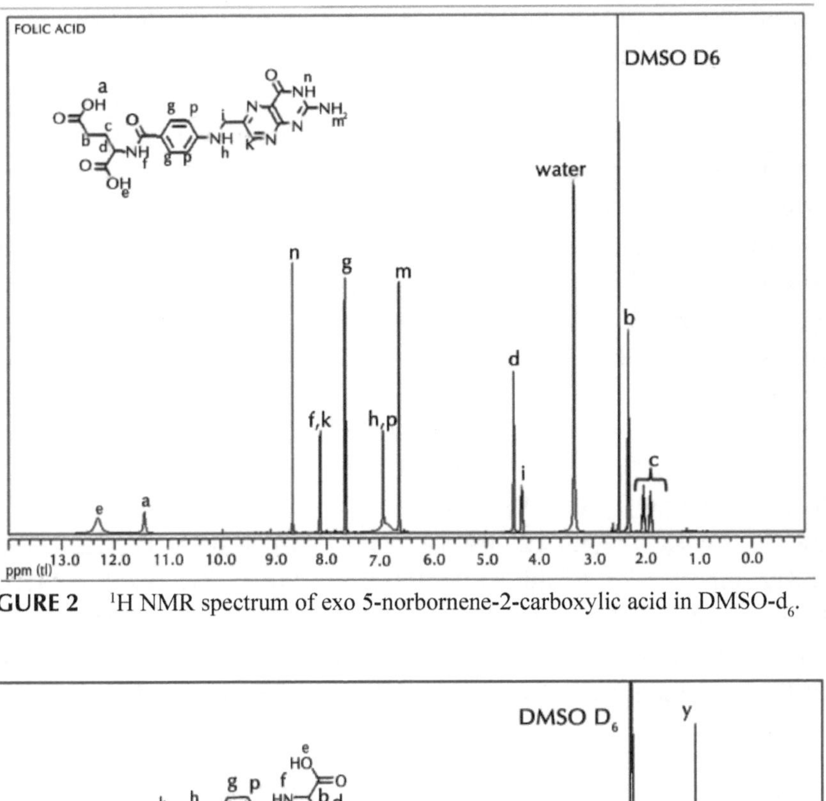

FIGURE 2 ^{1}H NMR spectrum of exo 5-norbornene-2-carboxylic acid in DMSO-d_6.

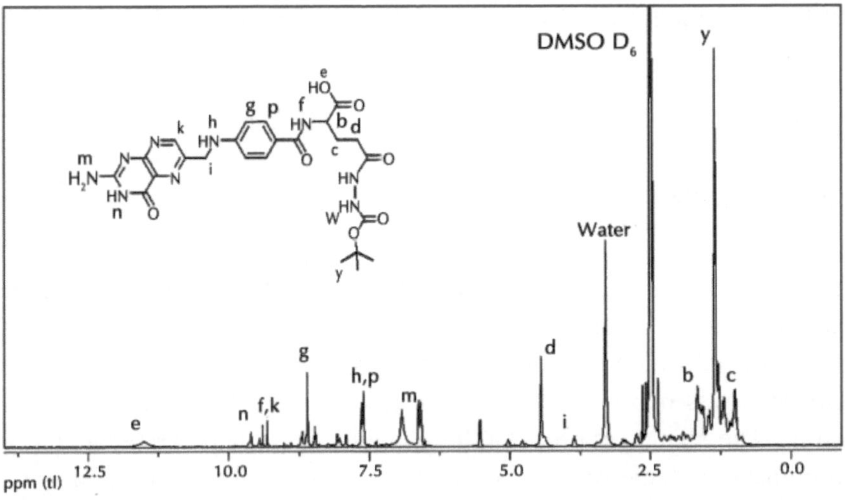

FIGURE 3 ^{13}C NMR spectrum of exo 5-norbornene-2-carboxylic acid in $CDCl_3$.

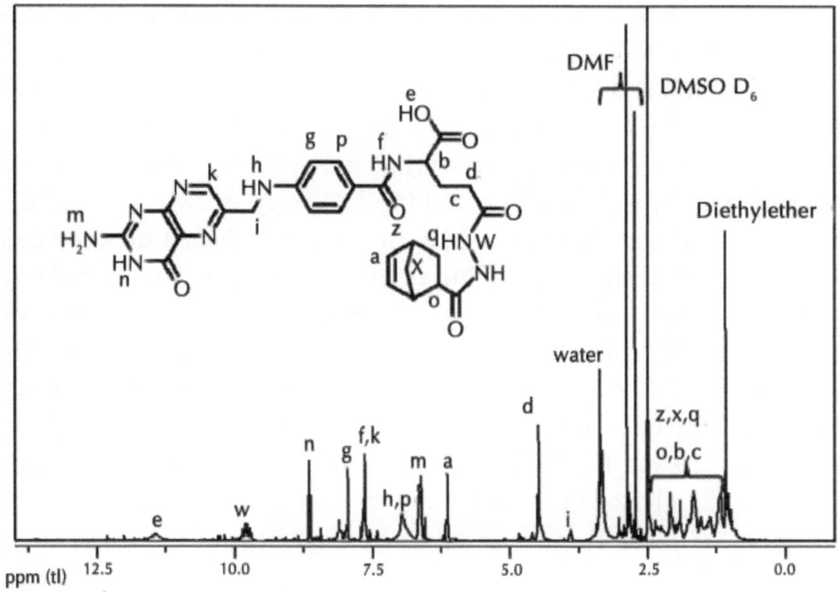

FIGURE 4 FT-IR spectrum of exo 5-norbornene-2-carboxylic acid.

Exo-oxabicylo-[2.2.1] hept-5-ene-2, 3-dicarboxylic anhydride was prepared following the reported process [13]. 3-(4-carboxyphenylcarbamoyl)-7-oxabicyclo[2.2.1]hept-5-ene-2-carboxylic acid was prepared by using 1 equivalent of 4-aminobenzoic acid and 5.1 equivalent of acetic anhydride in presence of 0.29 equivalent of sodium acetate in dimethyl formamide solvent. Product was confirmed by ^{1}H NMR and IR spectroscopy. The signal at δ 13.1 (bs, 1H) ppm and δ 176 ppm was responsible for carboxylic acid group (Figs. 5 and 6) δ 8.0–8.2 (m, 2H) and 7.4–7.5 (m, 2H) ppm were responsible for aromatic protons, while δ 6.6 (s, 2H), 5.34 (s, 2H) and 3.1 (s, 2H) ppm peaks were for oxo-norbornene protons. Broad stretching frequency at 3244 cm^{-1} was attributed to free carboxylic acid as shown in (Fig. 7a).

Tertiary butyl hydrazone nadic carboxylate was prepared by coupling of nadic acid and tertiary butyl carbazate by using 1.2 equivalent of N-(3-dimethylaminopropyl)-N'-ethylcarbodiimide hydrochloride (EDC) coupling reagent in presence of 1.2 equivalent of 1-hydroxybenzotriazole (HOBT) in dimethyl formamide solvent. Formation of monomer that was thoroughly characterized by ^{1}H NMR and IR spectroscopy. The signal δ

1.26 (s, 9H) ppm protons were responsible for tertiary butyl group, δ 10.30 (bs, 1H) and δ 8.96 (s, 1H) were for hydrazone protons. The signals at δ 7.8–7.9 (m, 2H) and δ 7.2–7.26 (m, 2H) were for benzene ring protons. The signals at δ 6.60 (s, 2H), 5.34 (s, 2H) and δ 3.11 (s, 2H) ppm were observed for oxo-norbornene protons and in ^{13}C NMR δ 156 ppm for Boc carbonyl (Figs. 8 and 9). The stretching frequency at 3244 cm^{-1} for free carboxylic acid was shifted to 3277 cm^{-1} which indicated the formation of amide group as shown (Fig. 7b). In the next step tert-butoxy carbonyl group was cleaved using TFA to yield the hydrazide. This product was confirmed by ^{1}H NMR and IR spectroscopy. The signal at δ 1.26 (s, 9H) ppm and in ^{13}C NMR δ 28 and δ 79 ppm for tertiary butyl group was absent in the product which indicated the complete deprotection of tertiary butyl group (Figs. 10 and 11). Shifting of stretching frequency from 3277 cm^{-1} to 3473 cm^{-1} indicated the formation of free amine group (Fig. 7c).

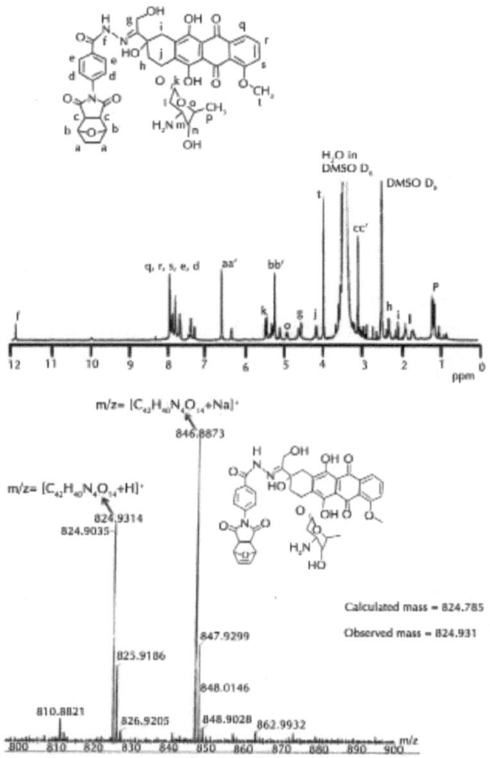

FIGURE 5 ^{1}H NMR spectrum of nadic acid in $DMSO\text{-}d_6$.

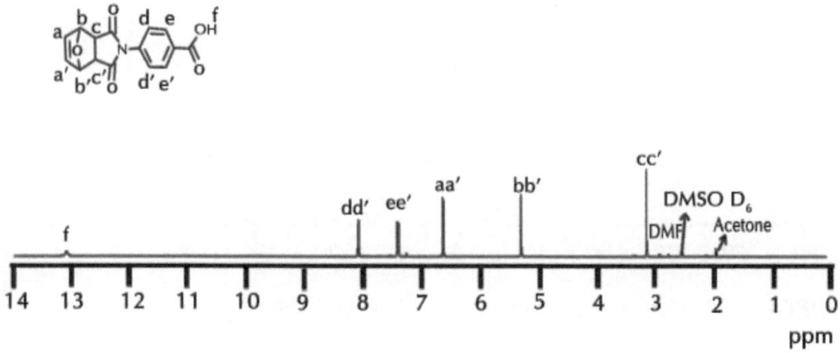

FIGURE 6 ^{13}C NMR spectrum of nadic acid in DMSO-d_6.

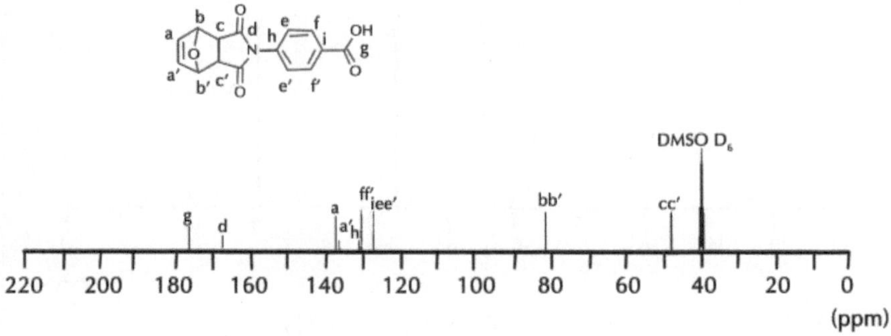

FIGURE 7 FT-IR spectra of (a) nadic acid (b) BOC-NADIC (c) nadic hydrazine.

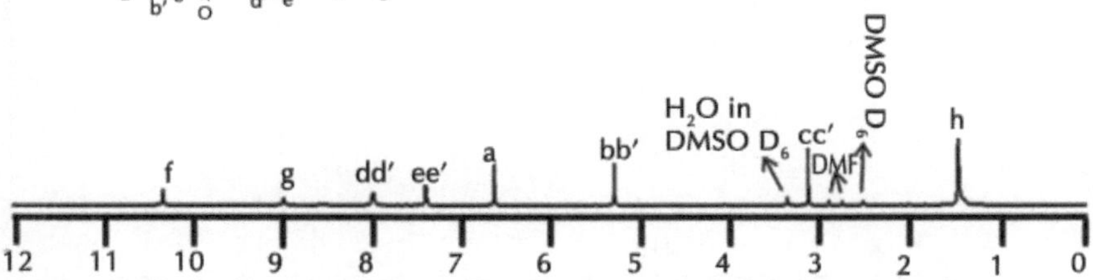

FIGURE 8 ^{1}H NMR spectrum of NADIC-BOC DMSO-d_6.

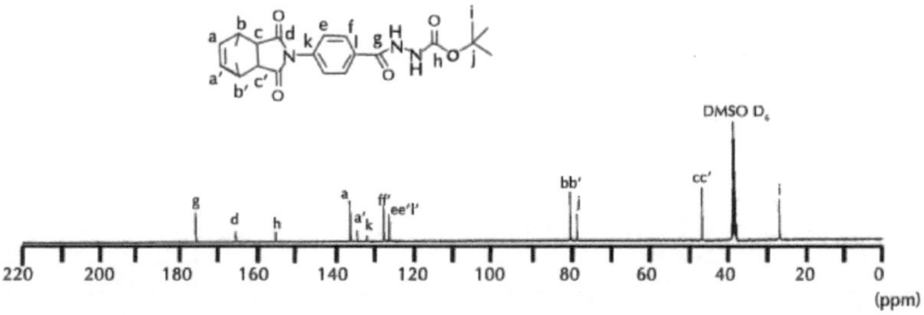

FIGURE 9 ^{13}C NMR spectrum of NADIC-BOC in DMSO-d_6.

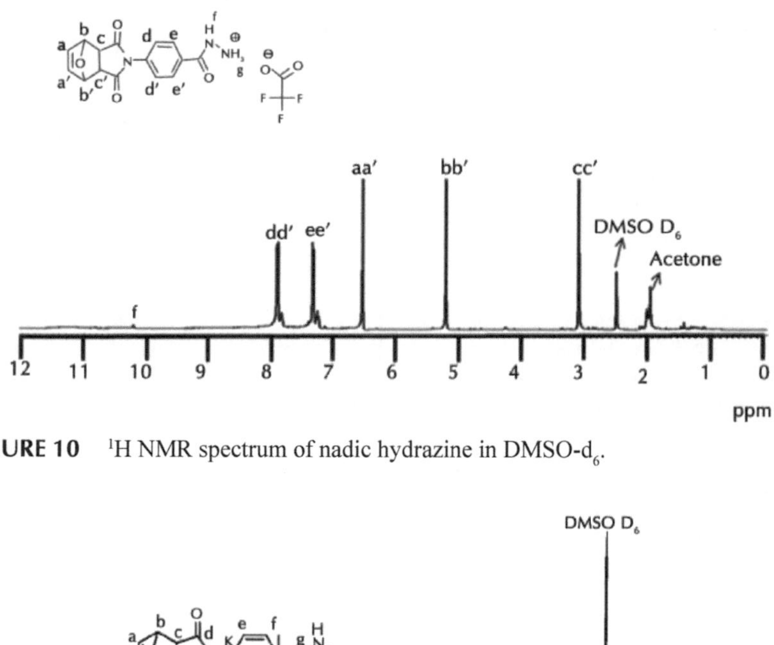

FIGURE 10 ^{1}H NMR spectrum of nadic hydrazine in DMSO-d_6.

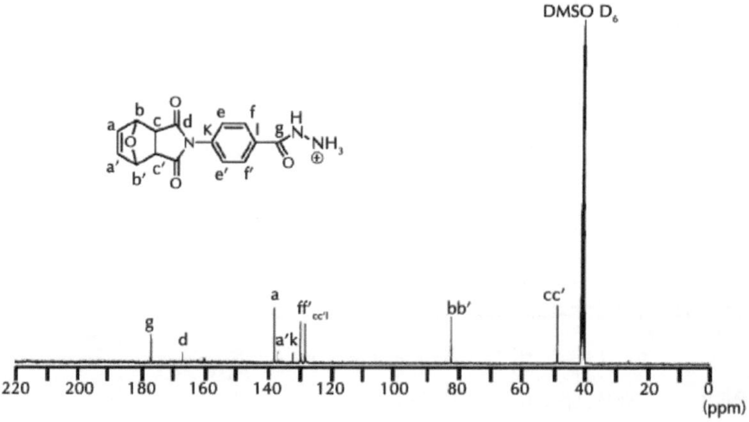

FIGURE 11 ^{13}C NMR spectrum of nadic hydrazine in DMSO-d_6.

Doxorubicin with hydrazone linker (Scheme 3) was prepared by addition of doxorubicin hydrochloride and norborne hydrazide in methanol in presence of TFA and the whole reaction set up was being protected from light for 24 h. The product was confirmed by ^{1}H NMR, IR spectroscopy and mass (Fig. 12b). The signals at δ 7.94–7.95 (m, 2H), 7.79–7.80 (m, 2H), 7.77–7.78 (m, 1H) and 7.36–7.39 (m, 1H) ppm were responsible for doxorubicin aromatic group protons. The signals at δ 6.62 (s, 2H) were observed for norbornene olefinic protons (Fig. 12a). In doxorubicin hydrazone linker the band at 1737 cm^{-1} (Fig. 13a) responsible for carbonyl-stretching frequency was absent that indicated the formation of hydrazone linker between ketone and hydrazine (Fig. 13b). All monomers were characterized by micromass spectrometer (Q-TOF), using methanol as solvent. Observed mass (m/z) and calculated mass for all monomers were in good agreement, which confirmed the formation of product.

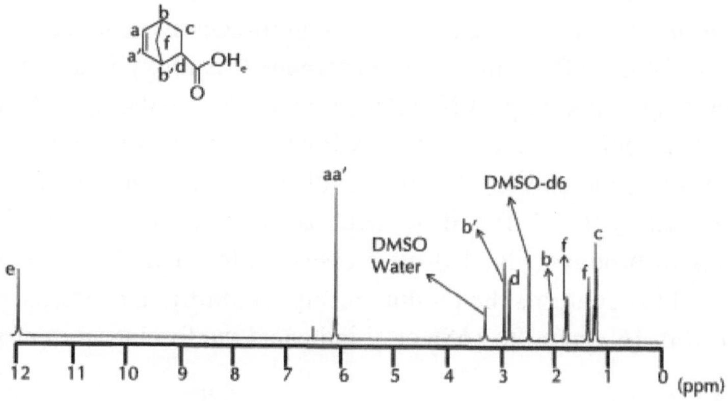

FIGURE 12 (a) ^{1}H NMR spectrum of mono 2; (b) mass spectrum of mono 2.

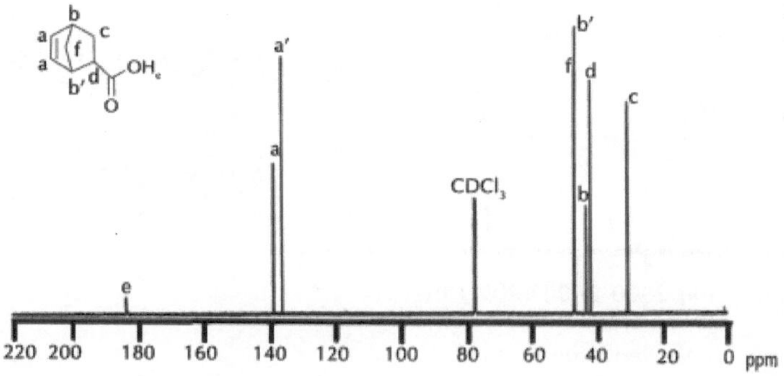

FIGURE 13 FT-IR spectra of (a) free doxorubicin and (b) mono 2.

Folic acid has two α-and γ-carboxylic acids (Fig. 14), but γ-carboxylic acid is more selectively activated due to its higher reactivity. First, the γ-carboxylic group of FA was activated by dicyclohexylcarbodiimide (DCC) and NHS (Scheme 2). Activated Folate-NHS ester was reacted with tertiary butyl carbazate in presence of pyridine to prepare BOC-FOLATE. The product was confirmed by ^{1}H NMR and IR spectroscopy. The signal at δ 1.30 (s, 9H) ppm protons (Fig. 15) were responsible for tertiary butyl group. Shifting of stretching frequency from 3377 cm^{-1} to 3392 cm^{-1} indicated the formation of amide group. The resulting residue (BOC-FOLATE) was dissolved in 20 ml of TFA to deprotect the BOC group. Deprotection was carried out at room temperature for 2 h. Deprotection was confirmed by ^{1}H NMR and IR spectroscopy. The signal at δ 1.30 (s, 9H) ppm for tertiary butyl group was absent in the product, which indicated the complete deprotection of tertiary butyl group. Shifting of stretching frequency from 3277 cm^{-1} to 3337 cm^{-1} indicated the formation of free amine group. Mono 1 (Scheme 2) was synthesized by coupling of deprotected BOC-FOLATE with exo 5-norbornene-2-carboxylic acid by using N-(3-dimethylaminopropyl)-N'-ethylcarbodiimide hydrochloride (EDC) coupling reagent in presence of 1-hydroxybenzotriazole (HOBT) in dimethyl formamide solvent. Formation of monomer that was thoroughly characterised by ^{1}H NMR and IR spectroscopy. The signal at δ 6.13-6.15 (s, 2H) ppm protons (Fig. 16) were responsible for norbornene olefinic protons, which confirms the product formation. Shifting of stretching frequency from 1696 cm^{-1} to 1686 cm^{-1} indicated the formation of product.

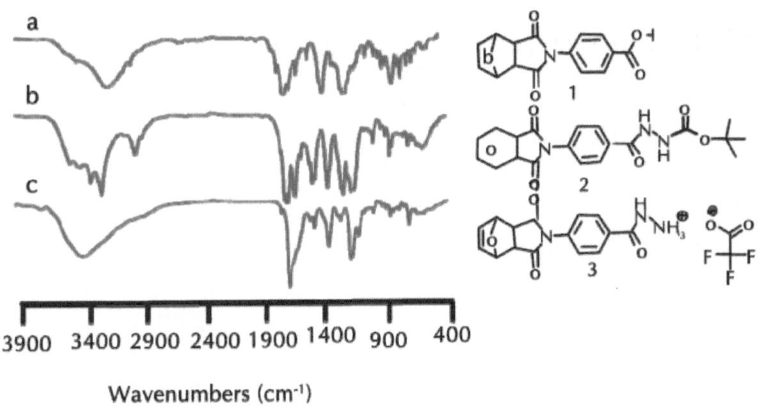

FIGURE 14 ^{1}H NMR spectrum of folic acid in DMSO-d_6.

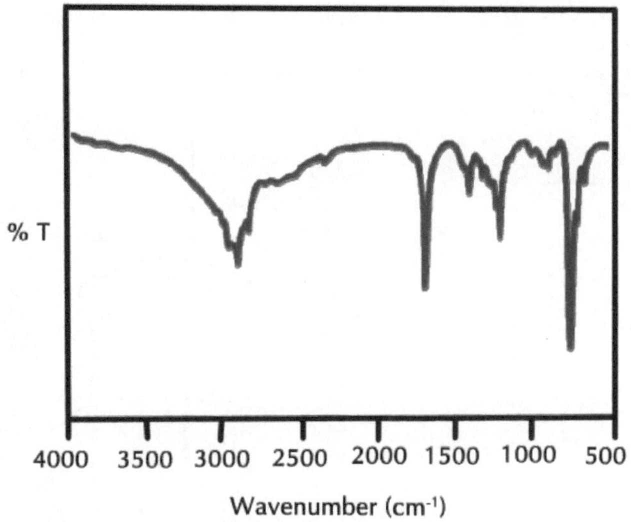

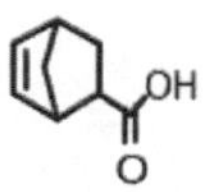

4

FIGURE 15 ^{1}H NMR spectrum of BOC-FOL in DMSO-d_6.

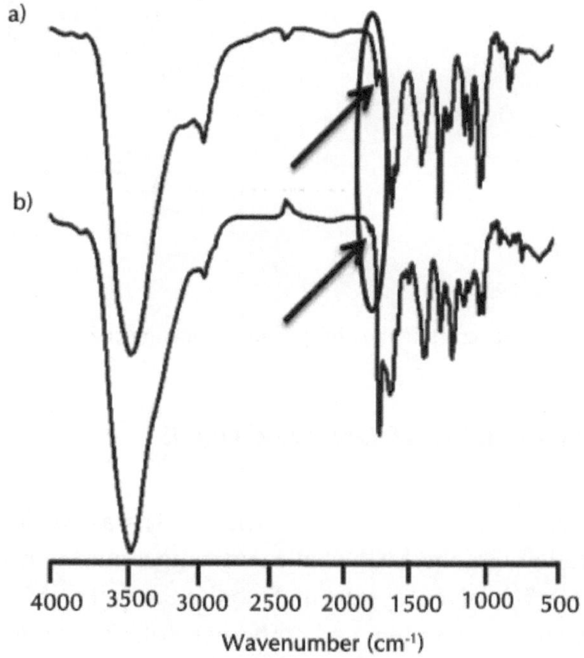

FIGURE 16 ^{1}H NMR spectrum of NOR-FOL in DMSO-d_6.

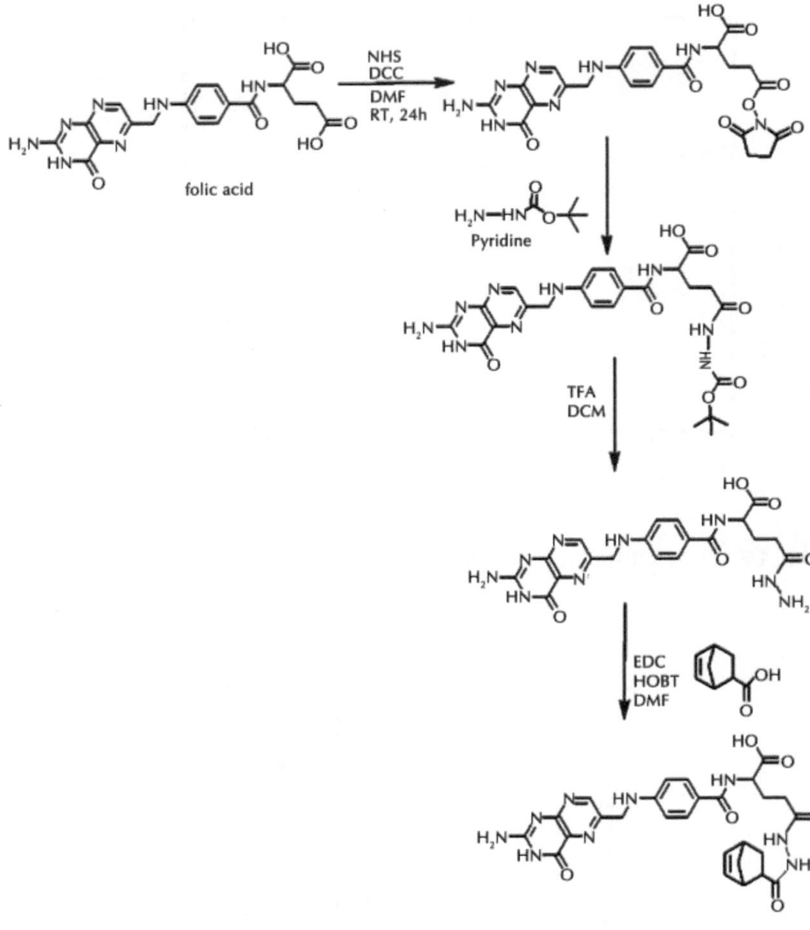

SCHEME 2 Schematic representation of synthesis of monomer 2.

5.2.3.2.2 POLYMERIZATION REACTIONS

Homopolymerization of monomer 2 (Scheme 3) was carried out by using 2nd generation Grubbs'catalyst at room temperature in dry $CDCl_3$ and CD_3OD (9:1 v/v %) solvent system and monitored by ^1H NMR. New peaks were observed at δ 5–5.3 ppm, and norbornene olefinic protons were disappeared at δ 6.6 ppm indicating the formation of the product. The stretching frequency at 1567 cm⁻¹ norbornene olifinic (C=C) was absent which confirms the homo polymer (Fig. 17).

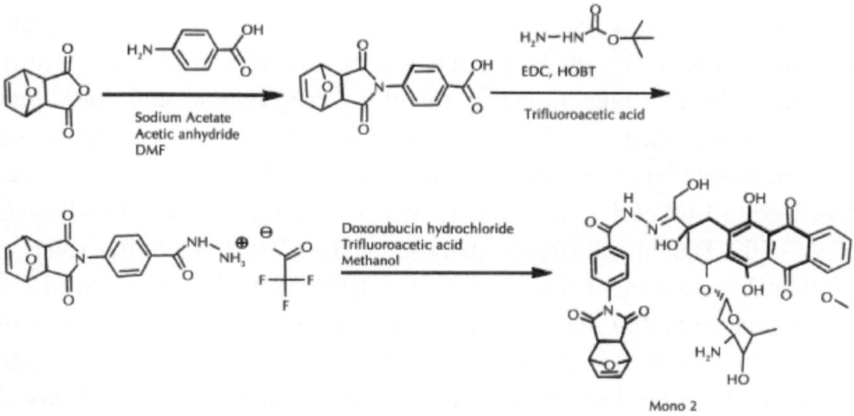

SCHEME 3 Schematic representation of synthesis of monomer 1.

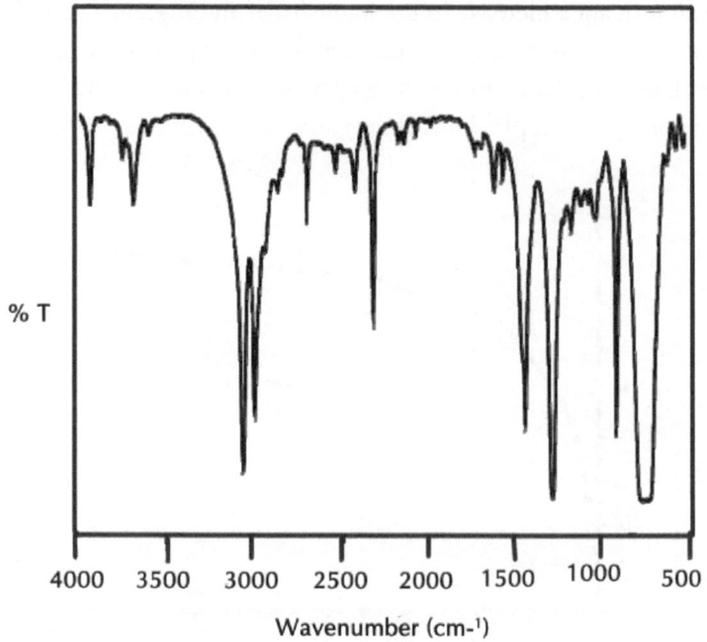

FIGURE 17 FT-IR spectra of poly 1.

5.2.3.3 DIALYSIS STUDY OF HOMOPOLYMER

For the drug release profile of homopolymer of DOX, pH 7.4 as well as acidic conditions was decided as the normal pH of blood for sustaining human life is about 7.4, while the endosomes and lysosomes of cells have a more acidic environment [11c]. Therefore, drug release study of homopolymer was carried out in pH 7.4 in phosphate buffer solution and pH 5.5 and 6 in acetate buffer solutions, respectively. A 1 mg of homopolymer was dissolved in 1 ml of distilled water (partially soluble) and loaded in dialysis tube (3,500; Dalton cut-off) and dialyzed against 70 ml of a buffer solution whose pH was maintained at 5.5. An aliquot of the sample was removed and its absorbance at 480 nm was measured as an indication of the release of doxorubicin. Fluorescence also recorded by exciting the solution at 510 nm. Emissions from the free DOX released from homopolymer were observed at 560 nm and 590 nm. The sample was then added back to the solution to maintain the volume of the solution. This procedure was repeated for every 1 h and results were observed for 48 h. It was observed that after 24 h, there was no significance increase in the intensity of fluorescence and UV intensity. The similar procedure was repeated to monitor the drug release at pH 6.0 as well as pH 7.4 and the results are shown in (Fig. 18). The DOX release from homopolymer at pH 7.4 was very minimal.

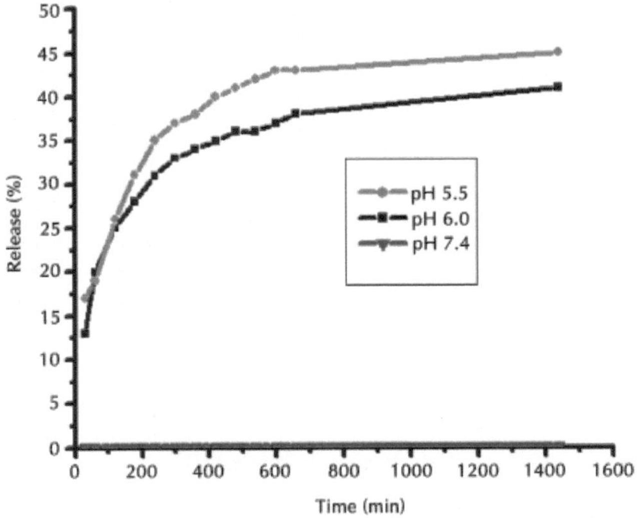

FIGURE 18 DOX Release Profile at 37°C in comparison with pH 5.5, 6 and 7.4.

Copolymerization of FOL-DOX have solubility issues. We are in the process of incorporating PEG moiety in FOL-DOX copolymer to make the copolymer's solubility in aqueous as well as biological medium. This synthetic attempt along with newly designed copolymer's biocompatibility studies will be our future report.

9.2.4 NON-COVALENT APPROACH

Self-assembling systems have a remarkable property of promptly responding to external stimuli, such as pH, solvent polarity, and temperature. So we are very specific among the structures like vesicles formed by the self-assembly of amphiphilic homopolymers. Because of their fundamental perspective, they have rich potential applications in drug or gene delivery, nanotechnology, and even they are considered to be model systems of biomembranes. These systems have demonstrated its versatility in the non-covalent drug carriers in the drug delivery therapy. Herein, we report a pH- bilayer vesicle formation from a unique design, an amphiphilic, norbornene derived barbiturate homopolymers, NDBH, is synthesized using the ring opening metathesis polymerization (ROMP). The barbiturate functionality of each monomer unit in NDBH can act as a hydrophilic head group, whereas the norbornene backbone can behave as a hydrophobic moiety (Scheme 4).

9.2.4.1 SYNTHESIS OF COMPOUND 1

A 10 g (102 mmol) of maleic anhydride and 8.32 g (122.4 mmol) of furan was dissolved in toluene. The reaction mixture was stirred at room temperature for 48 h. This product was isolated by filtration and washed with cold toluene and dried under vacuum to yield exo-7-oxabicyclo[2,2,1] hept-5-ene-2,3-dicarboxylic anhydride. ^{1}H NMR (500 MHz, $CDCl_3$): δ 6.56 (s, 2 H), 5.44 (s, 2H), 3.17(s, 2H); ^{13}C NMR (125 MHz, dmso-d_6): δ 171.54, 137.20, 81.63, 49.08; MS (ESI) calculated for $C_8H_6O_4$ (M^++H) 166.02, observed 166.

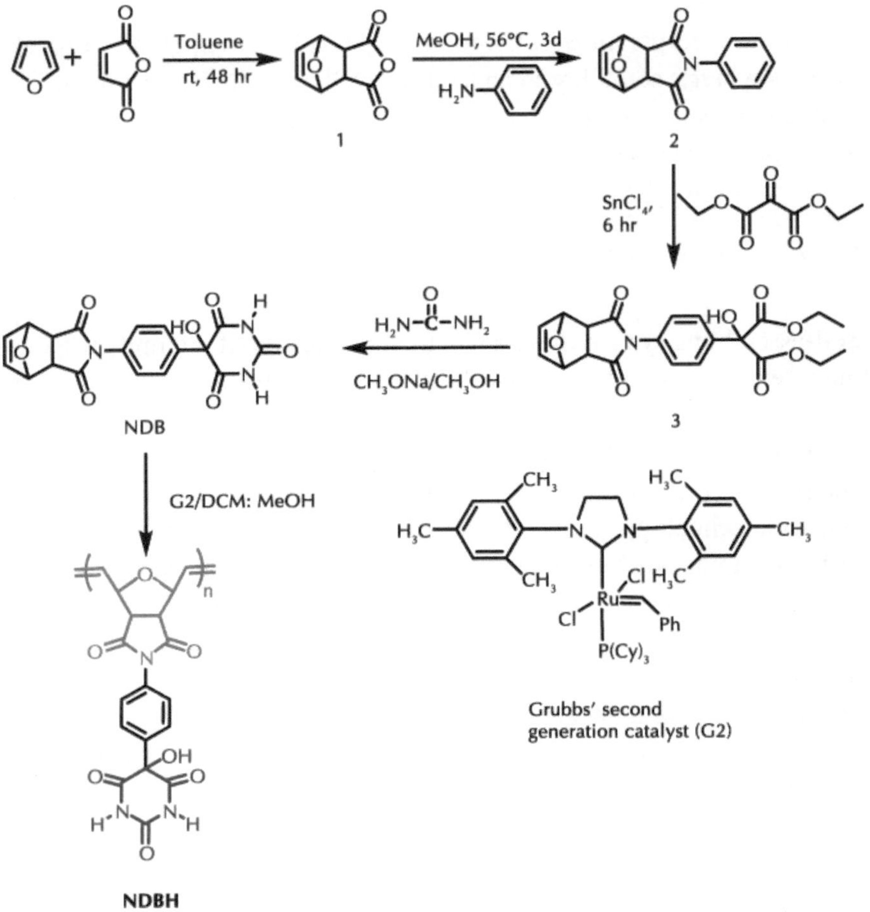

SCHEME 4 Synthesis of NDBH via ring opening metathesis polymerization.

9.2.4.2 SYNTHESIS OF COMPOUND 2

A 5 g (30.12 mmol) of compound 1 and 2.8 g (30.12 mmol) of aniline was dissolved in 50 ml methanol. The reaction mixture was stirred at 56°C for 72 h. This reaction mixture concentrated to the pasty mass and this pasty mass was dissolved in dichloromethane (50 ml), and washed with water (3 × 100 ml). The organic layer was dried over $NaSO_4$ and concentrated under vacuum to yield compound 2 as a brownish solid. ^{1}H NMR (500

MHz, dmso-d_6): δ 7.49 (s, 4H), 7.19 (s, 1H) 6.5 (s, 2 H), 5.2 (s, 2H), 3.0 (s, 2H); ^{13}C NMR (500 MHz, DMSO-d_6): δ 172.01, 137.31, 131.90, 128.01, 82.10, 49.53; MS (ESI) calculated for $C_{14}H_{11}NO_3$ (M^++H) 241.07, observed 241.62.

9.2.4.3 SYNTHESIS OF COMPOUND 3

A 200 mg of compound 2 was dissolved in 5 ml dichloromethane in a two necked round bottom flask fitted with $CaCl_2$ guard tube and the flask was kept in ice bath to avoid vigorous reaction. Diethyl ketomalonate (1 ml) was added via syringe followed by drop wise addition of 0.9 ml of stannous chloride. The reaction mixture was stirred at room temperature for 6 h. The reaction mass was concentrated under vacuum to yield brownish solid. ^{1}H NMR (500 MHz, DMSO-d_6): δ 7.49 (s, 2H), 7.19 (s, 2H), 7.14 (s, 1H), 6.5 (s, 2 H), 5.2 (s, 2H), 4.13 (m, 4H), 3.0 (s, 2H), 1.29 (t, 6H); ^{13}C NMR (500 MHz, DMSO-d_6): δ 172.01, 168.4, 137.31, 131.90, 128.01, 98.7, 82.10, 61.6, 49.53, 14.1. MS (ESI) calculated for $C_{21}H_{21}NO_8$ (M^++H) 415.12, observed 416.

9.2.4.4 SYNTHESIS OF NDB

50 mg of compound 3 was dissolved in 5 ml of dry tetrahydrofuran and freshly prepared sodium methoxide (50 mg) was dissolved in 20 ml methanol. Urea (50 mg was added to the reaction mixture. The reaction mixture was stirred for 24 h at room temperature. The product was isolated by precipitation using excess tetrahydrofuran. ^{1}H NMR (500 MHz, DMSO-d_6): δ 10.01 (s, 2H), 8.79 (s, 2H), 8.41 (s, 2H), 8.04 (s, 1H), 6.55 (s, 2 H), 5.38 (s, 2H), 3.2 (s, 2H); ^{13}C NMR (500 MHz, DMSO-d_6): δ 176.00, 168.4, 149.21, 137.31, 131.90, 128.01, 98.7, 82.10, 49.53; MS (ESI) calculated for $C_{18}H_{13}N_3O_6S$ (M^++H) 383.31, observed 384.17.

9.2.4.5 SYNTHESIS OF NDBH

Known amounts of NDB (50 mg) were weighed into separate Schlenk flasks, placed under an atmosphere of nitrogen, and dissolved in methanol and anhydrous dichloromethane (9:1 v/v %). Into another Schlenk flask, a desired amount of second generation Grubbs' catalyst 1.06 mg (100 mol %) was added, flushed with nitrogen, and dissolved in minimum amount of anhydrous dichloromethane. All two flasks were degassed three times by freeze-pump-thaw cycles. The NDB was transferred to the flask containing the catalyst via a cannula. The reaction was allowed to stir at room temperature until the polymerization was complete (24 h) before it was quenched with ethyl vinyl ether (0.2 ml). An aliquot was taken for GPC analysis, and the remaining product was precipitated from pentane, dissolved it again THF, passed it through neutral alumina to remove the catalyst and precipitated again from pentane to get a pure homopolymer. Gel permeation chromatography was done in tetrahydrofuran (flow rate = 1 ml/min). The molecular weight of polymer using was measured using polystyrene standards. M_n = 8300, PDI = 1.09. ^{1}H NMR (500 MHz, DMSO-d_6): δ 10.02(s, 2H), 8.32 (s, 2H), 8.19 (s, 2H), 7.54 (s, 1H), 5.4 (s, 2 H), 5.2 (s, 2H), 3.67(s, 2H).

9.2.4.6 RESULTS AND DISCUSSIONS

The Synthesis of exo-7-oxabicyclo [Section 7.2.2.1] hept-5-ene-2, 3-dicarboxylic anhydride (1) was previously reported in literature[16f–j]. From the ^{1}H NMR and ^{13}C NMR spectroscopy data, the new signal at δ 3.20 ppm and in δ 49.1 ppm confirmed the formation of compound 1 (Figs. 19 and 20). The compound 2 was synthesized by the coupling reaction of compound 1 with aniline, which was confirmed by ^{1}H NMR in which the aromatic proton signal of aniline was observed at δ 7.20 to δ 7.60 ppm. Also in ^{13}C NMR the signal at δ 176 ppm was observed due to the formation of cyclic amide (Figs. 21 and 22). The compound 3 was synthesized by Friedel-Crafts acylation reaction between compound 2 and diethyl ketomalonate by using stannous chloride as a catalyst. The formation of this product was confirmed by ^{1}H NMR. The signal at δ 7.39 ppm was responsible for the newly formed hydroxyl group where as the signal at δ 4.19 ppm and δ 1.20 ppm was responsible for OCH_2 and CH_3, respectively (Fig. 23). The ^{13}C NMR signal at δ 92 ppm was due to the newly formed hydroxyl group attached Carbon (Fig. 24). Formation of the compound 3 was also confirmed by FT- IR where the stretching frequency at 1708 cm^{-1} was due to ester carbonyl. Finally the NDB was prepared by coupling 3 with urea. The synthesis of NDB was confirmed by ^{1}H NMR and FT-IR. The ^{1}H NMR spectrum in DMSO-d_6 conveyed the absence of ester proton's signal at δ 4.19 ppm for the ester group and one new signal at δ 10.01 ppm which was responsible for amide –NH (Fig. 25). From the FT-IR spectra, the complete disappearance of ester carbonyl at 1708 cm^{-1} of compound 3 and the new stretching frequency at 1644 cm^{-1} for cyclic amide confirmed the formation of NDB (Fig. 26).

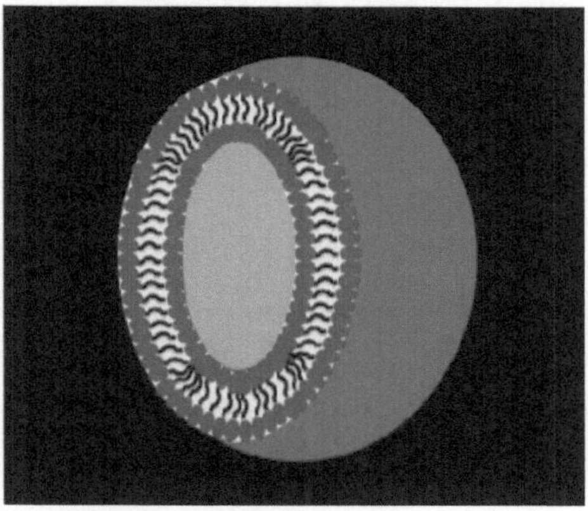

FIGURE 19 ^{1}H NMR of compound 1 in $CDCl_3$.

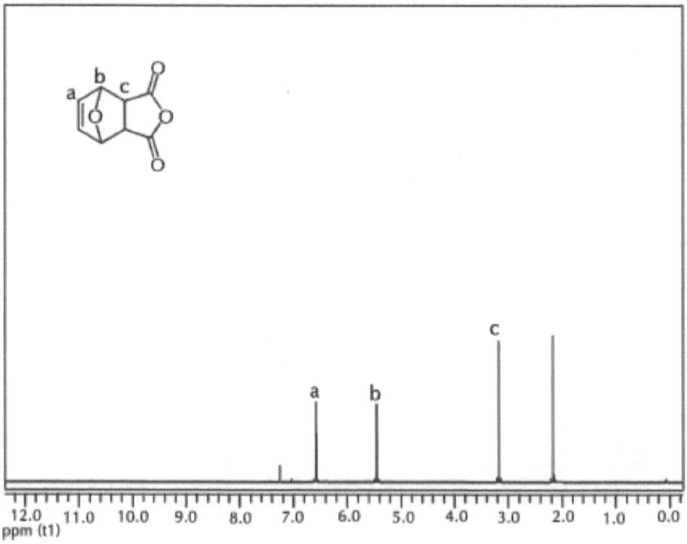

FIGURE 20 ^{13}C NMR of compound 1 in $CDCl_3$.

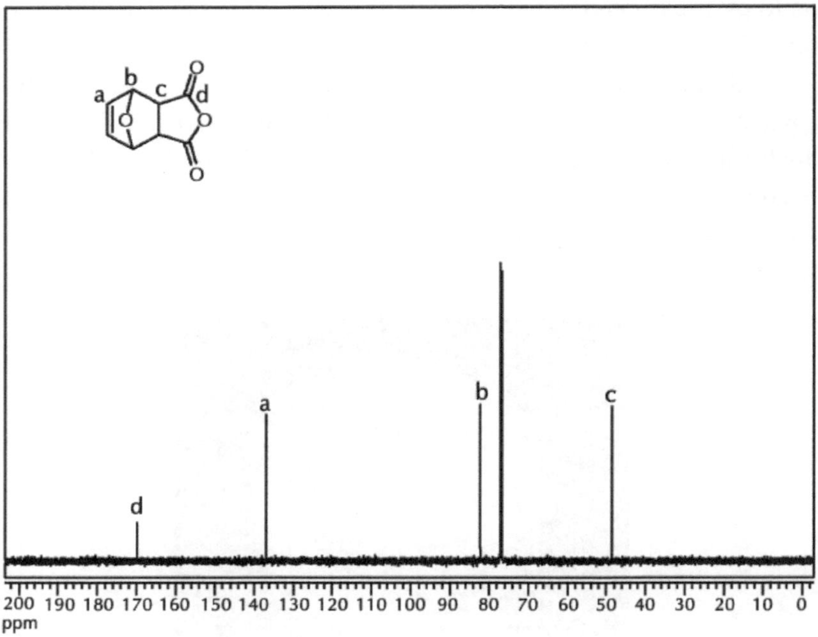

FIGURE 21 ^{1}H NMR of compound 2 in DMSO-d_6.

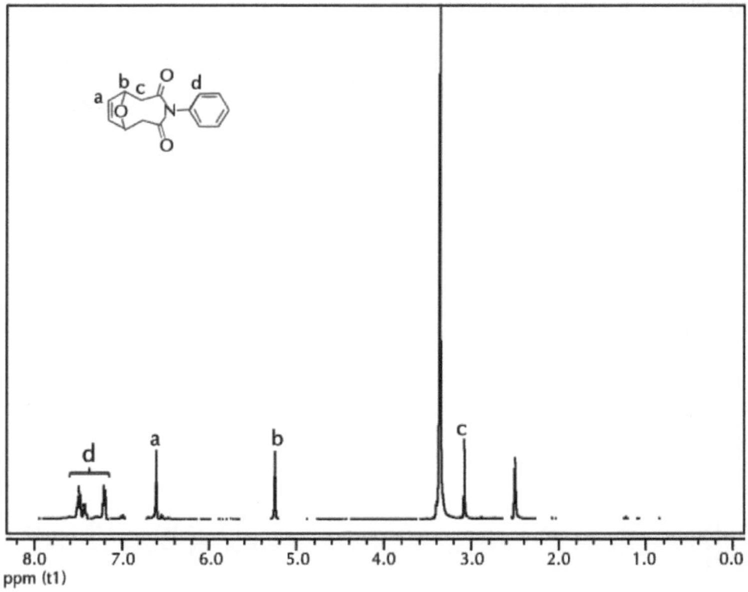

FIGURE 22 ^{13}C NMR of compound 2 in $CDCl_3$.

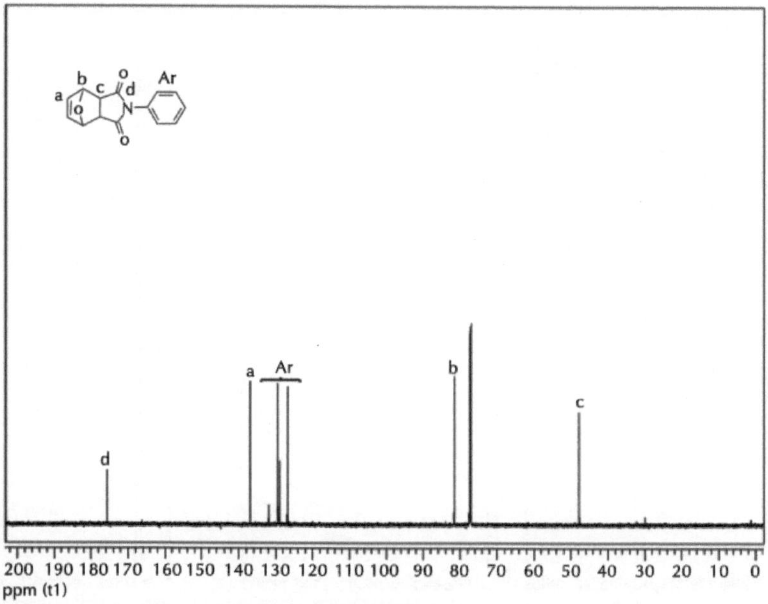

FIGURE 23 ^{1}H NMR of compound 3 in DMSO-d_6.

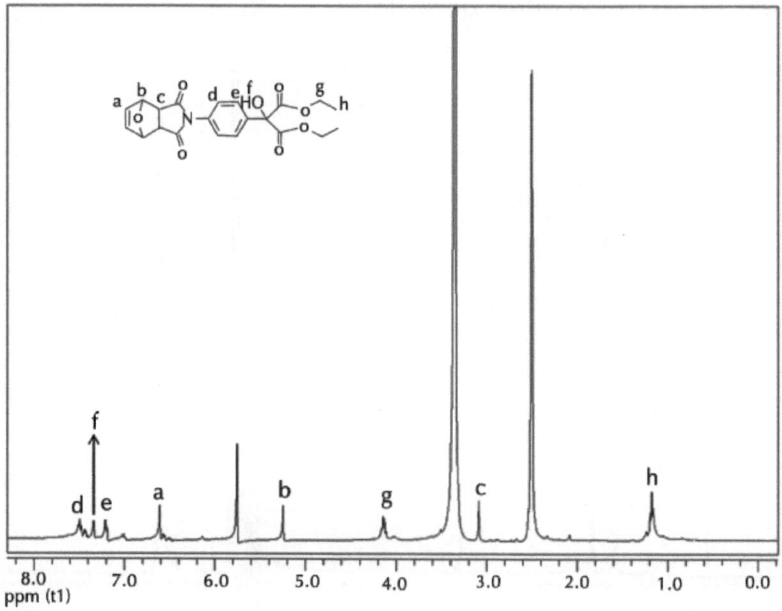

FIGURE 24 ^{13}C NMR of compound 3 in DMSO-d_6.

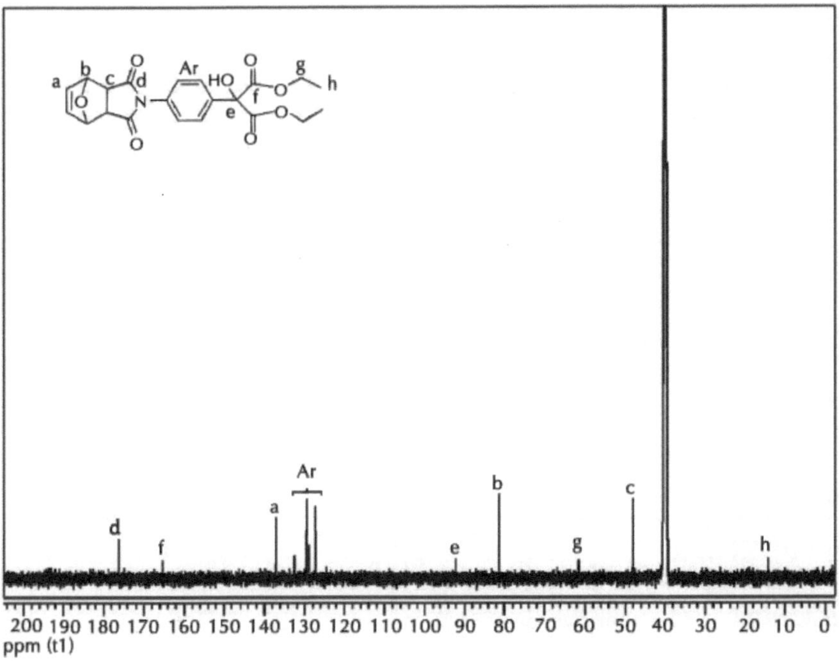

FIGURE 25 ^{1}H NMR of NDB in DMSO-d_6.

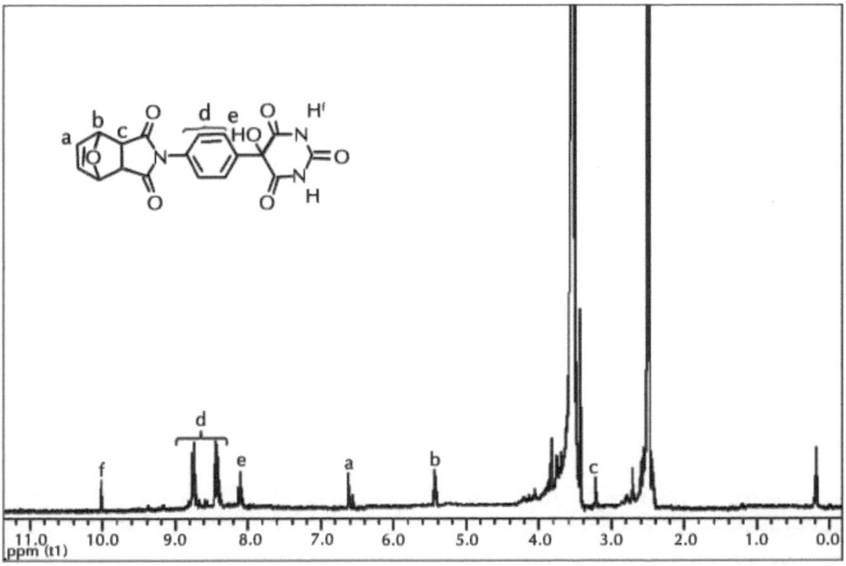

FIGURE 26 FT-IR spectroscopy of NDB.

Next, the homopolymerization of NDB (Scheme 4) was carried out by using 2nd generation Grubbs' catalyst at room temperature in dry dichloromethane and methanol (9:1 v/v %) solvent system and monitored by ^{1}H NMR. New signals were observed at δ 5.4 ppm, indicating the formation of the NDBH (Fig. 27). The molecular weight of the NDBH was determined by the gel permeation chromatography as shown in (Fig. 28). (M_n = 8300 with PDI 1.09). This NDBH polymersomes are expected to form a vesicle as we recently reported for the similar systems (Fig. 29). The barbituate functionality in the NDBH vesicles will be elaborately explored for its therapeutic nature. Also the drug encapsulated NDBH vesicles its use as delivery vehicle (Fig. 30).

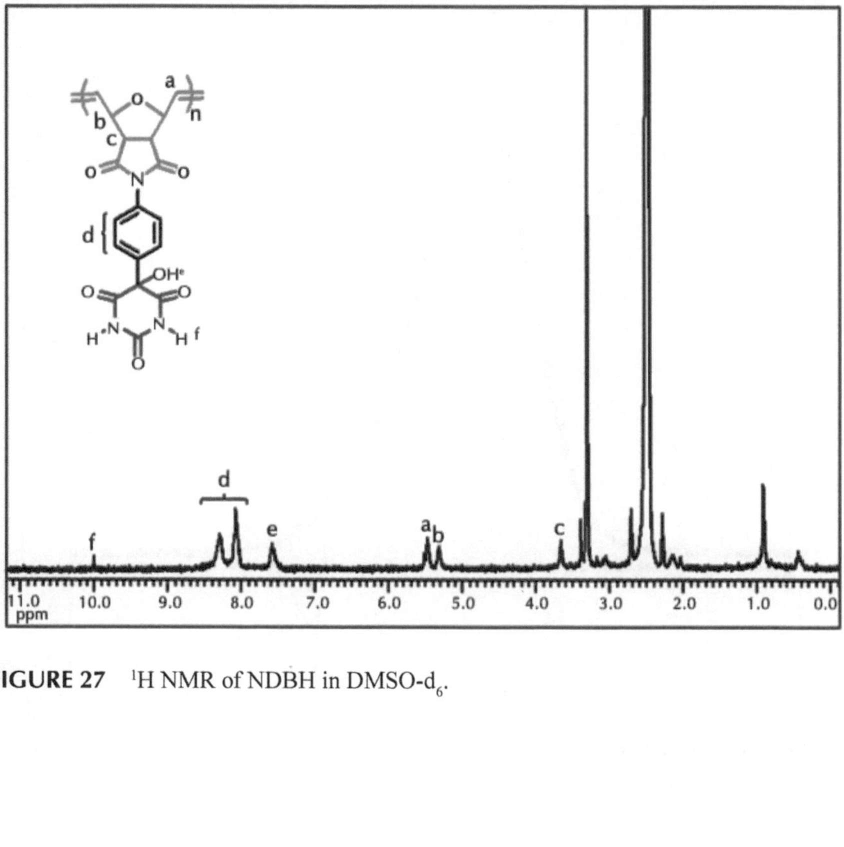

FIGURE 27 ^{1}H NMR of NDBH in DMSO-d_6.

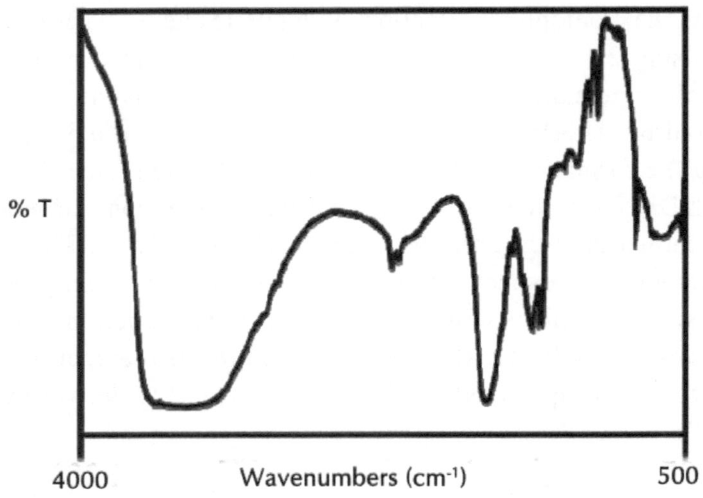

FIGURE 28 GPC traces of NDBH (M_n = 8300 Da, PDI = 1.09).

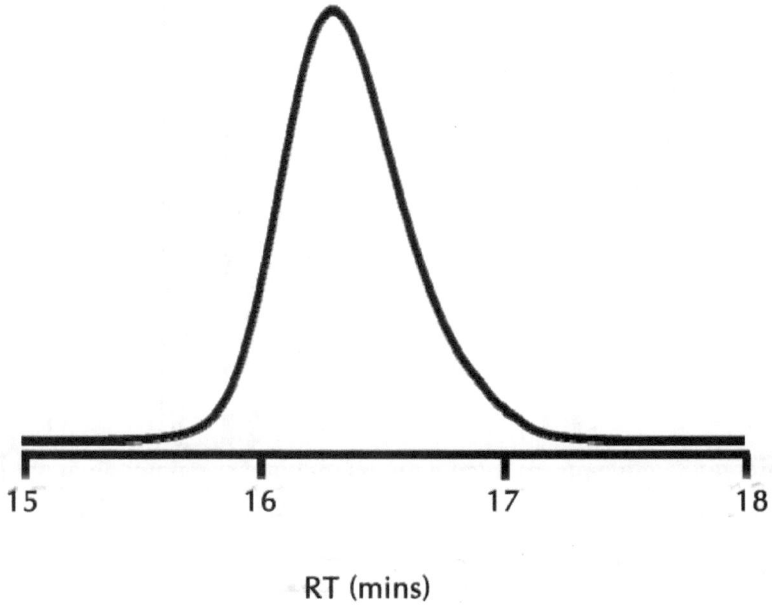

FIGURE 29 Cartoon representation of NDBH.

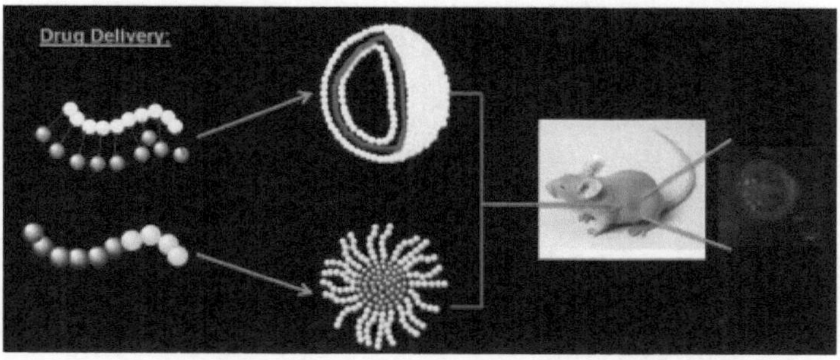

FIGURE 30 Cartoon representation of the overall biological evaluation of the polymer prodrugs (both covalent and non-covalent approach) discussed in this chapter.

9.3 CONCLUSION

Here, we report the design and synthesis of ROMP polymers that have potential application in cancer therapy. The synthesis and complete characterization of both norbornene derived doxorubicin (mono 1) and polyethylene glycol (mono 2) monomers are clearly described. The synthesis of the newly designed monomers is confirmed by NMR, IR, and mass spectroscopy. Secondly their copolymerization by ring-opening metathesis polymerization (ROMP) to get the block copolymer (COPY-DOX) is vividly elaborated. Similarly the NDB monomer as well as its homopolymer (NDBH) synthesis and complete characterization are thoroughly discussed. The newly synthesized polymers demonstrate both covalent and non-covalent approach to introduce the drug on to the nano-carriers. Overall, this present work clearly demonstrates the versatility of the ROMP technique in producing polymers, which have biological importance.

KEYWORDS

- **block copolymer**
- **doxorubicin**
- **Grubbs' catalyst**
- **homopolymer**
- **polyethylene glycol**
- **Schlenk flask**

REFERENCES

1. Fischbach, C.; Mooney, D. J. *Biomaterials,* **2007**, *28*, 2069.
2. Langer, R.; Tirrell, D. A. *Nature,* **2004**, *428*, 487.
3. Lutolf, M. P.; Hubbell, J. A. *Nat. Biotechnol,.* **2005**, *23*, 47–55.
4. Kiick, K. L. *Science,* **2007**, *317*, 1182–1183.
5. Duncan, R. *Nat. Rev. Cancer,* **2006**, *6*, 688–701.
6. Shuang L.; Ronak, M.; Kiick, K. L. *Macromolecules,* **2009**, *42*, 3–13.
7. a) Duncan, R.; *Nat. Rev. Drug Discovery,* **2003**, 2, 347–360. b) Duncan, R.; Kopecek, J. *Nat. Rev.,* **1984**, 57, 51–101.
8. a) Helminger, G.; Sckell, A.; Dellian, M.; forbes, N. S.; Jain, R. K. *Clic. Cancer Res.,* **2002**, 8, 1284–1291. b) Duncan, R. *Anti-Cancer Drugs,* **1992**, 3, 175–210. c) Duncan, R. *Biochem. Soc. Trans.,* **2007**, 35, 56–60.
9. a) Singla A. K.; Garg, A.; Aggarwal, D. *Int. J. Pharm.,* **2002**, 235, 179–192. b) Stukel, J. M.; Li, R. C.; Maynard, H. D.; Caplan, M. R. *Biomacromolecules,* **2010**, *11,* 160–167. c) Christman, K. L.; Broyer, R. M.; Schopf, E.; Kolodziej, C. M.; Chen, Y.; Maynard, H. D. *Langmuir,* **2011**, 27, 1415–1418.
10. Maeda, H.; Wu, J.; Sawa, T.; Matsumura, Y.; Hori, K. *J. Controlled Release* **2000**, 65, 271-284.
11. a) Ghosh, S.; Basu, S.; Thayumanavan, S. *Macromolecules,* **2006**, *39,* 5595–5597. b) Satav, S. S.; Bhat, S.; Thayumanavan, S. *Biomacromolecules,* **2010**, *11,* 1735–1740.
12. Gillies, E. R.; Frechet, J. M. J. *Bioconjugate Chem.*, **2005**, *16*, 361–368.
13. Gillies, E. R.; Frechet, J. M. J. *Chem. Commun.*, **2003**, *14*, 1640–1641.
14. a) Lehn, J.-M.; Eliseev, A. *Science,* **2001**, *291*, 2331–2332. b) Lehn, J.-M. *Chem. Soc. Rev.,* **2007**, *36*, 151–160; c) Jin, Y.; Song, L.; Su, Y.; Zhu, L.; Pang, Y.; Qiu, F.; Tong, G.; Yan, D.; Zhu, B.; Zhu, X. *Biomacromolecules,* **2011**, *12*, 3460–3468.
15. Gillies, E. R.; Frechet, J. M. *J. Pure Appl. Chem.*, **2004**, 76, 1295–1307.
16. a) Grubbs, R. H. Handbook of Metathesis; Wiley-VCH: New York, 2003; Vol. 3. b) Watson, K. J.; Park, S.-J.; Im, J.-H.; Nguyen, S. T. *Macromolecules,* **2001**, *34*, 3507. c) Mortell, K. H.; Gingras, M.; Kiessling, L.L. *J. Am. Chem. Soc.,* **1994**, *116*, 12053. d) Maynard, H. D.; Sheldon, Y. O.; Grubbs, R. H. *Macromolecules,* **2000**, *33*, 6239. e) Sterling, F. A.; Al-Badri, Z.M.; Madkour, A. E.; Lienkamp, K.; Tew, G. N. *J. Polym Sci., Part A: Polym.Chem.,* **2008**, *46*, 2640–2648. f) Mane, S. R., Rao, V. N., Shunmugam, R. *ACS Macro Lett.*, **2012**, *1*, 482–488. g) Mane, S. R.; Rao, V. N.; Chatterjee, K.; Dinda, H.; Nag, S.; Kishore, A.; Das Sarma, J.; Shunmugam, R. *Macromolecules,* **2012**, *45*, 8037–8042. h) Rao, V. N.; Mane, S. R.; Kishore, A.; Das Sarma, J.; Shunmugam, R. *Biomacromolecules,* **2012**, 13, 221–230. i) Mane, S. R.; Rao, V. N.; Chatterjee, K.; Dinda, H.; Nag, S.; Kishore, A.; Das Sarma, J.; Shunmugam, R. *J. Mater. Chem.,* **2012**, *22*, 19639–19642. j) Rao, V. N.; Kishore, A.; Sarkar, S.; Das Sarma, J.; Shunmugam, R. *Biomacromolecules,* **2012**, *13*, 2933–2944.
17. Alfred, S. F.; Lienkamp, K.; Madkour, A. E.; Tew, G. N. *J. Polym. Sci., Part A: Polym. Chem.,* **2008**, *46*, 6672.
18. Alfred, S. F.; Al-Badri, Z. M.; Madkour, A. E.; Lienkamp, K.; Tew, G. N. *J. Polym. Sci., Part A: Polym. Chem.*, **2008**, *46*, 2640.

19. Lienkamp, K.; Kins, C. F.; Alfred, S. F.; Madkour, A. E.; Tew, G. N. *J. Polym. Sci., Part A: Polym. Chem.,* **2009**, *47*, 1266.
20. Lienkamp, K.; Tew, G. N. *Chem. Eur. J.,* **2009**, *15*, 11784.
21. Lienkamp, K.; Kumar, K.-N.; Som, A.; Nuesslein, K.; Tew, G. N. *Chem. Eur. J.,* **2009**, *15*, 11710.
22. Lienkamp, K.; Madkour, A. E.; Kumar, K.-N.; Nuesslein, K.; Tew, G. N. *Chem. Eur. J.,* **2009**, *15*, 11715.
23. Lienkamp, K.; Madkour, A. E.; Musante, A.; Nelson, C. F.; Nusslein, K.; Tew, G. N. *J. Am. Chem. Soc.,* **2008**, *130*, 9836.
24. Gabriel, G. J.; Maegerlein, J. A.; Nelson, C. F.; Dabkowski, J. M.; Eren, T.; N€usslein, K.; Tew, G. N. *Chem. Eur. J.,* **2009**, *15*, 433.
25. Semra, C.; Tew, G. N. *Biomacromolecules,* **2009**, *10*, 353.
26. Al-Badri, Z. M.; Tew, G. N. *Macromolecules,* **2008**, *41*, 4173.
27. Madkour, A. E.; Koch, A. H. R.; Lienkamp, K.; Tew, G. N. *Macromolecules,* **2010**, *43*, 4557–4561.
28. Matson, J. B.; Stupp, S. I. *Chem. Commun.,* ***2011***, *47 (28)*, 7962–7964.
29. Bae, Y.; Fukushima, S.; Harada, A.; Kataoka, K. *Angew. Chem., Int. Ed.*, **2003**, *42*, 4640–4643.

CHAPTER 10

CORRELATION OF THE EXPERIMENTAL AND THEORETICAL STUDY FOR THE CYCLISATION REACTION OF MESOIONIC HETEROCYCLE IMIDAZO[1,5 A]PYRIDINES IN TERMS OF THE DENSITY FUNCTIONAL DESCRIPTORS

S. ROY, S. DHAIL, S. NASKAR and T. CHAKRABORTY

CONTENTS

10.1 INTRODUCTION

Recently, Roy et al., published the synthetic procedure of biologically active mesoionic heterocycles. They have observed a peculiar mechanistic feature that substituted mesoionic derivatives do not undergo cyclisation whereas the unsubstituted one readily participates into the cyclisation. It is also revealed from their experimental study that sulphur containing mesoionic heterocycles are less reactive as compared to the oxygen containing derivatives. In this report, we have evaluated theoretical DFT based local reactivity parameters like fukui functions, local softness and local philicity indices to validate the above mentioned experimental facts. We have also calculated the global reactivity parameters to explain the reactivity differences between the substituted and unsubstituted mesoionic derivatives. The nice correlation between the experimental and theoretical counterpart supports our venture.

The central idea of chemistry is that the physical and chemical properties of molecules are determined by its geometrical as well as electronic structures. It is expected that all the chemical, biological and physical properties of a molecule must be coded in its structural formula. The challenge of fundamental science is to correlating structure with property. The drug is a molecule and bacteria are also bio-molecules and the interaction between the two is a chemical process and will occur in accordance with the fundamental laws of chemistry. It is reasonable to assume that nature has given a clue of the drug activity of molecules in their structure and it is the task of a scientist to locate and identify it and to link it with the process of curing, i.e., the healing mechanism. The action of the potent carcinogens lies in their chemical formula and the electronic structures.

The branch of theoretical science, which is engaged in studying the relationship between structure and property is labeled as Structure Activity Relationship (SAR) and Structure Property Relationship (SPR). Thus the jargon of the trade of SAR and SPR is the correlation of properties with structure in terms of some descriptors–global and local. The quantum mechanical descriptors that are useful in SAR/SPR studies are as follows: (i) charge densities on atomic sites, (ii) the dipole moment, (iii) eigen values of the frontier orbitals, the HOMO and LUMO, (iv) the HOMO-LUMO gap, (v) the chemical potential (μ) or electronegativity (χ), (vi) the global hardness (η), (vii) the global softness (S), (viii) electrophilicity index (ω). Of these, electrophilicity index (ω), has special relevance in the present context because ligands binding phenomena are of general interest in catalysis and drug design and protein and DNA functioning. In addition to this list, some density functional theoretic local descriptors can also be invoked in this study. These are fukui functions and local softness and local philicity indices. The descriptors mentioned above [1–8] play a critical role in correlating the structure with reactivity and the site selectivity of various bio-active molecules [9, 10]. The local philicity index is by far the most powerful concept of reactivity and selectivity among the descriptors invoked in this type of study.

Imidazo[1,5-a]pyridines are an important class of mesoionic heterocyclic compounds, some of which are found utility in a number of research areas including potential applications in organic light-emitting diodes (OLED) [11–13] and thin-layer field effect transistor (FET) [14]. The unique biological and photo-physical properties of nonionic imidazo[1,5-a]pyridines also led to a wide variety of its applications; such as inhibitors of several enzymes (HIV protease, thromboxane synthetase and Mek Kinase) [15–17], antibiotics [18], fluorophores [19, 20]. Furthermore, the mesoionic heterocycles are being used now-a-days as a useful synthon for many valuable organic compounds [21].

The recent advent of the copper-catalyzed cyclo addition reaction makes available Imidazo[1,5-a]pyridines by Roy et al., although efficiency and reliability is matched with the expected reaction procedure [22]. In the proposed catalytic cycle, after metal reduction nucleophilic attack by the pyridyl-N of the pmtpm ligand to the C-atom of the SCN part which is equatorially aligned in the Cu(II)-complex [23] (Scheme 1). This nucleophilic attack is the important part to form the final product (3) but substituted pmtpm derivatives at the C2 and C6 position of the pyridine ring with different substituents make it least nucleophilic center (Scheme 2). While investigating the effects of O-containing schiff base ligands rather than S-containing is more pronounced for the faster cyclo-addition reaction (Scheme 3), which is highly supported by the experimental fact.

In this manuscript, we report the feasibility of the cyclo-addition reaction of the final product Imidazo[1,5-a]pyridines (3) through a Cu(II)-intermediate (2) by theoretical calculation.

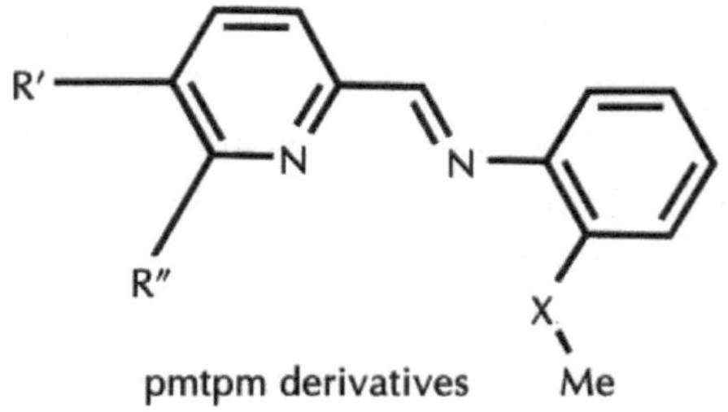

Set-1: X=S, R′= H, R″= -Br, -Br, $-CH_3$, $-C_6H_4$-Cl, $-OCH_3$
Set-2: X=S, R′= Br, R″= -H
Set-3: X= O, R′ = H, R″ = -Br, $-CH_3$, $-C_6H_4$-Cl,OCH_3
Set-4: X= O, R′ = Br, R″ = -H

SCHEME 1 Selected thioether and ether containing Schiff base ligands.

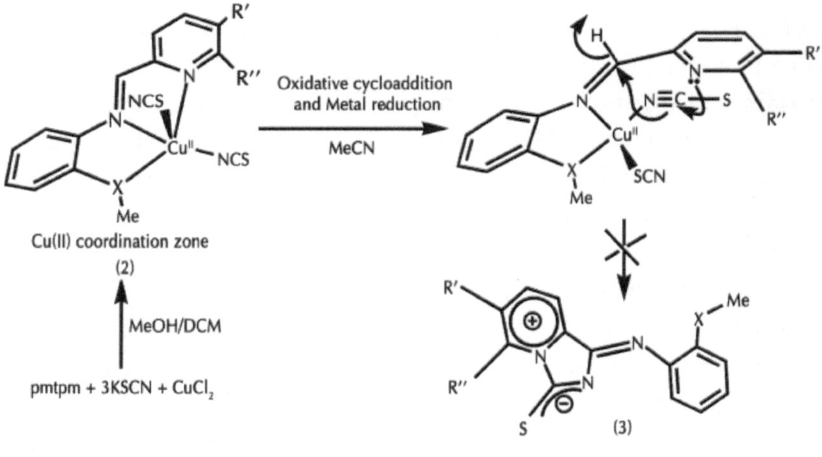

SCHEME 2 Possible mechanistic pathway for the cycloaddition reaction.

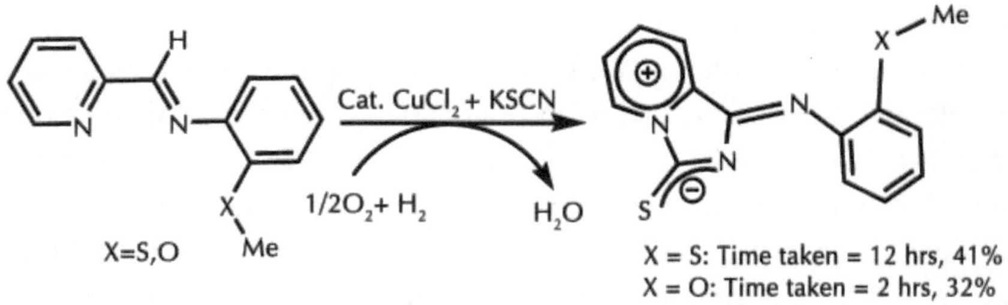

SCHEME 3 Copper catalyzed cycloaddition of unsbstituted Schiff base ligand.

10.2 THEORETICAL BACKGROUND

Recent day, quantum mechanics becomes very popular to explain the mechanistic features of bio-active molecules. There are several quantum chemical descriptors through which we can predict reaction mechanism and as well as structure activity relationship of numerous bioactive molecules. A number of excellent reviews have been published on the application of quantum chemical descriptors in SAR/SPR studies [24–26]. To determine the equilibrium geometry, the molecular force field and to compute the quantum mechanical descriptors of the drug molecules, some suitable quantum mechanical method are invoked [27].

10.2.1 THE GLOBAL DESCRIPTORS

Parr et al. [1] for the first time put the qualitative concept, electronegativity, on a sound quantum mechanical basis when they defined, within the paradigm of density functional theory of Hohenberg and Kohn [28], that the electronegativity is the negative of a partial derivative of energy (E) of an atomic or molecular system with respect to the number of electrons N, at a constant external potential v(r).

$$\mu = -\chi = -(\partial E/\partial N)_{v(r)} \tag{1}$$

The operational and approximate formula was also suggested by them invoking finite difference approximation as

$$\chi = -\mu = (I.E. + E.A.)/2 \tag{2}$$

where I.E. and E.A. represent ionization energy and electron affinity, respectively.

The absolute hardness is defined as [2]

$$\eta = \tfrac{1}{2}\,[\partial\mu/\partial N]_{v(r)} = \tfrac{1}{2}\,[\partial^2 E/\partial N^2]_{v(r)} \tag{3}$$

Parr and Pearson [2] invoked finite difference approximation to provide approximate and operational definitions of the hardness as under

$$\eta = \tfrac{1}{2}\,(\text{I.E.} - \text{E.A}) \quad (4)$$

Pearson [29] further connected electronegativity and hardness to the SCF-MO theory and invoked Koopmans' theorem to compute the ionization potential (IE) and the electron affinity (EA) as

$$\text{IE} = -\,\varepsilon_{\text{HOMO}} \quad (5)$$

$$\text{EA} = -\,\varepsilon_{\text{LUMO}} \quad (6)$$

where $\varepsilon_{\text{HOMO}}$ and $\varepsilon_{\text{LUMO}}$ are the orbital energies of the highest occupied and the lowest unoccupied orbitals.

From the above, it can be rearranged that

$$\chi = -\,\tfrac{1}{2}\,(\varepsilon_{\text{LUMO}} + \varepsilon_{\text{HOMO}}) \quad (7)$$

$$\eta = \tfrac{1}{2}\,(\varepsilon_{\text{LUMO}} - \varepsilon_{\text{HOMO}}) \quad (8)$$

Softness is another property of molecular system used in structure activity relationship studies. It is a reactivity index and is defined as the reciprocal of hardness

$$S = (1/\eta) \quad (9)$$

Parr et al. [6] defined one more global reactivity descriptor of atoms and molecules. The parameter was labeled as global electrophilicity index, (ω). This has special reference to correlate the reactivity of the reference molecules, a drug, with a biomolecule when the reaction becomes either electrophilic or nucleophilic. The proposed ansatz electrophilicity index (ω) is defined as

$$\omega = (\mu)^2 / (2\eta) \quad (10)$$

where μ is the chemical potential and η is the hardness of the system.

10.2.2 THE LOCAL DESCRIPTORS

Parr et al. [3–5] related the density functional theory with the frontier orbital theory of Fukui [30, 31]. In frontier orbital theory, the two orbitals,

the HOMO and the LUMO, are the most important in correlating the molecular reactivity and suggesting orientation of a group in the molecule. It is a fact that the reaction takes between two reactants locally and not globally. Ingenious work of Parr et al. [3, 5] identified the theoretical parameter to suggest site selectivity in a reaction, which is defined as fukui function. The ansatz which have been used to define the fukui function is as follows:
for governing electrophilic attack,

$$f^{-}(r) = [\partial\rho(r) / \partial N]^{-}_{v(r)} \tag{11}$$

for governing nucleophilic attack ,

$$f^{+}(r) = [\partial\rho(r) / \partial N]^{+}_{v(r)} \tag{12}$$

for governing radical attack ,

$$f^{0}(r) = [\partial\rho(r) / \partial N]^{0}_{v(r)} \tag{13}$$

Fukui function measures the response of the electron density at every point and the sites with the largest value for the fukui functions are those with the largest response, and as such the most reactive sites within the molecule. According to the "frozen core" approximation, the operational algorithms proposed [3] to calculate the fukui functions are as follows

for governing electrophilic attack,

$$f^{-}(r) \approx \rho_{HOMO}(r) \tag{14}$$

for governing nucleophilic attack,

$$f^{+}(r) \approx \rho_{LUMO}(r) \tag{15}$$

for governing radical attack,

$$f^{0}(r) \approx ½ [\rho_{HOMO}(r) + \rho_{LUMO}(r)] \tag{16}$$

Though local hardness is not straight forward, local softness can be defined following the algorithm of Parr et al. [3–5].

The local softness parameters with their specific use are as follows:

The s^- (r) is for governing electrophilic attack,

$$s^-(r) = S\, f^-(r) \tag{17}$$

The s^+ (r) is for governing nucleophilic attack

$$s^+(r) = S\, f^+(r) \tag{18}$$

The s^0 (r) is for governing radical attack

$$s^0(r) = S\, f^0(r) \tag{19}$$

But site selectivity is more transparent in the local philicity indices [7]. The ω^- is for governing electrophilic attack,

$$\omega^- = \omega\, f^-(r) \tag{20}$$

The ω^+ is for governing nucleophilic attack

$$\omega^+ = \omega\, f^+(r) \tag{21}$$

The ω^0 is for governing radical attack

$$\omega^0 = \omega\, f^0(r) \tag{22}$$

10.3 METHODS OF COMPUTATION

In this report we have studied twelve mesoionic derivatives, which have biological activity. The parent structure of mesoionic has been presented in the Fig. 1. The derivatives of bioactive mesoionic that have been used in this study are arranged in the Table 1. The structures of all mesoionic derivatives have been created by ISIS Draw 2.2 software and 3D modeling of these bioactive compounds has been preformed with the help of Argus Lab [32].

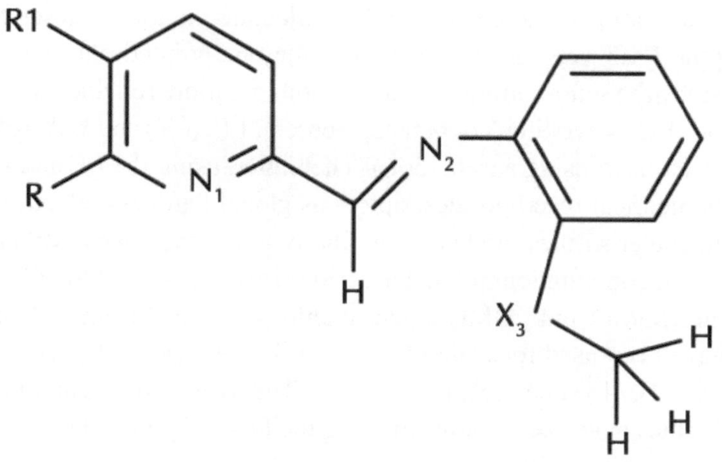

FIGURE 1 Basic graphical structure of the parent mesionic compound.

TABLE 1 Description of the site substitution in the basic structure of Mesoionic Heterocycle.

Compound No.	R	R1	X
1	H	H	S
2	$-CH_3$	H	S
3	$-C_6H_5Cl$	H	S
4	Br	H	S
5	$-O-CH_3$	H	S
6	H	Br	S
7	H	H	O
8	$-CH_3$	H	O
9	$-C_6H_5Cl$	H	O
10	Br	H	O
11	$-O-CH_3$	H	O
12	H	Br	O

We have evaluated global descriptors by using SCF–MO methods.

Argus Lab [32] have been used to calculate the global descriptors by using the PM3 version of SCF MO. It may be pointed out that in SAR/SPR study the semi-empirical SCF methods are more reliable than abinitio methods [33]. Invoking Koopmans' theorem, I.E. (eV) and E.A. (eV) have been calculated using ansatz (5, 6). Thereafter, using the I.E.and E.A, the global chemical reactivity descriptors as global hardness, electronegativity, molecular softness and electrophilicity index have been evaluated.

For the computation of local descriptors, we have used the PM3 method. The HOMO and LUMO eigen functions obtained from PM3 calculation have been used for computing fukui function (f^-, f^+, f^0) with the help of Eqs. (14), (15) and (16), respectively. The local softness and local philicity values have been evaluated using the Eqs. (17), (18), (19), (20), (21) and (22), respectively.

The calculated global reactivity parameters for twelve mesoionic derivatives have been presented in the Table 2. The DFT based local reactivity parameters of twelve instant heterocycles along their computed global electrophilicity index have been arranged in the Table 3.

TABLE 2 Calculated global reactivity descriptors; global hardness (η), global softness (S), electrophilicity index (ω), electronegativity (χ) of the mesoionic heterocyclic compounds.

Molecule	η	χ	S	ω
1	0.1026	0.1703	9.746589	0.1414
2	0.1369	0.1961	7.304602	0.1404
3	0.9077	0.1755	1.101686	0.1378
4	0.9045	0.1981	1.105583	0.1076
5	0.1079	0.1669	9.267841	0.1291
6	0.9055	0.1968	1.104362	0.1028
7	0.1012	0.1639	9.881423	0.9046
8	0.1282	0.1555	7.800312	0.9022
9	0.9065	0.1874	1.103144	0.9001
10	0.9032	0.1817	1.107174	0.1529
11	0.1032	0.1654	9.689922	0.9015
12	0.9024	0.1782	1.108156	0.1064

TABLE 3 Calculated local reactivity parameters; fukui functions (f^-, f^+, f^0) , local softness(s^-, s^+, s^0) and local philicity indexes (ω^-, ω^+, ω^0) along with the global electrophilicity indices (ω) of the mesionic derivatives.

Reactive Centers	f^-	f^+	f^0	s^-	s^+	s^0	ω^-	ω^+	ω^0	ω (Global)	Compound
N1	0.0094	0.0687	0.0391	0.0916	0.6697	0.3807	0.0013	0.0097	0.0055	0.1414	1
N1	0.0075	0.0002	0.0039	0.0550	0.0014	0.0282	0.0010	0.00003	0.0005	0.1404	2
N1	0.0106	0.0477	0.0292	0.1089	0.4889	0.2989	0.0017	0.0075	0.0046	0.1378	3
N1	0.00002	0.0109	0.0054	0.0003	0.1149	0.0576	0.000005	0.0238	0.0119	0.1076	4
N1	1.6421	0.0518	0.8620	15.2169	0.7585	7.9877	0.2121	0.0106	0.1113	0.1291	5
N1	0.00003	0.0596	0.0298	0.0003	0.6244	0.3124	0.000007	0.0121	0.0060	0.1028	6
N1	0.1190	0.2985	0.2087	0.8389	2.1035	1.4712	0.0113	0.0282	0.0197	0.9046	7
N1	0.1036	0.0570	0.0953	0.8757	0.7352	0.8054	0.0106	0.0089	0.0097	0.9022	8

TABLE 3 *(Continued)*

Reactive Centers	f^-	f^+	f^0	s^-	s^+	s^0	ω^-	ω^+	ω^0	ω (Global)	Compound
N1	0.0173	0.3739	0.1956	0.1777	3.8379	2.0078	0.0031	0.0674	0.0352	0.9001	9
N1	0.1061	0.1369	0.1215	0.7257	0.9363	0.8310	0.0120	0.0154	0.0137	0.1529	10
N1	0.2995	0.0043	0.1519	2.0174	0.0291	1.0232	0.0276	0.0004	0.0140	0.9015	11
N1	0.1552	0.3436	0.2494	1.0406	2.3040	1.6723	0.0165	0.0366	0.0265	0.1064	12

10.4 RESULTS AND DISCUSSION

The theoretical correlation and identification of the probable site of reaction in the bio-active molecules is possible in terms of the local density functional descriptors. Such selectivity in drug research is the jargon of the trade. The molecular reactivity is expected due to locally but not globally. We have mentioned earlier that the theoretical descriptors effective in site selectivity are local softness, fukui functions and local philicity index. We have computed the electrophilicity index, a global electron seeking parameter, for all the structures and have already noted that the values of the electrophilicity indexes and the bioactivity of the mesoionic decrease hand in hand. Further, the electrophilicity index is a descriptor of electron donating and electron accepting power of the molecules. The global electrophilicity index together with the fukui function, local softness and local electrophilicity index will decide whether a molecule will preferably enter a nucleophilic or electrophilic substitution reaction at the reactive site of a molecule.

Recently, Roy et al. [23] reported the synthesis of mesoionic derivative and probable mechanistic feature of its derivatives. In this report we have tried to establish the experimental observation by theoretical model. We have first tried to establish the mechanistic pathway of the cyclo-addition procedure in terms of the computed local reactivity descriptors like fukui functions, local softness and local philiciy indices.

Table 3 explains all the local activity descriptors for predicting the probable site of nucleophilic reaction, of nitrogen atom at the position 1 of the compound 1 are consistently higher than the corresponding values of nitrogen at the position 1 of other similar type of compounds. Therefore, the computed values of the local reactivity descriptors unequivocally predict that the Unsubstituted mesoionic derivative undergoes necleophilic reaction readily. Relying upon the relative magnitudes of fukui functions and local softness and local philicity values, we may predict that, if the compound 1undergoes the nucleophilic reaction, the preferred attacking site is on the nitrogen atom of position 1by a bio-molecule.

Roy et al. [23] have reported that the compound 1 undergoes cyclo-addition when nitrogen atom at position 1 behaves as a potent nucleophile. They also reported that substituted mesoionic derivatives do not undergo the cyclisation mechanism and nucleophilic behavior of the nitrogen atom at position 1of the substituted mesoionic derivatives are much lower as compared to that unsubstituted one. In this report, we have supported this mechanistic feature of mesoionic compounds in terms of our computed local DFT based reactivity parameters. From our computed local reactivity parameters, it is clear that nitrogen atom position 1 is more nucleophilic for the unsubstituted compound and as the mesoionic derivatives are substituted, the nucleophilicity of nitrogen atom decreases.

From the Table 2, it is clear that the unsubstituted mesoionic compounds (Compounds 1 and 7) are more reactive than their substituted derivatives. Global electrophilicity indexes of the instant molecules run hand-in-hand with their activities. From, the global reactivity parameters, it can be concluded that unsubstituted mesoionic compound containing oxygen atom at position 3 (Compound 7) is more reactive as compared to the unsubstituted mesoionic compound containing sulphur atom at position 3 (Compound 1). From the computed global DFT based reactivity parameters, the overall reactivity order of mesoionic compounds are nicely observed.

Another interesting fact reveals from our theoretical calculation. The mesoionic derivatives containingoxygen atom at position 3 are more reactive than compounds containing sulphur atom at position 3.

Roy et al. already observed experimentally that mesoionic derivatives containing oxygen atom at position 3 undergo complete cyclisation in two hours, whereas the same type of compounds containing sulphur atom in place of oxygen at position 3, take a long time, near about 12 h, to undergo complete cyclisation. From our calculated local reactivity descriptors, we may also conclude that the mesoionic compounds containing oxygen atom at position are more reactive as compared to the same compounds containing sulphur atom in place of oxygen at position 3.

10.5 CONCLUSION

In the present report, we have studied the reactivity orders and mechanistic pathway of the substituted and unsubstituted mesoionic derivatives in terms of the global and local reactivity descriptors under the paradigm of SPR/SAR study. The global descriptors nicely correlate the variation of activity with structure of the biologically active mesoionic derivatives. We have also explored the type of reactions, which the biomolecules will prefer while entering into reaction with the instant heterocycles in terms of the local descriptors.

We have launched a search of identifying the preferred site of attack by the bio-molecules (bacteria) on the mesoionic derivatives. We have also tried to predict the nature of the reaction that will be preferred by the biologically active mesoionic compounds molecules while attaching with the biomolecules in terms of local quantum mechanical descriptors. Thus we hopefully predict the reactive site in the mesoionic heterocycles and also the mechanism of the reaction that will be preferred by these compounds while entering into reaction with the biomolecules.

KEYWORDS

- **eigen function**
- **frozen core**
- **fukui function**
- **Koopmans' theorem**

REFERENCES

1. Parr, R. G.; Donnnelly, R. A.; Levy, M.; Palke, W. E. *J. Chem. Phys.,* **1978**, *68*, 3801.
2. Parr, R. G.; Pearson, R. G. *J. Am. Chem. Soc.,* **1983**, *105*, 7512.
3. Parr, R. G.; Yang, W. *J. Am. Chem. Soc.,* **1984**, *106,* 4049.
4. Yang, W.; Parr, R. G. *Proc. Natl. Acad. Sci. USA.,* **1985**, *82*, 6723.
5. Parr, R. G.; Yang, W. *Density Functional Theory of Atoms and Molecules*; Oxford University Press: New York, 1989.
6. Parr, R. G.; Szentpaly, L. V.; Liu, S. *J. Am. Chem. Soc.* **1999**, *121*, 1922.
7. Chattaraj, P. K.; Maiti, B.; Sarkar, U. *J. Phys. Chem. A.* **2003**, *107*, 4973.
8. Chattaraj, P. K.; Roy, D. R. *Chem. Rev.,* **2007**, *107*, PR46.
9. Chatterje, A.; Balaji, T.; Matsunaga, H.; Mizukami, F. *J. Mol. Graph. Model.,* **2006**, *25*, 208.

10. Roos, G.; Loverix, S.; De. Proft, F.; Wyns, L.; Geerlings, P. *J. Phys. Chem. A* **2003**, *107*, 6828.
11. Nakatsuka, M.; Shimamura, T. Jpn. Kokai Tokkyo JP 2001035664, 2001; *Chem Abstr.* **2001**, *134*, 170632.
12. Tominaga, T.; Kohama T.; Takano, A. Jpn. Kokai Tokkyo JP 2001006877, 2001; *Chem Abstr.* **2001**, *134*, 93136.
13. Kitasawa, D.; Tominaga, T.; Takano, A. Jpn. KokaiTokkyo JP 2001057292, 2001; *Chem, Abstr.* **2001**, *134*, 200276.
14. Nakamura, H.; Yamamoto, H. PCT Int. Appl. WO 2005043630, 2005; *Chem, Abstr.* **2005**, *142*, 440277.
15. Kim, D.; Wang, L.; Hale, J. J.; Lynch, C. L.; Budhu, R. J.; MacCoss, M.; Mills, S. G.; Malkowitz, L.; Gould, S. L.; DeMartino, J. A.; Springer, M. S.; Hazuda, D.; Miller, M.; Kessler, J.; Hrin, R. C.; Carver, G.; Carella, A.; Henry, K. Lineberger, J.; Schleif, W. A.; Emini, E. A. *Bioorg. Med. Chem. Lett.*, **2005**, *15*, 2129.
16. Ford, F. F.; Browne, L. J.; Campbell, T.; Gemenden, C.; Goldstein, R.; Gude, C.; Wasley, W. F. *J. Med. Chem.*, **1985**, *28*, 164.
17. Price, S.; Heald, R.; Savy, P. P. A. WO2009085980, 2009, *Chem. Abstr.*, **2009**, *151*, 123970.
18. Yoshimura, Y.; Miyake, A.; Nishimura, T.; Kawai, T.; Yamaoka, M. J. *Antibiotics*, **1991**, 44, 1394.
19. Shibahara, F.; Sugiura, R.; Yamaguchi, E.; Kitagawa, A.; Murai, T. *J. Org. Chem.*, **2009**, *74*, 3566.
20. Shibahara, F.; Yamaguchi, E.; Kitagawa, A.; Imai, A.; Murai, T. *Tetrahedro,* **2009**, *65*, 5062.
21. Earl, J. C.; Mackney, A. W. *J. Chem. Soc.*, **1935**, *53*, 899.
22. Roy, S.; Mitra, P.; Patra, A. K. *Inorg. Chim. Acta.*, **2011**, *370*, 247.
23. Roy, S.; Javed, S.; Olmstead, M. M.; Patra, A. K. *Dal. Trans.*, **2011**, *40*, 12866.
24. Franke, R.*Theoretical Drug Design Methods*, (Elesvier, Amsterdam, 1984).
25. Gupta, S. P.; Singh, P.; Bindal, M. C. Chem. Rev. **1983**, *83*, 633.
26. Gupta, S. P. *Chem. Rev.*, **1991**, *91*, 1109.
27. Allinger, N. L. In *Advances in Physical Organic Chemistry*; Elesvier, 1976.
28. Hohenberg, P.; Kohn, W. *Phys. Rev.*, **1964**, *136*, B864.
29. Pearson, R. G. *Proc. Natl. Acad. Sci.*, **1986**, *83*, 8440.
30. Fukui, K. In *Theory of Orientation and Stereoselection*; Springer-Verlag: Berlin, 1973.
31. Fukui, K. *Science*, **1982**, *218*, 747.
32. Argus Lab. Version 4.0, Planaria Software LLC.
33. Karelson, M. In *Molecular Descriptors in QSAR/QSPR*; Wiley-Interscience: New York, 2000.

CHAPTER 11

QUANTUM BREATHERS AND PHONON BOUND STATE IN THE FERROELECTRIC SYSTEM

A. K. BANDYOPADHYAY, S. DAS, and A. BISWAS

CONTENTS

11.1 INTRODUCTION

11.1.1 FERROELECTRICITY

Lithium niobate and lithium tantalate are technologically very important ferroelectric materials with a low switching field that have several applications in the field of nonlinear photonics and memory switching devices. In a Hamiltonian system, such as dipolar system, the modal dynamics of such ferroelectrics can be well-modeled by a nonlinear Klein-Gordon (K-G) equation. Due to strong localization coupled with discreteness in a nonlinear K-G lattice, breathers and multi-breathers manifest in the localization peaks across the domains in polarization-space-time plot. Due to the presence of impurities in the structure, dissipative effects are observed. To probe the quantum states related to discrete breathers, the same K-G lattice is quantized to give rise to quantum breathers (QBs) that are explained by a periodic boundary condition (Bloch state). The gap between the localized and delocalized phonon-band is a function of impurity content that is again related to the effect of pinning of domains due to antisite defects, i.e., a point of easier switching, which is related to Landau coefficient (read, nonlinearity). Secondly, in a non-periodic boundary condition approach, the temporal evolution of quanta shows a 'critical' time of redistribution of quanta that is proportional to QB's lifetime in femtosecond having a possibility for THz and other applications in quantum computation. Hence, the importance of both of the methods for characterizing quantum breathers is shown in these perspectives.

In the field of solid state physics, one of the most investigated materials is ferroelectric, which has important applications as memory switching [1–4], nonlinear optical communications [5], non-volatile memory devices [6, 7], and many others [8, 9]. Ferroelectrics have also emerged as important materials as: (a) piezoelectric transducers, (b) pyroelectric detectors, (c) surface acoustic wave (SAW) devices, and (d) four-phase mixing doublers. Both lithium tantalate and lithium niobate appear to be promising candidates as the key photonic materials for a variety of devices: (a) optical parametric oscillators, (b) nonlinear frequency converters, (c) second-order nonlinear optical material, and (d) holography, etc. Many of such devices include important nano-devices [9–11].

Ferroelectricity is an electrical phenomenon and also an important property in solids. It arises in certain crystals in terms of spontaneous dipole moment below Curie temperature [1]. The direction of this moment can be switched between the equivalent states by the application of an external electric field [2–4]. It is observed in some crystal systems that undergo second-order structural changes below the Curie temperature, which results in the development of spontaneous polarization. This can be explained by Landau-Ginzburg free energy functional [3, 4, 9]. The ferroelectric behavior is commonly explained by the presence of domains with uniform polarization. This behavior is nonlinear in terms of hysteresis of polarization (P) and electric field (E) vectors. Phenomenological models of ferroelectrics have been developed for engineering computation and for various applications.

From the physics standpoint, ferroelectric behavior is commonly explained by the rotation of domains and domain walls [5, 6]. This behavior is due to the collections of domains, where the ferroelectric domains, as with ferromagnetic domains, are created and oriented by a need to minimize the fields as well as the free energy of the crystal. The bulk properties and domain structure of these materials have been extensively studied [12–16]. However, recently they have gained renewed interest for potential applications in nano science and the design of nano devices, where the focus is on properties exhibited at small length scales. The new applications are based both on classic ferroelectric properties and features that are particular to nano-structured arrays and include applications to [10]: (a) integrated or surface mounted capacitors, (b) electromechanical sensors, (c) actuators, (d) transducers, (e) infrared sensors, (f) tunable thermistors, and (g) nonlinear dielectric materials. Due to these current interests, we look for details of the dynamical properties of domain arrays, which may become significant features at some of the length scales of interest, and this aspect is definitely very important.

11.1.2 FERROELECTRIC DOMAINS

Several studies have been made by Vanderbilt et al. [17] on the ab initio calculation of energy of the domain wall having a narrow width of the order of one lattice spacing and on defect pinning of domain wall in lead titanate [18]. By NMR experiments on intrinsic defects of ferroelectrics, Yatsenko et al. [19] studied domain dynamics of lithium and antisite niobium defects structure. Quantum chemical calculations were undertaken by Stashans et al. [20] on the oxygen vacancy defects in lead titanate crystals. The relaxor ferroelectrics and intrinsic inhomogeneity were studied by Bussman-Holder et al. [21] for dielectrically soft matrix.

Phenomenological level of description has been used in many previous theoretical and experimental investigations of ferroelectric domain walls, particularly by Scott and coworkers [6, 7], and by many others on important experimental and theoretical investigations [22–25]. First principle calculations have also been performed on ferroelectrics by Klotins et al. [26] in nano polar regions. A study on the domain structure by dynamic contact 'electrostatic force microscopy' revealed the distribution of polar nano-regions [27]. Many authors have modeled the structure of domain walls by using a Landau-Ginzburg type of continuum theory, notably by Zhirnov et al. [28], In the more recent work, Lee et al. [29], using density functional theory (DFT) and molecular dynamics simulations, showed that the 180° domain walls in lead titanate and lithium niobate have mixed wall character and this can be dramatically enhanced in nanoscale thin film heterostructures where the internal wall structure can form polarization vortices.

In tip-induced nanodomain formation in scanning probe microscopy (SPM) of ferroelectric films and crystals, using the analytical Landau-Ginzburg-Devonshire approach and phase-field modeling, Morozovska et al. [30] showed that for infinitely thin domain wall, the depolarization field outside the semi-ellipsoidal domain tip is always higher than the intrinsic coercive field that must initiate the local domain breakdown through the sample depth, while the domain length is finite in the energetic approach (LM paradox). The relative usefulness of the above two approaches was also shown compared to those in Refs. [17] and [28]. The behavior of Er defects in lithium niobate by energetics and stability considerations by DFT approach combined with thermodynamic calculations were also shown recently [31], which yields exclusively the role of defects on the charge balance and not direct relevance to the ferroelectric domains or domain walls.

More recently, Vasudevarao et al. [32] on "interface dynamics" showed that the nucleation bias is found to increase from a 2D nucleus at the (softer) wall to 3D nucleus in the (harder) bulk and the wall is also found to greatly affect switching on micrometer length scales. The mechanism of correlated switching and domain wall energy were analyzed using DFT and phase-field modeling to ascribe the long range effect to wall bending under the influence of a tip with bias that is well below the bulk nucleation level at large distances from the wall. These studies, compared to those mentioned in Ref. [15], obviously provide an experimental link between the macroscopic and mesoscopic physics of domain walls in ferroelectrics and atomistic models of 2D nucleation. Here, it may be mentioned that XRD work of Floquet and Valot [33] showed a much larger domain wall width, but this was related to the presence of "twin walls" to different concentrations of point defects in barium titanate that was not backed up by theoretical calculations. An in-depth analysis based on the comparison of different techniques could be useful, but it is beyond the scope of this chapter.

The above description shows the importance of domain wall in ferroelectrics. The preparation techniques of different ferroelectric materials may also affect the properties in terms of distribution of defects in the domain walls [34] that was shown to be important by Scott et al. [35, 36]. In a recent work by Bandyopadhyay et al. for the effect of damping on switching behavior, the importance of domains and domain walls was also highlighted [11]. Some of the above investigations have been made to get an overview of ferroelectrics in terms of "smaller length scales" in which the excitations can exist. As the domains are small and discrete in the nano range, a description is given later on localization in ferroelectrics to understand the meaning of classical and quantum breathers.

Moreover, having explained the inter-relation between domains and discrete breathers, it should be pointed out that as the rotation of domain and domain wall is important for switching and again, as switching is influenced by pinning of defects on the domain walls in ferroelectrics, it was considered useful to deal with discrete breathers through our discrete Hamiltonian, as detailed later. This Hamiltonian gives rise to nonlinear Klein-Gordon (K-G) equation with a damping term to take care of dissipation, as the defects also tend to promote the nature of dissipation in the ferroelectric system. Nonlinear K-G equation was particularly developed for ferroelectric materials in our previous work [9]. A particular facility in these treatments is that this is a well-known equation of mathematical physics, which exhibits a wide variety of interesting properties and has applications in different physical systems. In the present work, we numerically investigate the existence and stability of both dissipative and energy conserved breathers in discrete and periodic arrays of ferroelectric domains. In the next section, let us discuss about classical discrete breathers.

11.1.3 DISCRETE BREATHERS

The term "breather" has a historical significance. It can be created in translationally invariant nonlinear lattice models. It is in contrast to soliton solutions, which are moving wave-packets, i.e., nonlinear localized traveling waves that are robust and propagate without change in shape, giving the polarization profile and the distribution of the elastic strain across the domain wall [37]. Its temporal evolution has always been an area of intensive research. On the other hand, "breathers" are discrete solutions, periodic in time and localized in space, and whose frequencies extend outside the phonon spectrum, i.e., breathers are localized and time-periodic wave packets [38, 39]. The first discovered breather was in a sine-Gordon (integrable) system, which is also analytically tractable. There are various methods to characterize discrete breathers. Classical breathers can also be obtained in a non-integrable K-G system, as done numerically by well-known technique such as spectral collocation method, which involves minimum errors in the analysis of different breather modes. Due to the localization, the length scale of such excitation assumes more significance that obviously drives us to the nano domain, whose importance in the field of solid-state physics cannot be denied.

Discrete breathers (DB) of the domain array, also known as intrinsic localized modes (ILM), are nonlinear excitations that are produced by the nonlinearity and discreteness of the lattice. These excitations are characterized by their long time oscillations. These are highly localized pulses in space that are found in the discrete nonlinear model formulation. Unlike the plane wave like modes, DBs have no counterparts in the linear system, but exist only because of the system nonlinearity in a periodic lattice [9]. They are formed as a self-consistent interaction or coupling between the mode and the system nonlinearity. In this way, DB modifies the local properties of the system at the DB peak, and the modified local properties of the system provide the environment for the DB to exist. As the continuum limit formulation cannot be applied to their study, the present formulation on discrete domains is appropriate to highly localized pulses having widths that are not large compared to the domain of interest. Hence, the question is about the appropriate length scale, which drives us to the nanorange of domains in ferroelectric system. Therefore, localization assumes more significance.

Localization is an important aspect for applications in a variety of devices under the broad field of solid-state physics. It plays a crucial role in qualifying and quantifying a systems' operations. The extent of localization in the quantum regime assumes more significance for very small-structures, e.g., for nano devices. Now, the question comes: how do we get localization in a system or a lattice? Localization is evolved mainly either by "disorder" in the lattice [40] or by the systems' interplay of nonlinearity and discreteness [41], i.e., our attention is diverted towards DB. The first one, i.e., Anderson localization, has been implemented in details in many types of devices. As nonlinearity arises in ferroelectrics in terms of *P–E* hysteresis due to the rotational movement of the discrete domains and domain walls, they could also give rise to the localization. Here, we shall mainly discuss about the localization due to nonlinearity by adding some nonlinear components in the governing equation and discreteness. DBs seem to be quite versatile in managing localized energy, i.e., in targeted energy transfer or trigger mechanism [42]. They can transport this energy efficiently by engaging the lattice in their motion after DBs are formed, and moreover, under specific circumstances they can transfer this energy in selected lattices [43]. Combining these facts from model and some general studies, it can be said that DBs in ferroelectrics could in principle act as an able energy managers. Hence, the above explanations are given to relate the localized waves of discrete breathers and domain walls in ferroelectric materials.

On the application front, DBs have been investigated in several systems: (a) solid state mixed-valence transition metal complexes [44], (b) quasi-one dimensional antiferromagnetic chains [45], (c) array of Josephson junctions [46], (d) micromechanical oscillators [47, 48], (e) optical waveguide systems [49], and (f) proteins [50]. DBs modify system properties viz. lattice thermodynamics and introduce the possibility of non-dispersive energy transport [51, 52], because of their potential application for translatory motion along the lattice [53]. As already indicated, DBs are observed both in integrable (viz. sine-Gordon equation) and non-integrable systems (viz. Klein-Gordon equation).

However, integrability imposes a criterion for obtaining DBs analytically. DBs are obtained analytically for integrable systems, while for non-integrable systems it is obtained by various numerical methods viz. spectral collocation method, finite-difference method, finite element method, Floquet analysis, etc. As evident from many numerical experiments, DBs mobility is achieved by an appropriate perturbation [42]. From the practical application perspective, dissipative DBs are more relevant than their Hamiltonian counterparts. The latter with the character of an attractor for different initial conditions in the corresponding basin of attraction may appear whenever power balance, instead of energy conservation, governs the nonlinear lattice dynamics. The attractor character for dissipative DBs allows for the existence of quasi-periodic and even chaotic DBs [54, 55].

Without going into the history, it can be said that there is a considerable amount of research activity on DBs since the work of Sievers and Takeno [56] was published in 1988. Their existence has been theoretically proposed in several discrete many-body systems, and observed experimentally in different systems [57]. Thus, large volume of analytical and numerical studies has revealed the existence and properties of DBs in various nonlinear systems [39]. Detailed discussions of DBs have been reviewed extensively in the works of Sievers et al. [39, 56, 57] and Segev et al. [58]. Flach et al. [59, 60] studied the subject and, also Mackay and Aubry [61, 62] that was also followed by a presentation on "what we know about discrete quantum breathers" by Fleurov [63]. Here, another review by Flach and Gorbach [38] also needs a mention that contains almost all the relevant references on DBs.

It is pertinent to mention that a richer variety of K-G equation was also derived in other important nonlinear optical materials such as split-ring-resonator based metamaterials, where DB pulse has been observed [64,

65]. The Fano resonance due to DBs has also been described on two-channel ansatz in K-G lattice by various workers [66–68]. Various parameters, such as dielectric permittivity, coupling, focusing-defocusing nonlinearity, are included in the Hamiltonian that is used to describe the nonlinear modes in metamaterials. Hence, the Fano resonance due to DBs has also been explored in terms of these material parameters [69] for applications in the (a) biosensing technology, (b) spectral selectivity, (c) beam filtering, etc.

Furthermore, as the discreteness is found to trap the breathers, the moving breathers are non-existent in highly discrete nonlinear system that has been presented by Bang and Peyrard in the context of a K-G model [70]. These authors [71] did a numerical study on the "exchange of energy and momentum" between the colliding breathers to describe an effective mechanism of "energy localization" in K-G lattice, arising out of the discreteness and non-integrability of the system, Here, the bright soliton solutions have been used for nonlinear dynamics of DNA molecules to demonstrate the generation of highly localized modes. It is known that Quasi-phase matching (QPM) is an important issue in a quadratic nonlinear photonic crystal (QNPC) or photonic band-gap materials with tunability. In an interesting work on QPM by Kobyakov et al. [72], the influence of an induced cubic nonlinearity on the amplitude and phase modulation was analytically studied to predict an efficient all-optical switching. Further, in the application front for a QNPC, a stable soliton solution was shown by Corney and Bang [73] for cubic nonlinearity and QNPC was found to support both dark and bright solitons even in the absence of quadratic nonlinearity, and the "modulation instability" in such systems was also shown [74]. Trapani et al. [75] studied focusing and defocusing nonlinearities in the context of parametric wave mixing.

The literature on soliton is so vast that it is very difficult to mention all the references. However, the Ref. [37] is very useful for further references. For soliton propagation, in many optical systems, it is a common practice to use nonlinear Schrodinger equation (NLSE). However, we have recently shown that the NLSE can be derived through perturbation on K-G equation when progressive wave passes through nonlinear medium, such as lithium niobate ferroelectrics, where dispersion may take place, and discrete energy levels due to dipole-dipole interaction were estimated via hypergeometric function [76].

As also indicated in Ref. [37], the ferro-para phase change occurs through a global and coordinate displacement of the ions. Hence, the presence of solitons is due to the Landau double well potential in which the pentavalent metal (niobium or tantalum) ions are sitting with their coupling that is strong enough to lead to cooperative effects. A two-well potential has been used to derive kink solution of the nonlinear propagating waves in ferroelectrics in the context of a diatomic chain model [77–79]. The impact of this potential in case of a discrete system has been quite extensively studied by Comte [80]. However, it should be noted that these motions become spatially localized due to nonlinearity and discreteness along with the pinning effect of the ferroelectric domains and domain walls, which are typically in the "nano-range." Therefore, the essence of nonlinearity and discreteness paves the way for the nano-ferroelectric devices [10, 11]. Then, there are other aspects, such as quantum computation and targeted energy transfer (TET) by using a new concept through DBs [43]. After discussing about classical breathers, next let us look for quantum breathers.

11.1.4 QUANTUM BREATHERS

In this new approach, for the characterization of DBs or classical breathers [81], the bulk system was the right tool, but when we are dealing with smaller systems, we have to use quantum physics, which brings us to the quantum breathers (QBs) [82, 83]. Once generated, QBs modify system properties such as lattice thermodynamics and introduce the possibility of non-dispersive energy transport, as generally described for DBs [84]. These are observed in many systems: (a) ladder array of Josephson junction for superconductors [85], (b) BEC in optical lattices or nonlinear photonic lattices [86], (c) interacting optical waveguides [87], cantilever vibrations in micromechanical arrays [88], macromolecules such as DNA [89], split-ring-resonator (SRR) based meta-materials in antenna arrays [90], two-magnon bound states in anti-ferromagnets [91], two-phonon bound states (TPBS), i.e., quantum breathers due to charge defects in ferroelectrics, as investigated by Fourier Grid Hamiltonian method [92].

The next question arises as: how do we characterize discrete breathers? This is analogous to its classical counterpart, where some higher harmonics lie outside the linear spectrum (continuum) of the system [38]. When some states of the quantum mechanical system lie outside the continuum of the eigen-energies, there exists a quantum breather. The branching out of the quantum breather state from the single-phonon continuum is quite noteworthy in systems with charge defects [92]. A brief account is given here on phonon bound state or breather state. Despite our work on discrete breathers [9, 13], so far the pinning has been explained classically, thereby prompting us to think about quantum explanations that have been briefly explored for ferroelectrics. Although impurity data for lithium tantalate are not available, we could work through Landau coefficient or nonlinearity route to explore if there is also pinning in such systems (see Section 3.2). For quantum breathers, it is important to consider detailed information on phonons and their bound state concept, which is sensitive to the degree of nonlinearity. In the eigen-spectrum or more traditionally E_k vs k plot, a quantum breather band separates itself from the delocalized phonon band. Or in other words, it is the hopping tendency of the phonons that describe the quantum breathers.

So, let us consider that the phonons in one sublattice may hop from one domain to another adjacent domain. This hopping might have some consequences with the change of nonlinearity or switching field that is again related to the impurity in the lattice, thereby the "hopping strength" can be directly related to this phenomenon. It is determined by finding the phonon-band energy gap (i.e., the energy gap between the delocalized and localized phonons) in the usual eigen-spectrum [38]. In quantum mechanics, single phonons are considered delocalized. If two-phonons are bound, then it can be in a localized state that is a necessity for the formation of quantum breathers. For these two types of phonons, as the energy is different, the two-phonon localized band has to separate out of the single phonon continuum. As the phonons are quantized vibrations, the breathers thus formed through spatial localization are called "quantum breathers" (QBs) [82, 83]. Therefore, the two-phonon bound states are considered as signature of QBs.

In a non-exhaustive literature search, a brief account is given here on the study of phonon and its vibrations. Corso et al. did an extensive study of density functional perturbation theory for lattice dynamics calculations in a variety of materials including ferroelectrics [93]. They employed a nonlinear approach to mainly evaluate the exchange and correlation energy, which were related to the non-linear optical susceptibility of a material at low frequency [94]. The phonon dispersion relation of ferroelectrics was also studied extensively by Ghosez et al. [95, 96]; these data were, however, related more with the structure and metal-oxygen bonds rather than domain vibrations or soliton motion. In a very interesting work, a second peak in the Raman spectra was interpreted by Cohen and Ruvalds [97] as evidence for the existence of "bound state" of the two phonon system and the repulsive anharmonic phonon-phonon interaction which splits the bound state off the phonon continuum was estimated for diamond.

A femtosecond time-domain analog of light-scattering spectroscopy called impulsive stimulated Raman scattering (ISRS) is a very useful technique [98]. This has been extensively used by Nelson et al. [99] in dealing with the anharmonic vibrations in both lithium niobate and lithium tantalate crystals. Stone and Dierolf [100] did some work on *the influence of domain walls on the Raman scattering process in lithium niobate and lithium tantalate.* A powerful technique, such as molecular-dynamics simulations of vibrational wave packets, was used by Phillpot et al. [101] to study the scattering of longitudinal-acoustic modes and predicted that the presence of gaps in the phonon spectrum of thin high-symmetry nanowires will result in a complete reflection of phonons at the interfaces.

The goal of this chapter is to explore whether switching is easier at particular point of "nonlinearity" and also "coupling" within the domains by quantum calculations on ferroelectrics, such as lithium tantalate. Therefore, the hopping strength of phonon and thus the phonon energy gap have been derived from the quantized model of the ferroelectric system. The calculations of various TPBS parameters are made against Landau coefficients (read, nonlinearity) to highlight the quantum origin of pinning in lithium tantalate with important consequences for various nano-ferroelectric devices. So far, we have described a situation for TPBS route by a periodic boundary condition involving Bloch function. In the next section, let us also discuss about QB in a non-periodic boundary condition approach.

11.1.5 NON-PERIODIC BOUNDARY CONDITION

Quantum breathers can be characterized by various methods [102] such as: (1) Splitting and correlations by computing the nearest neighbor energy spacing (tunneling splitting) between pairs of symmetric-anti-symmetric eigen-states, (2) Fluctuation of total number of quanta: by measuring the relative contribution of each basis state, (3) Entanglement: by minimizing the distance of a given state to the space of product states of the many body problem, (4) Temporal evolution of initially localized excitations in the system-time evolution of total number of quanta or phonons, (5) Avoided crossings and degenerate eigen-states in the system: the exponentially small weight of the tunneling pair states in the dynamical barrier region. Some of the above methods involve matrix diagonalization and Feynman path integral technique, as presented by Schulman to describe quantum tunneling and the stability of DBs [103]. In this section, out of the above five methods, our main focus will be on point number (4), i.e., on the temporal variation of the number of quanta, as it is convenient to characterize QBs by this method.

Now the question comes: what are the applications of QBs or, does it help in making a device? What is the advantage of a system, which has QB over a system, which doesn't have it? Once generated, QBs modify system properties such as lattice thermodynamics and introduce the possibility of non-dispersive energy transport, as generally described for DBs [51]. Then, there comes the aspect of quantum computation and targeted energy transfer (TET) [43]. When we use the combination of multiple qubits (quantum bits) to encode a signal instead of combination of classical bits (viz. 1, 0) we are in the regime of quantum computation. The hardware is made from various materials. However, for the case of quantum computers, the material selection [104, 105] still remains a debatable issue, which can be solved by using an anharmonic model such as K-G model, in which the energy levels are non-equidistant and hence giving us the capability to properly harness the physical and also computational behaviors. Thus, these non-equidistant anharmonic potential wells could have important implications from the application point of view, and this is the reason why we are working on K-G lattice.

Althouh QBs are characterized by various methods [82, 102], here our main focus will be on the temporal evolution of the number of quanta, as it is possible to find out the "critical time of redistribution" that is proportional to QB's lifetime in femtoseconds, which might be useful for quantum computation application. QBs have been studied for dimer and trimer cases, and that also by (mainly) periodic boundary condition approach. However, a real material consists of many subunits, i.e., thousands of domains make ferroelectrics, each acting as sites and phonons act here as bosons or quanta. Again, how the increase of number of sites and bosons affects a system can also be regarded as an interesting topic. Hence, it drives us to a study that considers more number of sites and quanta. This is also the main aim of this chapter.

Quantum localization behavior in K-G lattice has been studied by many researchers in terms of four atom lattice with periodic function, notably by Proville [106], delocalization and spreading behavior of wave-packets by Flach et al. [38], dimer case for targeted energy transfer by Aubry et al. [43]. Here, we present a generalized method for any number of sites and quanta without periodic boundary condition to show the QB states. In K-G lattice, it is important to calculate the "critical time" of redistribution of quanta under various physical conditions. It is the "time" when the temporal evolution of the number of quanta first meets or tends to meet.

For strengthening our focus on temporal evolution of quanta, it has to be noted that the application of QB in ferroelectrics consists of many different fields of technology, namely "phase-coherent optical pulse synthesis" [107], "parametric light generation" [108], and "ultrafast spectroscopy" [109]. On the latter application, Nelson et al. [98] studied both lithium tantalate and lithium niobate. The critical point at which the temporal evolution of quanta meets or tends to meet may be directly related to ferroelectric switching phenomenon, which if tailored well could lead to any of the above applications. Hence, viewing ferroelectricity in terms of phonons and its study via temporal evolution under various controlling parameters assume significance.

The chapter is organized as follows: in the Section 7.2, we first present some details of spectral collocation method to develop space-time evolution of polarization plots for overall view of classical breathers in Section 7.2.1, then we present the mathematical model for TPBS parameters after second quantization in Section 7.2.2.1 and finally second quantization on K-G lattice is done with Bosonic field operators in Section 7.2.2.2. In Section 7.3, the results and discussion are also presented in three parts for the above three cases. In Section 7.4, the conclusions are given.

11.2 THEORETICAL DEVELOPMENT

Here, we discuss the theoretical aspect of our discrete Hamiltonian to show nonlinear K-G equation and then deal with quantum mechanical approach for phonon bound state or QB state to explain the possible dependence of criticality on the Landau coefficient through quantum route.

11.2.1 CLASSICAL BREATHERS

Let us consider an idealized one-dimensional array of N identical ferroelectric domains layered along the x-direction. The domains are considered to be rectangular parallelepipeds. For simplicity, the polarizations in each domain are oriented in the z-direction and translationally invariant in the y-direction. Between the neighboring domains, there is domain wall and here we consider nearest neighbor coupling between the domains. The domain arrangement has been shown in Ref. [9]. In a previous treatment, a time-dependent formulation for the dynamics of the domain array was obtained as a generalization of the Landau-Ginzburg free energy functional involving polarization (P) and electric field (E) vectors: The nearest neighbor domains [i.e., the polarization in the ith domain (P_i) with that in the (i–1)th domain (P_{i-1})] were taken to interact by a harmonic potential with a phenomenological spring constant (k) so that the resulting Hamiltonian for the polarization is given by [9, 13]:

$$H=\sum_{i=1}^{N}\left(\frac{1}{2m_d}\right)p_i^2+\sum_{i=1}^{N}\frac{k}{4}(P_i-P_{i-1})^2+\sum_{i=1}^{N}\left(\left(-\frac{\alpha_1}{2}P_i^2+\frac{\alpha_2}{4}P_i^4\right)-EP_i\right) \quad (1)$$

The momentum (p_i) can be defined in terms of order parameter (P_i) in relation with inertial constants (m_d and Q_d).

Eq. (1) gives a good general treatment of the mode dynamics in the array, particularly for modes, which are strongly localized over a small number of the domains in the array. For extended modes and modes which are localized, and slowly range over a large number of consecutive domains, Eq. (1) can be approximated by a continuum treatment by Taylor expansion. In this limit, expressed in dimensionless units, Eq. (1) yields a nonlinear K-G equation with a damping term [9, 13] as:

$$\frac{\partial^2 P}{\partial t^2}+\bar{\gamma}\frac{\partial P}{\partial t}-\bar{k}\left(\frac{\partial^2 P}{\partial x^2}\right)-(\bar{\alpha}_1 P-\bar{\alpha}_2 P^3)-E_0\cos(\omega t)=0 \quad (2)$$

for the dynamics of polarization $P(x,t)$. Here, Eq. (2) with an ac driver contains all the non-dimensional terms as: $P'=P/P_s$, where P_s is the saturation polarization in C/m^2 (typical value for lithium tantalate ferroelectrics as 0.55 C/m^2 [3]), $E'=E/E_c$, where E_c is the coercive field (when $P'=0$) in kV/cm in the usual non-linear hysteresis curve of P vs. E with a typical value for the same material as 13.9 kV/cm [3], $t'=t/t_c$, where t_c is considered as the critical time scale for polarization to reach a saturation value, i.e., at or near the domain walls that are of importance to our study with a typical value of 10 ns for a switching time of (say) 200 ns for a damping value $\bar{\gamma}$=0.50 [15] that are based on the above data, and $x'=x/W_L$, where W_L = domain wall width of the order of a few nm. Eq. (2) is obtained after dropping the prime notation, and by taking $a_1=a_2/P_s^{\,2}$ and $\bar{\alpha}_1=\bar{\alpha}_2=\bar{\alpha}=(a_1P_s)/E_c$ [3, 9]. Here, the interaction and damping terms are defined as: $\bar{k}=\frac{kP_s}{2E_c}$ and $\bar{\gamma}=\frac{\gamma P_s}{t_cE_c}$, where γ is a decay constant relating the loss of polarization due to internal friction during its motion in the system of domains, which is important for the motion of localized traveling waves (i.e., soliton). Although discrete breathers rides on the background of phonon, the phonons are themselves dispersive in nature and the nonlinear systems are also dispersive. Moreover, the ferroelectric switching involves the domain rotation that gives rise to dispersion. This is the reason of including the "damping term" in our spectral collocation method of describing 3-D figures of polarization profile.

Among all the known methods of numerical simulation, we use the most versatile method of spectral collocation to analyze classical breathers in our system of ferroelectrics. This method is not only the latest numerical technique with ease of implementation, but also gives rise to a minimum of errors in the analysis. Spectral methods are a class of spatial discretizations for differential equations. In order to prepare the equation for numerical solution we introduce the auxiliary variable: $Q_i = \dot{P}_i = \frac{\partial P_i}{\partial t}$. This reduces the second order Eq. (2) to the first order system:

$$\dot{P} = Q, \text{ and } \dot{Q} = \bar{k}DP - \bar{\gamma}Q + \bar{\alpha}(P - P^3) + E_0 \cos(\omega t).$$

An appropriate banded matrix D has to be selected and then we solve our system of the first order differential equations that can be written as:

$$\begin{pmatrix} \dot{P} \\ \dot{Q} \end{pmatrix} = \begin{pmatrix} I & 0 \\ -\bar{\gamma} & \bar{k}D \end{pmatrix} \begin{pmatrix} Q \\ P \end{pmatrix} + \bar{\alpha} \begin{pmatrix} 0 \\ (P - P^3) \end{pmatrix} + \begin{pmatrix} 0 \\ U \end{pmatrix} E_0 \cos(\omega t) \tag{3}$$

Here, U is a column vector whose elements are unity. We have used well-known 4th order Runge-Kutta method for the system (3). The numerical simulation for DBs is shown in Section 3.1.

11.2.2 QUANTUM BREATHERS

11.2.2.1 PERIODIC BOUNDARY CONDITION APPROACH

Our discrete Hamiltonian gives a general treatment of the mode dynamics in the array, particularly for modes, which are strongly localized over a small number of domains in the array. For such modes, Eq. (1) can be split as:

$$\tilde{H} = H_0 + H_1 \tag{4}$$

where,

$$H_0 = \sum_i \frac{p_i^2}{2} - \frac{(\alpha_1 + \lambda)P_i^2}{2} + \frac{\alpha_2 P_i^4}{4} - EP_i \tag{5}$$

$$H_1 = -\frac{\lambda}{2}\sum_i P_i P_{i-1} \tag{6}$$

Then, a general "basis" can be written for *n* particles. The numerical analysis was carried out with Fourier grid Hamiltonian method [84] with 1000 grids and 0.006 spacing to calculate various eigenvalues and eigenvectors. We restrict ourselves to two phonon states, since at the working temperature the number of phonon is smaller. In order to reduce the computer memory requirement, we take the advantage of translational invariance by periodic Bloch wave formulation, as detailed in Ref. [92]: $|\psi> = \sum_j \nu_j |\phi_2^j>$. Due to translational invariance, the eigenstates of *H* are also eigenstates of the translation operator (*T*) where: $\tau = \exp(iq)$ is its eigenvalue with $q = \frac{2\pi\nu}{f}$ being allowed Bloch wave number and $\nu \in \left[-\left(\frac{f-1}{2}\right), \left(\frac{f-1}{2}\right)\right]$. Here, τ is the eigenvalue of *T*, and *f* is taken as a renormalizing constant. Thus, we can construct the Bloch states. For a given basis of number operators, we can get the eigenvalues for each Block wave number (*q*). For a two-phonon case, the non-zero hopping coefficients (D_{mn}) are: $D_{01} = D_{10}, D_{12} = D_{21}$. The energy gap between the single phonon continuum and a bound state is given by:

$$E_g = E_2 - E_0 - 2(E_1 - E_0) \tag{7}$$

where E_0, E_1 and E_2 are three eigenvalues at different points of wave vector (*k*) that are calculated from our computation to generate *E(k)* vs. *k* curve, which gives the signature of QB in terms of two-phonon bound state. The width of the single-phonon in the eigen-spectrum is given by the magnitude of 4s, i.e., the width of the single delocalized phonon-band, where s is expressed as:

$$\sigma = -\frac{\lambda}{2} D_{01}^2 \tag{8}$$

where l is an interaction term (i.e., equivalent to *k*). D_{01} represents the coefficient for zero to single phonon generation. The variation of the single phonon spectrum width (W_{ph}) represents (through $D_{01} = D_{10}$) the creation of a new phonon or annihilation of an existing phonon. Again, the hopping coefficient for a single phonon to become a two-phonon bound state is given by:

$$\mu = -\frac{\lambda}{2} D_{01} D_{12} = -\frac{\lambda}{2} D_{10} D_{21} \tag{9}$$

All the above calculations were done for 51 sites or domains and l = 10. More data points could be used in our present simulation, but here we are primarily focused to study nonlinearity or impurity induced critical behavior of DB motion and its quantum origin. To treat the problem analytically we take the help of second-quantization method. By quantizing the Hamiltonian in Eq. (1) in a number conserving quantized form with $N \rightarrow \infty$ [38] with $\lambda_1 = \frac{\lambda}{(\lambda - \alpha)}$, $\eta = \frac{2\alpha}{(\lambda - \alpha)}$ and $E = 0$ in Eq. (1) leads to the equation for two-phonon amplitudes as:

$$\hat{H} = \sum_n a_n^+ a_n + \frac{3}{8} \eta a_n^{+2} a_n^2 + \frac{\lambda_1}{2} \left\{ a_n^+ \left(a_{n+1} + a_{n-1} \right) + h.c. \right\} \tag{10}$$

where a_n^+ and a_n are the creation and annihilation operators with *hc* as Hermitian conjugate and *c* is the velocity of light; the terms h and 1_l are already defined above.

Hence, for a fixed total quasi-momentum *K*, the critical α-value or Landau parameter may be calculated as:

$$\alpha_{TPBS} = \lambda \left[\frac{Cos \frac{KR}{2}}{E_V - 1} - 1 \right] \tag{11}$$

Here *R* is a lattice parameter and E_v is the eigenvalue representing the bound state. Hence, for lithium tantalate type ferroelectrics, after α_{TPBS}, a branch is separated from the continuum due to corresponding value of impurity content (read, nonlinearity) and has got a critical point for pinning transition. Now, for bound state $m = n$ and the bound state energy (E_{BS}) can be derived as:

$$E_{BS} = 1 + \frac{3\eta}{8} - 4\lambda_1 Cos\frac{KR}{2} \tag{12}$$

Substituting the values of h and λ_1 and taking the eigenvalue equal to the energy of the bound state, and also taking the cosine term as equal to 1, the critical α-value or Landau parameter may be related as:

$$E_{BS} = \frac{\alpha}{\alpha - \lambda} \tag{13}$$

The results on the calculation of various TPBS parameters for lithium tantalate are shown with nonlinearity in Section 7.3.2. In the next section, let us deal with the non-periodic boundary condition approach for QBs.

11.2.2.2 NON-PERIODIC BOUNDARY CONDITION APPROACH

The general Hamiltonian for the Klein-Gordon equation for order parameter (y_n) at *n*th site is written as:

$$H = \sum_n \frac{p_n^2}{2m} + \frac{A}{2}y_n^2 + \frac{B}{4}y_n^4 + k\left(y_n - y_{n-1}\right)^2 \tag{14}$$

The first term is momentum at *n*th site (p_n), the second and third terms are nonlinear potential formulation and the last term contains an interaction constant (k). From the above Eq. (14), after deducing the classical equation of motion and rescaling of time, we get:

$$\tilde{H} = \sum_n \frac{1}{2}p_n^2 + \frac{1}{2}y_n^2 + \eta y_n^4 + \lambda\left(y_n - y_{n-1}\right)^2 \tag{15}$$

where $\eta = \frac{B}{4A}$ $\lambda = \frac{k}{2A}$, A and B are two constants. Now, let us use creation and annihilation Bosonic operators at the *n*th site and the above Hamiltonian [Eq. (15)] is quantized. In an important work done by Proville [106] the non-number conserving methods for four sites and an arbitrary number of particles are shown. In distinction to other methods, the above analysis gives a generalized way to solve the system for arbitrary number of particles on arbitrary number of sites.

After second quantization, a general "basis" is then created. For the characterization of QBs, we need to make the Hamiltonian time-dependent. Let us take the help of temporal evolution of number of bosons at each site of the system:

$$< n_i > (t) = < \Psi_t \,|\, \hat{n}_i \,|\, \Psi_t > \tag{16}$$

We take i-th eigen-state of the Hamiltonian, and then we make it time dependent as:

$$| \Psi_i(t) > = \sum_i b_i \exp(-iE_i t/\hbar) \,|\, \psi_i > \tag{17}$$

where ψ_i and E_i is the i-th eigenvector and eigenvalue respectively, and t is time. The Planck's constant (h) is taken as unity and $b_i = \langle \psi_i | \psi(0) \rangle$ for each site i and for a given range of t, where $\psi(0)$ stands for the initial state.

It is pertinent to mention that in contrast with the Discrete Non-Linear Schrodinger equation, where complete energy transfer takes place [43], in case of nonlinear K-G lattice, complete energy transfer does not take place between the anharmonic oscillators and there is a critical time of redistribution for the quanta [38]. This is an important point to be noted. With the above methodology, we can now proceed to deal with the applications of non-periodic boundary condition approach in lithium tantalate.

11.3 RESULTS AND DISCUSSION

The results are again presented in two subsections. In Section 7.3.1, classical breathers are shown in 3D figures. In Sections 7.3.2.1 and 7.3.2.2, the data on QBs are shown for periodic and non-periodic boundary conditions respectively. First, we present the figures for classical breathers, and then we present the figures for QBs for better understanding.

11.3.1 SPECTRAL COLLOCATION ANALYSIS

Typical 3D polarization diagrams for lithium tantalate are shown in Figs. 1 and 2, respectively, that are considered as manifestations of DBs in our system due to localization, as our analysis is based on discrete domains. All the units in these figures are dimensionless. For a case with zero field and no damping, i.e., Hamiltonian breathers, symmetric breathers are normally observed in 3D figures (not shown here). It is noted that in the simulation of 2D figures of polarization (P) with site index (n), i.e., distance, the peaks have been found to be symmetric Gaussian bands and even in 3D pictures the same type of symmetric breather bands are observed. However, for a sample with poling field of 13.9 kV/cm with a finite value of field (E = 0.01) and a moderate damping (0.50) with low level of interaction constant (0.50), it is seen from Fig. 1 that the dissipation starts to visibly show up and the symmetric dissipative DBs are still observed. If the damping value is further increased to a high value of 0.9, the intensity almost decays to zero that is expected from dissipative breathers. With an increasing value of interaction, i.e., at a very high value of 50, as taken in our previous work on ILM [64], it is seen from Fig. 2 that the "tri-breathers" are formed, as also observed in the case of lithium niobate [81].

In this case, the importance of the coupling parameter is also observed in creating multi-breathers in lithium tantalate system. However, in our numerical simulation, bi-breathers have already been observed at a lower value of coupling and there is no formation of multi-breathers up to a coupling value of 5. As coupling increases further, multi-breathers start forming. There are important observations made on the appearance of multi-solitons by controlling various parameters by several authors with numerical solutions for optical communication devices [100, 101]. Hence, 3D figures of classical DBs reveal important information against different values of damping and coupling within the system of ferroelectrics that has implication for application as switching materials with low switching field.

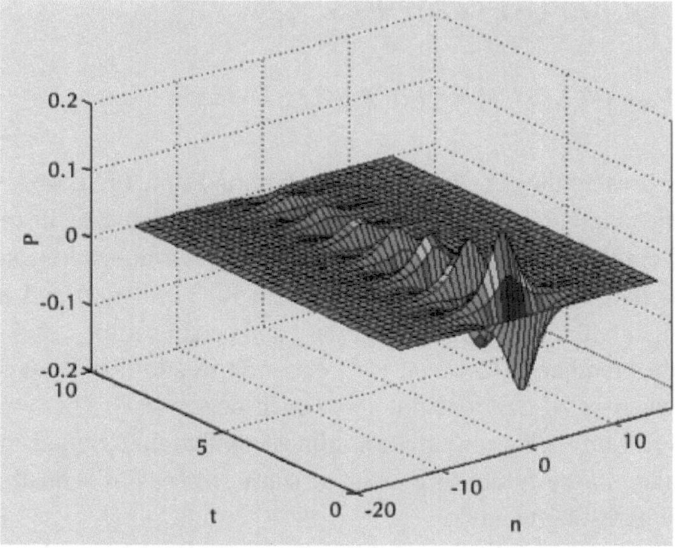

FIGURE 1 3D polarization-time-space plot of classical discrete breathers in lithium tantalate system with poling field = 13.9 kv/cm, non-dimensional electrical field = 0.01 and damping = 0.50 and a low level of coupling constant = 0.50. Symmetric breathers are still observed at this damping value that decays further on increasing the damping.

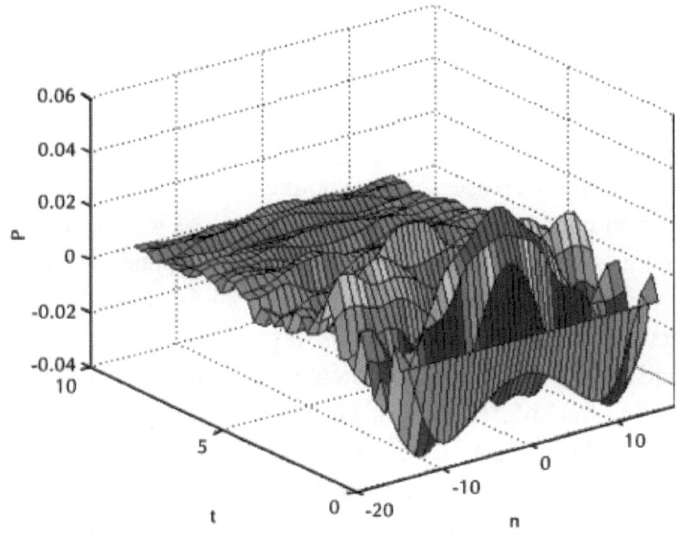

FIGURE 2 3D polarization-time-space plot of classical discrete breathers with poling field = 13.9 kv/cm, non-dimensional electrical field = 0.01 and damping = 0.50 and a very high level of coupling constant = 50. Multi-breathers are formed with increasing coupling value showing the effect of coupling on lithium tantalate system with low poling field.

11.3.2 QUANTUM BREATHERS

11.3.2.1 TWO-PHONON BOUND STATES

Here, we present the data for two-phonon bound state (TPBS) or QB state in lithium tantalate system. It is pertinent to mention that in our recent work on another important nonlinear optical material, such as split-ring-resonator (SRR) based meta-material, both K-G equation and nonlinear Schrodinger equation (NLSE) show dark and bright solitons, and also dark and bright breathers. However, K-G equation in addition shows breather pulses, whereas NLSE does not show such pulses [64]. This work could also be relevant for an important nonlinear optical material, such as lithium tantalate in the future directions of study, where K-G equation seems to show interesting behavior.

The eigen-spectra showing TPBS in lithium tantalate are shown in Fig. 3 at a switching field value of 13.90 kV/cm. The coupling value is taken as constant at 12 for the spectra. The Landau parameter (a) is inversely proportional to the impurity content or switching field. In the absence of the impurity data, we would operate through Landau coefficient (read, nonlinearity), which should also give a correct picture of QBs in terms of various TPBS parameters, as nonlinearity and discreteness give rise to quantum localization or rather to the formation of QBs.

It is seen from various eigen-spectra including that shown in Fig. 3 that as the nonlinearity decreases, i.e., the poling field increases, the shape of the single-phonon continuum also changes in terms of its width that could be considered as significant indicating a functional dependence on nonlinearity. This width is measured by W_{ph} as shown in Eq. (8) which indicated a slight decrease towards lower a values (a = 514 to 421; i.e., average value = 467) and then it drops quite drastically towards lower a values, i.e., the highest poling field of 210 kV/cm. This is in contrast to that in lithium niobate system, where it sharply drops towards a = 471 and then it sharply increases towards lower a values [92]. This may be due to the different levels of the degree of polarization in the respective systems [112]. It also indicates that in lithium niobate, as the nonlinearity decreases from 1767 towards 471, the difference in values of W_{ph} is about 19.1 in absolute term, whereas that for lithium tantalate system, it is only about 1.1 for nonlinearity values from 4427 to 467 (average). For this phenomenal change despite having a higher level of variation in nonlinearity, the drop in values of the width in the single-phonon continuum is extremely small implying that the effect of nonlinearity is relatively smaller. It is quite significant for lithium tantalate, which is gaining popularity as an important candidate as devices that are also complimented by its lower value of switching field. This might indicate that whatever be the impurity content in lithium tantalate as antisite tantalum defects, there does not seem to be an appreciable effect of pinning in the system. It should be emphasized here that "pinning" is a macroscopic phenomena involving the charge defects in the system [6, 112], whereas TPBS parameters are microscopic in nature that are obtained after detailed quantum calculations. However, defect structure of ferroelectrics has also to be understood. Another powerful theory, such as density function theory (DFT) as done by some authors [31] and also first principle calculations [17] could be useful, but it is beyond the scope of this chapter.

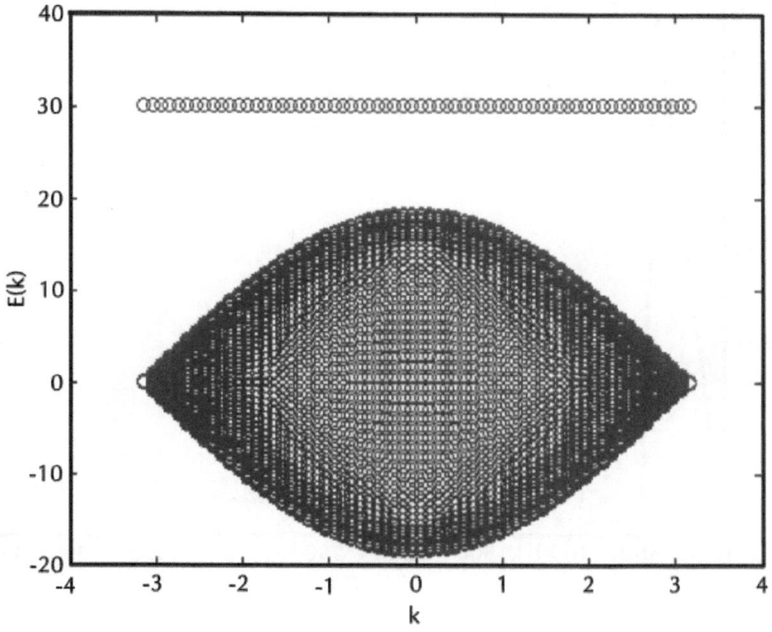

FIGURE 3 Eigenspectra of two-phonon bound state under periodic boundary condition approach for lithium tantalate system with poling field = 13.69 kV/cm and coupling = 12. The upper spectrum is the signature of quantum breather that separates out of the localized spectrum.

Next, let us look into the other TPBS parameters, namely the energy gap (E_g) that should indicate a different trend to that shown by phonon band width. The relation between the energy gaps that is calculated from Eq. (7) indicates some general trends. It has been observed that the energy gap decreases drastically towards an average value of nonlinearity of 467 and thereafter it drops relatively less towards lower a values, i.e., toward the highest poling field of 210 kV/cm, as it was expected. This might signify that as the nonlinearity decreases in this system, the formation of QB becomes more and more difficult and up to a certain value, the switching will also be much easier in the system and thereafter it becomes relatively more difficult, as revealed by quantum calculations of TPBS parameters. It also indicates a possibility of pinning transition, even if the effect may be relatively smaller. It is seen from Fig. 4 that the hopping coefficient (m) that is calculated from Eq. (9) against nonlinearity, there is a sharp transi-

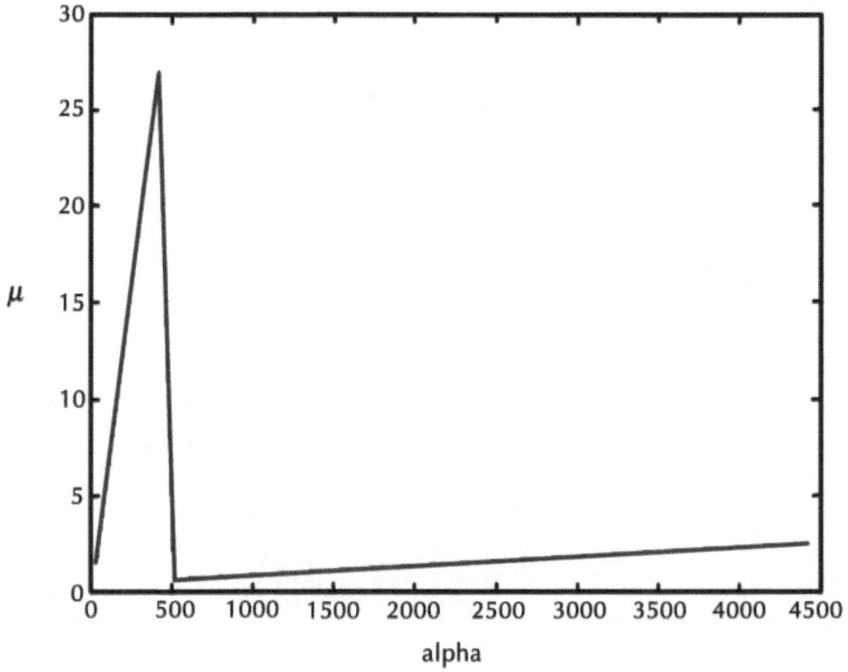

FIGURE 4 Phonon hopping coefficient that is calculated from Eq. (9) against nonlinearity parameter showing a sharp transition around poling field value of 17 kV/cm giving directions towards future application of lithium tantalate ferroelectrics.

tion at a poling field of 17 kV/cm with a nonlinearity value of 421. For QB under periodic boundary condition, this behavior of the micro level properties (such as TPBS parameters) can be considered quite significant for future applications. This supports our quantum calculations in that the tendency of phonon hopping to the second excited state may be considered to be showing a direction towards interpretation of some physical behavior of lithium tantalate for some applications in devices.

11.3.2.2 LIFETIME OF QUANTUM BREATHERS

For a non-periodic boundary condition approach, a typical simulation curve for a smaller value of coupling of 0.9 and nonlinearity of 421 is shown in Fig. 5. This simulation was done with 6 particles on 3 sites with initial condition as: $|\psi(0)\rangle = |5,1,0\rangle$. Here, the initial localization is mainly at the first site and then there is a fast redistribution of quanta between the other two sites until they become equal or almost equal, and the critical time for redistribution (t_{re}) is calculated at the crossing point of three quanta in femtosecond at this coupling value. The curve with higher intensity that represents the first site is clearly visible and they evolve due to the faster process of redistribution of quanta in lithium tantalate system that was not observed in lithium niobate.

Here, the critical time for redistribution (t_{re}) that is proportional to QB's lifetime is around 1359 fs. As the number of quanta increases from 6 to 12, the QB's lifetime decreases significantly to: 421, 207, 127 and 93 fs, respectively. It is noteworthy that for 6 quanta in lithium niobate, the QB's lifetime is higher at 441 fs for the same value of coupling of 0.9, whereas that for lithium tantalate under the same condition it is 421 fs. A lower lifetime of QBs in the latter case might make it suitable candidate for certain devices.

It has been observed that as the number of quanta increases, the "critical time of redistribution" continuously decreases. This effect is more pronounced in case of lithium tantalate than that observed in lithium niobate. Moreover, at each value of quanta, the QB's lifetime decreases as the coupling increases, even between 0.1 and 0.9. This change is particularly noticeable for lower quanta at 4 and 6, respectively. From 8 quanta onwards, the effect seems to be relatively less. Hence, it can be stated that the QB's lifetime is relatively more sensitive in case of lithium tantalate than that in lithium niobate [113]. It has to be noted that there is a qualitative difference of the QBs lifetime data between ferroelectric system and another important nonlinear optical material, i.e., split-ring-resonator based metamaterials for antenna application in the THz regime that has been investigated quite extensively [65, 114].

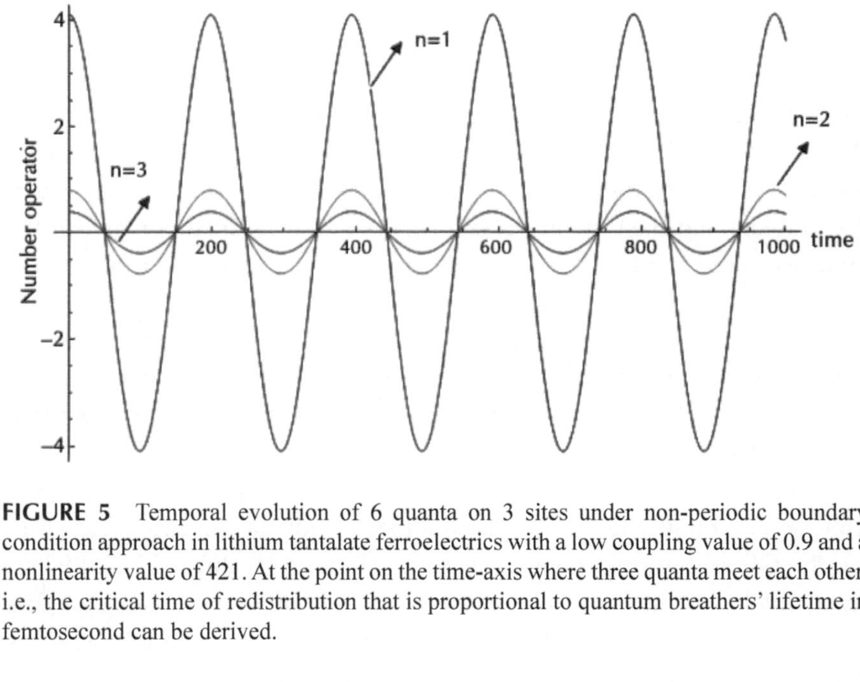

FIGURE 5 Temporal evolution of 6 quanta on 3 sites under non-periodic boundary condition approach in lithium tantalate ferroelectrics with a low coupling value of 0.9 and a nonlinearity value of 421. At the point on the time-axis where three quanta meet each other, i.e., the critical time of redistribution that is proportional to quantum breathers' lifetime in femtosecond can be derived.

It is pertinent to mention that for a given macromolecule, Tretiak et al. [115] observed that the lifetime of discrete breathers increases as the crystals become more and more defective. As the defect or disorder could create localization, it is suitable for the formation of QB giving us insight on quantum localization. The above discussion includes the effect of impurities, as it depends on nonlnearity parameter, which in turn is embodied in our Hamiltonian. Therefore, this is discussed in Section 7.3.2.1 in terms of TPBS parameters. Finally, it should be mentioned that the application of QB, particularly in terms of temporal evolution of quanta, could assume significance if novel technologies and methods are to be adopted, as discussed earlier.

11.4 CONCLUSIONS

On the evolution of polarization with both time and space, the present study shows non-linear Klein-Gordon equation that is based on our discrete Hamiltonian in a typical array of ferroelectric domains. By spatial discretization technique via spectral collocation method, the Klein-Gordon equation shows the behavior of classical breathers for an overall 3D view that are sensitive towards higher damping, or rather the oscillations of the symmetric breathers seem to be quite stable up to a relatively moderate value of damping (~0.50). In a periodic boundary condition approach for quantum breathers, the second quantization gives rise to some interesting values of TPBS parameters against nonlinearity within the system. This shows that after a value of nonlinearity (» 421), i.e., equivalent to a switching field of about 14 kV/cm, the formation of quantum breathers starts becoming relatively more difficult. The behavior of eigen-energy again nonlinearity or Landau coefficient was in contrast to that shown in lithium niobate system. The hopping coefficient shows a sharp transition at this point indicating some sort of pinning in this system as well. Without the impurity data in this system, it is difficult to specify the role of antisite tantalum defects. For a non-periodic boundary condition approach, the temporal evolution spectra show that with increasing number of quanta, there is a decrease of time of redistribution that is proportional to the QB's lifetime in fs. It also shows a strong dependence on the coupling. This is considered useful for a future study in this new field of investigation of quantum breathers in ferroelectrics and other applications of QBs in important nonlinear optical materials.

KEYWORDS

- **Bloch state**
- **breather**
- **energy localization**
- **hopping strength**
- **Klein-Gordon**
- **Landau-Ginzburg type**
- **quantum breathers**

REFERENCES

1. Fu, H.; Cohen, R. E. *Nature* **2000,** *403*, 281–283.
2. Lines, M. E.; Glass, A. M. In *Principles and Applications of Ferroelectrics and Related Materials.* Clarendon: Oxford, 1977.
3. Kim, S.; Gopalan V.; Gruverman, A. *Appl. Phys. Lett.* **2002,** *80*, 2740–2742.
4. Bandyopadhyay, A. K.; Ray, P. C. *J. Appl. Phys*. **2004,** *95*, 226–230.
5. Gahagan, K. T. et al., *Appl. Opt.* **1999,** *38*, 1186–1190.
6. Gopalan, V. *Ann. Rev. Mater. Res.,* **2007,** *37*, 449–489.
7. *Dawber, M.; Rabe, K. M.; Scott, J. F. Rev. Mod. Phys.* **2005,** *77, 1083–1130.*
8. Catalan, G.; Schilling, A.; Scott J. F.; Greg, J. M. *J. Phys: Cond. Matter* **2007,** *19*, 132201–132207.
9. Bandyopadhyay, A. K.; Ray, P. C.; Vu-Quoc, L.; McGurn, A. R. *Phys. Rev.* B **2010,** *81*, 064104–064114.
10. Rainer Waser. In *Nanoelectronics and Information Technology*, Wiley: Weinhiem, 2005.
11. Giri, P.; Ghosh, S.; Choudhary, K.; Alam, Md.; Bandyopadhyay, A. K.; Ray, P. C. *Phys. Scr.* **2011,** *83*, 015702–015706.
12. Srinivas, V.; Vu-Quoc, L. *Ferroelectrics* **1995,** *163*, 29–57.
13. Bandyopadhyay, A. K.; Ray, P. C.; Gopalan, V. *J. Phys: Cond. Matter* **2006,** *18*, 4093–4099.
14. Bandyopadhyay, A. K.; Ray, P. C.; Gopalan, V. *Euro. Phys. J.-B* **2008,** *65*, 525–531.
15. Bandyopadhyay, A. K.; Ray, P. C.; Gopalan, V. *J. Appl. Phys.* **2006,** *100*, 114106–114109.
16. Scrymgeour, D. A.; Gopalan, V.; Itagi, A.; Saxena A.; Swart, P. J. *Phys. Rev. B* **2005,** *71*, 184110–184122.
17. Meyer B.; Vanderbilt, D. *Phys. Rev. B* **2002,** *65*, 104111–104121.
18. He L.; Vanderbilt, D. *Phys. Rev. B* **2003,** *68*, 134103–134109.

19. Yatsenko, A. V.; Ivanova-Maksikova, M. H.; Sergeev, N. A. *Physica B* **1998,** *254*, 256–259.
20. Stashans, A.; Serrano, S.; Medina, P. *Physica B* **2006,** *381*, 82–89.
21. Bussman-Holder, A.; Bishop, A. R.; Egami, T. *Europhys. Lett.* **2000,** *71*, 249–255.
22. Erhart, J.; Cao, W. *J. Appl. Phys.* **1999,** *86*, 1073.
23. Emelyanov, A Y.; Pertsev, N. A.; Salje, E. H. K. *J. Appl. Phys.* **2001,** *89*, 1355.
24. Wittborn, J.; Canalias, C.; Rao, K. V.; Clemens, R.; Karlsson, H.; Laurell, F. *Appl. Phys. Lett.* **2002,** *80*, 1622.
25. Wu, Q.; Huang, N. D.; Liu, Z. R. et al., *J. Appl. Phys.* **2007,** *101*, 014112.
26. Klotins, E. *Physica E* **2010,** *429*, 614–617.
27. John, N. S.; Saranya, D.; Parui, J.; Krupanidhi, S. B.*J. Phys. D: Applied Phys.* **2011,** *44*, 415401.
28. Zhirnov, V. A. *Sov. Phys. JETP* **1959,** *8*, 822.
29. Lee, D.; Behera, R. K.; Wu, P.; Xu, H.; Li, Y. L.; Sinnott, S. B.; Phillpot, S. R.; Chen, L. Q.; Gopalan, V. *Phys. Rev. B* **2009,** *80*, 060102–060105.
30. Morozovska, A. N. et al., *Phys. Rev. B* **2009,** *80*, 214110.
31. Xu, H.; Lee, D.; Sinnott, S. B.; Gopalan, V.; Dierolf, V.; Phillpot, S. R. *Phys. Rev. B* **2009,** *80*, 144104–144114.
32. Vasudevarao, A. *Phys. Rev. B* **2010,** *82*, 024111.
33. Floquet, N.; Valot, C. *Ferroelectrics* **1999,** *234*, 107.
34. Tsai, D.; Suresh, M. B.; Chou, C. C. *Phys. Scr.* **2007,** *T129*, 175.
35. Scott, J. F. *J. Phys.: Cond. Matter 18,* **2006,** *R361.*
36. *Dawber, M.; Gruverman, A.; Scott, J. F. J. Phys: Cond. Matter 18,* **2006,** *L71.*
37. Dauxois, T.; Peyrard, M. In: *Phys. of Solitons*, Cambridge University Press: Cambridge, 2006; pp. 211.
38. *Flach*, S.; Gorbach, A. V. *Phys. Rep.* **2008,** *467*, 1–116.
39. Sato, M.; Hubbard, B. E.; Sievers, A. J. *Rev. Mod. Phys.* **2006,** *78*, 137–157.
40. Anderson, P. W. *Phys. Rev. Lett.* **1958,** *109*, 1492.
41. Campbell, D. K.; Flach, S.; Kivshar, Y. S. *Phys. Today*, pp. 43, Jan. 2004.
42. Chen, D.; Aubry, S.; Tsironis, G. P. *Phys. Rev. Lett.* **1996,** *77*, 4776–4779.
43. Maniadis, P.; Kopidakis, G.; Aubry S. *Physica D* **2004,** *188*, 153–177.
44. Swanson, B. I.; Brozik, J. A.; Love, S. P.; Strouse, G. F.; Shreve, A. P.; Bishop, A. R.; Wang, W. Z.; Salkola, M. I. *Phys. Rev. Lett.* **1999,** *82*, 3288–3291.
45. Schwarz, U. T.; English, L. Q.; Sievers, A. J. *Phys. Rev. Lett.* **1999,** *83*, 223–226.
46. Trıas, E.; Mazo, J. J.; Orlando, T. P.*Phys. Rev. Lett.* **2000,** *84*, 741.
47. Sato, M.; Hubbard, B. E.; Sievers, A. J.; Ilic, B.; Czaplewski, D. A.; Graighead, H. G. *Phys. Rev. Lett.* **2003,** *90*, 044102–044105.
48. Dick, A. J.; Balachandran, B. C.; Mote, D. Jr., *Nonlinear Dynamics* **2007,** *54,* 13.
49. Eisenberg, H. S.; Silberberg, Y.; Morandotti, R.; Boyd, A. R.; Aitchison, J. S. *Phys. Rev. Lett.* **1998,** *81*, 3383–3386.
50. Edler, J.; Pfister, R.; Pouthier, V.; Falvo, C.; Hamm, P. *Phys. Rev. Lett.* **2004,** *93*, 106405–106408.
51. Tsironis, G. P. *Chaos* **2003,** *13*, 657–666.
52. Kopidakis, G.; Aubry, S.; Tsironis, G. P. *Phys. Rev. Lett.* **2001,** *87*, 165501–165504.
53. Flach, S.; Kladko, K. *Physica D* **1999,** *127*, 61.
54. Martınez, P. J.; Meister, M.; Floria, L. M.; Falo, F. *Chaos* **2003,** *13*, 610.

55. Martınez, P. J.; Flor′ıa, L. M.; Falo, F.; Mazo, J. J. *Europhys. Lett.* **1999,** *45*, 444.
56. Sievers, A. J.; Takeno, S. *Phys. Rev. Lett.* **1988,** *61*, 970–973.
57. Sievers, A. J.; Page, J. B. In: *Dynamical Properties of Solids*, Vol. 7, Horton, G. K.; Maradudin, A. A. Ed.; North-Holland: Amsterdam, 1995.
58. Fleischer, J. W.; Segev, M.; Elfremidis, N. K.; Christodoulides, D. N. *Nature* **2003,** *422,* 147–150.
59. Flach, S. *Phys Rev E* **1995,***51*, 1503
60. Flach, S.; Willis, C. R. *Phys. Rep.* **1998,** *295*, 181.
61. Mackay, R. S.; Aubry, S. *Nonlinearity* **1994,** *7*, 1623.
62. Aubry, S. *Physica D* **1997,** *103*, 201.
63. Fleurov, V. *Chaos* **2003,** *13*, 676.
64. Giri, P.; Choudhary, K.; Sengupta, A.; Bandyopadhyay, A. K.; McGurn, A. R. *Phys. Rev. B* **2011,** *84*, 155429–155437.
65. Mandal, B.; Adhikari, S.; Basu, R.; Choudhary, K.; Mandal, S. J.; Biswas, A.; Bandyopadhyay, A. K.; Bhattacharjee, A. K.; Mandal, D. *Phys. Scr.* **2012,** *86*, 015601–015610.
66. Flach, S.; Miroshchinko, A. E.; Fleurov, V.; Fistul, M. V. *Phys. Rev. Lett.* **2003,** *90*, 084101–084104.
67. Miroshnichenko, A. E.; Flach, S.; Kivshar, Y. S. *Rev. Mod. Phys.* **2010,** *82*, 2257.
68. Kim, S. W.; Kim, S. *Phys. Rev. B* **2001,** *63*, 212301.
69. Choudhary, K.; Adhikari, S.; Biswas, A.; Ghosal, A.; Bandyopadhyay, A. K. *J. Opt. Soc. Am.* B **2012,** *29*, 2414.
70. Bang, O.; Peyrard, M. *Physica D* **1995,** *81*, 9.
71. Bang, O.; Peyrard, M. *Phys. Rev. E* **1996,** *53*, 4143.
72. Corney, J. F.; Bang, O. *Phys. Rev. E* **2001,** *64*, 047601.
73. Kobyakov, A.; Lederer, F.; Bang, O.; Kivshar, Y. S. *Opt. Lett.* **1998,** *23*, 506–508.
74. Corney, J. F.; Bang, O. *Phys. Rev. Lett.* **2001,** *87*, 133901–133904.
75. Di Trapani, P.; Bramati, A.; Minardi, S.; Chinaglia, W.; Conti, C.; Trillo, S.; Kilius, J.; Valiulis, G. *Phys. Rev. Lett.* **2001,** *87*, 183902–183905.
76. Giri, P.; Choudhary, K.; Dey, A.; Biswas, A.; Ghosal, A.; Bandyopadhyay, A. K. *Phys. Rev. B* **2012,** *86*, 184101–184108.
77. Gonzalez, J. A.; Guerrero, I.; Bellorin, A. *Phys.Rev. E* **1996,** *54*, 1265.
78. Holyst, J. A. *Phys. Rev. E* **1998,** *57*, 4786.
79. Benedek, G.; Bussmann-Holder, A.; Bilz, H. B. *Phys. Rev. B* **1987,** *36*, 630.
80. Comte, J. C. *Phys. Rev. E* **2002,** *65*, 046619.
81. Giri, P.; Choudhary, K.; Sengupta, A.; Bandyopadhyay, A. K.; Ray, P. C. *J. Appl. Phys.* **2011,** *109*, 054105–054112.
82. Scott, A. C.; Eilbeck, J. C.; Gilhøj, H. *Physica D* **1994,** *78*, 194–213.
83. R. A. Pinto, M. Haque, S. Flach, *Phys. Rev. A* **2009,** *79*, 052118–052125.
84. Nguenang, J. P.; Pinto, R. A.; Flach, S. *Phys. Rev. B* **2007,** *75*, 214303–214308.
85. Johnson, P. R.; Strauch, F. W.; Dragt, A. J.; Ramos, R. C.; Lobb, C. J.; Anderson, J. R.; Wellstood, F. C. *Phys. Rev. B* **2003,** *67*, 020509(R).
86. Eiermann, B.; Anker, Th.; Albiez, M.; Taglieber, M.; Treutlein, P.; Marzlin, K. P.; Oberthaler, M. K. *Phys. Rev. Lett.* **2004,** *92*, 230401–230404.
87. Fleischer, J. W.; Segev, M.; Efremidis, N. K.; Christodoulides, D. N. *Nature* **2003,** *422*, 147–150.

88. Sato, M.; Hubbard, B. E.; Sievers, A. J.; Ilic, B.; Czaplewski, D. A.; Craighead, H. G. *Phys. Rev. Lett.* **2003,** *90,* 044102–044105.
89. Satarik, M.; Zdravkovic, S.; Tuszynski, J. *BioSystems* **1999,** *49,* 117. 90. A. L. Rakhmanov, A. M. Zagoskin, S. Savel'ev, F. Nori, *Phys. Rev. B* **2008,** *77,* 144507.
91. Zvyagin, S. A.; Wosnitza, J.; Batista, C. D.; Tsukamoto, M.; Kawashima, N.; Krzystek, J.; Zapf, V. S.; Jaime, M.; Oliveira Jr., N. F.; Paduan-Fihlo, A. *Phys. Rev. Lett.* **2007,** *98,* 047205–047208.
92. Biswas, A.; Choudhary, K.; Bandyopadhyay, A. K.; Bhattacharjee, A .K.; Mandal, D. *J. Appl. Phys.* **2011,** *110,* 024104–111.
93. Baroni, S.; Gironcoli, S. D.; Corso, A. D. *Rev. Mod. Phys.* **2001,** *73,* 515.
94. Lazzeri, M.; Gironcoli, S. D. *Phys. Rev. Lett.* **1998,** *81,* 2096–2099.
95. Ghosez, Ph.; Gonze, X.; Michenaud, J. P. *Ferroelectrics* **1998,** *206–207,* 205.
96. Ghosez, Ph.; Michenaud, J. P.; Gonze, X. *Phys. Rev. B* **1998,** *52,* 6224.
97. Cohen, M. H.; Ruvalds, J. *Phys. Rev. Lett.* **1969,** *23,* 1378–1381.
8. Dougherty, T. P.; Wiederrecht, G. P.; Nelson, K. A.; Garrett, M. H.; Jensen, H. P.; Warde, C. *Science* **1992,** *258,* **5083.**
99. Brennan, C. J.; Nelson, K. A. *J. Chem. Phys.* **1997,** *107,* 9691–9694.
100. *Stone, G.; Dierolf, V. Opt. Lett.* **2012,** *37, 1032–1034.*
101. Becker, B.; Schelling, P. K.; Phillpot, S. R. *J. Appl. Phys.* **2006,** *99,* 123715–123719.
102. Pinto, R. A.; Ricardo, A. *Phys. Rev. B* **2008,** *77,* 024308–024310.
103. Schulman, L. S.; Tolkunov, D.; Mihokova, E. *Chem. Phys.* **2006,** *322,* 55–74.
104. Leggett, A. et al., Eds.; In: *Quantum Computing and Quantum Bits in Mesoscopic Systems,* Kluwer Academic/Plenum Publishers: New York, 2004.
105. Gershenfeld, N. A.; Chuang, I. L. *Science* **1997,** *275,* 350.
106. Proville, L. *Phys. Rev. B* **2005,** *71,* 104306–104311.
107. Shelton, R. K.; Ma, L. S.; Kapteyn, H. C.; Murnane, M. M.; Hall, J. L; Ye, J. *Science* **2001,** *293,* 1286–1289.
108. Ebrahimzadeh, M. *Phil. Trans. Royal Soc. A* **2003,** *361,* 2731–2750.
109. Rousse, A.; Rischel, C.; Gauthier, J. *Rev. Mod. Phys.* **2001,** *73,* 17–31.
110. Ablowitz, M. J.; Biondini, G. *Opt. Lett.* **1998,** *23,* 1668–1670.
111. Gabitov, I.; Indik, R.; Mollenauer, L. F.; Shkarayev, M.; Stepanov, M.; Lushnikov, P. M. *Opt. Lett.,* **2007,** *32,* 605–607.
112. Yang, T. J.; Gopalan, V.; Swart, P. J.; Mohideen, U. *Phys. Rev. Lett.,* **1999,** *82,* 4106–4109.
113. Mandal, S. J.; Choudhary, K.; Biswas, A. Bandyopadhyay, A. K.; Bhattacharjee, A. K. Mandal, D. *J. Appl. Phys.* **2011,** *110,* 124106–116.
114. Mandal, et al., B. *Quant. Phys. Lett.* **2012,** *1,* 59–68.
115. Tretiak, S.; Saxena, A.; Martin, R.L.; Bishop, A.R. *Phys. Rev. Lett.* **2002,** *89,* 97402–97405.

CHAPTER 12

TRIPLE DENSITY QUANTUM SIMILARITY MEASURES AND THE TENSORIAL REPRESENTATION OF MOLECULAR QUANTUM OBJECT SETS

RAMON CARBÓ-DORCA*[1]

CONTENTS

*[1]E-mail: quantumqsar@hotmail.com

12.1 INTRODUCTION

In this chapter, molecular quantum similarity (QS) measures involving three density functions are studied, providing the necessary algorithms and programming sources for application purposes. Triple density representation of some known molecular quantum object set (MQOS) permits to express each element as a symmetric matrix with dimension equal to the MQOS cardinality. Such matrix formulation appears instead of the vector representations founded on double density QS (DDQS) measures or as a result of the usual classical descriptor parameterization. The whole triple density quantum similarity (TDQS) measures description of a given MQOS corresponds to a third order hypervector or tensor, whose elements are symmetric matrix representations of every molecular structure belonging to the MQOS. Such tensorial representation permits to set up an extended set of procedures in order to study the relationships between the MQOS elements, beyond the usual vector description. For the sake of completeness the quantum mechanical origin of the TDQS integrals is sketched as an introduction. The three p-type Gaussian orbitals are employed along the theoretical development to illustrate, via a concrete and particularly simple case example, the structure of the definitions encountered along the discussion, as well as the ability of QS measures to discriminate between DF belonging to degenerate wave functions. The present study can be considered in this way a first step towards the general theory and computational feasibility of a hypermatricial or tensorial representation of molecular structures associated to any MQOS. Generalized Carbó similarity indices (CSI) are also studied as a way to manipulate the TDQS measures for easy interpretation. Besides the appropriate description of the programs associated to this paper, here are given several application examples, based essentially on the same background philosophy as the usual DDQS measures of previous papers.

Quantum similarity (QS), applied to molecular structures or briefly molecular QS (MQS) which started in our laboratory with a 1980 pioneering paper [1], has followed a steady road, see for example a large résumé study on QS theory and applications [2], several theoretical studies [3–10] as well as a recent theoretical description [11] leading towards the initial paper of this series [12], where an as complete as possible bibliography of the QS evolution is given.

Actually, several MQS papers [12–14] associated to theoretical questions studied from the computational side*2, constitute three steps towards the goal to give, not only a modern up to date account of the background of MQS measures, but the basic computational tools and examples for general applications, like molecular set ordering [14] as well as to new insights on several MQS computational problems [12, 13] and questions attached to quantum quantitative structure-properties relationships (QQSPR) [15, 16].

However, despite of the previous endeavor, while writing the manuscript of the present work, the author has felt that some basic problems related with the MQS measures in general and with triple density QS (TDQS) measures in particular were lacking of some detailed study into the literature. To include this large amount of information within the present discussion will certainly enlarge the volume of the present contribution. So, it was opted prior the ending of the actual discussion to publish separately such material elsewhere in several independent papers. In this way, readers interested deepening the understanding of the mathematical and theoretical aspects of QS, can have a look at a general feature of the products of Gaussian functions [17], follow a discussion on the discrete general representation of molecular sets [18], peruse the mathematical possibilities of function basis sets transformation [19], inquire about the nature of the scaling of Euclidian distances and compare them with more flexible and general similarity indices [20], survey a discussion about the nature of the generalized Carbó similarity indices (CSI) [21], have handy a résumé of the principal mathematical definitions associated to QS [22], realize the holographic nature of the QS matrix representations of molecular structures [23] and finally analyze the geometric nature of the quantum object sets (QOS), see Ref. [22] for detailed updated definitions, density functions (DF) tags relationships [24].

---

*2The previous papers of this series will be named into the text generically as communications on quantum similarity (CQS). Every particular member of the collection will be referred as CQS1 [12], CQS2 [13] and CQS3 [14], respectively. The present chapter will be sometimes mentioned as CQS4.

12.1.1 TRIPLE DENSITY QS (TDQS) MEASURES OUTLOOK

With this set of previous studies, TDQS measures can be further considered from a new vantage point, despite they have been already described at the initial stages of the MQS development; see, for instance, the following previous work references, where not only the definitions and theoretical background were presented in a schematic manner [25], but some not so simple applications have been also put forward [26].

In any case, the appearance of TDQS into the MQS literature has been somehow scarce, for multiple reasons but essentially because of the computational demand of the associated integral measures. However, the development of the modern computational facilities permits to use with certain ease the present kind and even higher order tensors chosen amid the MQS tools.

Among other characteristics, TDQS measures defined over molecular QOS (MQOS)[*3], can provide another sort of discrete description of molecular structures carrying extended information if compared with the usual double density QS (DDQS) measures. Moreover, TDQS measures open the way to higher order QS measures [27], and can help to present them as a richer new set of discrete molecular descriptions, differing both in generality, operator bias and information content, from the classical usual vector parameterization employed in discrete molecular description. These previous considerations about TDQS measures might justify the contents of this study.

12.1.2 PRESENT WORK PLANNING

As in the previous papers, this one will furnish not only the basic theory and practical formulation used, but a full description of the algorithms

[*3]MQOS are a particular class of tagged sets. MQOS are molecular sets, where each element is described via a quantum mechanical density function. The set of molecular structures is termed the *object set* and the set of attached density functions is the *tag set*. The couples: (submicroscopic element; density function tag) are termed quantum objects. See below in the section: *The mathematical structure of the collection of TDQS integrals, when computed over a molecular quantum object set (MQOS),* for further detailed definitions and references. See also reference Ref. [22] for an up to date account of terms, references and definitions.

employed as well. Essentially, among other interesting issues, the question of the triple molecular superposition (MS) problem will be deeply analyzed. Reproducibility and computational accessibility are stressed in the preceding set of previous papers, which will be called: CQSn(n=1,3). As the present contribution can be considered in fact CQS4, then it will also follow these working lines; therefore, a web page containing the program codes and several input–output examples will be also furnished and made openly available for download to any potential user.

Besides the continuation of the CQS*n* studies, the basic computational background of the present work also leans on previous mathematical development on the nature of GTO's [28] and the HEDT [29], which constitutes a recent research effort, dealing about the so-called General Gaussian Product Theorem (GGPT) [17]. There, the structure and computational characteristics of the product of any number of GTO's, arbitrarily centered in three dimensional space, has been presented. Once the useful mathematical tools furnished by the GGPT are known, it is a straightforward matter to compute the QS integrals needed for the development of this paper and the computation of higher orders densities QS measures. Because of this previous knowledge, no extended information is presently given about the deduction of the involved TDQS integral form.

Accordingly, this contribution will be essentially centered into several arguments by means of the following scheme. First, it will be given some simplified data about the needed basic integrals formulation of TDQS measures over 1s type GTO's. The structure of the TDQS hypermatrices (TDQSH) will be discussed next in a second step. Among the possible applications of the TDQS measures several manipulations, similar to the ones given in the double density QS (DDQS) previous work as shown in CQS3 will be also presented as a third part. A discussion will be also given afterwards, describing the general extended theoretical background of MQOS, the TDQS tensorial description and interrelationships among its elements. Such an analysis will be done to evidence the powerful source of molecular structure information contained in TD and within higher order QS measures. Finished this extended discussion, a short argumentation connecting TDQS measures and quantum quantitative structure-properties relationships (QQSPR) will be given in order to have a complete overview of the present theoretical panorama. The computational details with the set of programs specially constructed to enhance the present work will constitute the final main section of CQS4, which will end with several application examples. A glossary has been added at the end describing the acrostics appearing in the present CSQ4 discussion.

Finally, it can be stressed that along the theoretical development, when necessary to illustrate the structure of the proposed TDQS measures and the associated similarity indices, a schematic example will be given. This illustrative example will be carried out along the paper development. It has been chosen based on three degenerate p-type GTO's. The corresponding density functions attached to these three p-type orbitals are coincident with the three diagonal elements of d-type GTO's. With these DF, the TDQS measures can be analytically described in a complete and easy manner. From the tensor structure associated to the TDQS measures of the p-type GTO DF, all further described manipulations of the TDQS tensors within the text are also particularly performed with the p-type GTO's TDQS measures.

12.2 TRIPLE 1S GTO INTEGRALS

The computational background where the present CQS4 paper will be developed does not differ abruptly from the previous CQS papers and is basically grounded in the well-known atomic shell approximation (ASA) [30–34] in order to express, within an advanced polarized promolecular framework, the quantum mechanical molecular density functions (DF). To obtain additional information on recent ASA application development, see for example the previous CQS chapters [12–14] and some application to describe DF for large macromolecular structures [35, 36] to see the adequate behavior for ASA DF.

While this paper was developed several applications and studies of ASA behavior have been described: DF representation [41] and molecular electrostatic potentials [37, 38]. In order to have a detailed view of ASA DF performance separate work path studies have been written. First, a pair of papers [37, 38] studied the behavior of the electrostatic molecular potential (EMP) [39] based upon ASA DF in a promolecular and polarized framework. Second, inspired by Mezey's holographic electronic density theorem [40], another work [41], describes a novel manner of imaging quantum chemical functions using stereographic representations [42] and the chosen examples are also based on ASA DF shapes. Finally, a new way to compute soft EMP has been described [43].

Within the ASA context the important and basic function needed is the simple and well-known 1s GTO, therefore the involved integrals of the present CQS4 paper will be associated to this kind of familiar functions, as it has been in the previous elements of the CQS series. This section development will give the basic information about this subject, which will be noticeably related to TDQS measures and the subsequent algorithms.

12.2.1 NORMALIZED 1S GTO

Suppose that a 1s GTO is written as:

$$\gamma_I\left(\mathrm{r}-\mathrm{R}_I\,\middle|\,\alpha_I\right)=N\left(\alpha_I\right)e^{-\alpha_I|\mathrm{r}-\mathrm{R}_I|^2} \quad (1)$$

where a normalization factor is defined as the usual inverse square root Euclidian norm:

$$N\left(\alpha_I\right)=N_I=\left\langle e^{-2\alpha_I|\mathrm{r}-\mathrm{R}_I|^2}\right\rangle^{-\frac{1}{2}}=\left(\iiint_{-\infty}^{+\infty} e^{-2\alpha_I|\mathrm{r}-\mathrm{R}_I|^2}d\mathrm{r}\right)^{-\frac{1}{2}}=\left(\frac{2\alpha_I}{\pi}\right)^{\frac{3}{4}}$$

using the well-known integral [44]:

$$I_0\left(\alpha\right)=\int_{-\infty}^{+\infty}e^{-\alpha x^2}dx=\left(\frac{\pi}{\alpha}\right)^{\frac{1}{2}} \quad (2)$$

12.2.2 TRIPLE 1S GTO INTEGRAL

Then, a triple 1s GTO overlap integral can be defined as:

$$S_3 = \langle \gamma_I \gamma_J \gamma_K \rangle = N_I N_J N_K \iiint_{-\infty}^{+\infty} e^{-\left(\alpha_I |\mathrm{r}-\mathrm{R}_I|^2 + \alpha_J |\mathrm{r}-\mathrm{R}_J|^2 + \alpha_K |\mathrm{r}-\mathrm{R}_K|^2\right)} d\mathrm{r} \quad (3)$$

It is trivial to see that the three normalization factors can be expressed as a constant under a unique symbol:

$$N_{IJK} = N_I N_J N_K = N_{IJ} N_K = \left(\frac{8\alpha_I \alpha_J \alpha_K}{\pi^3}\right)^{\frac{3}{4}}$$

Thereafter, in order to obtain an analytic expression for the integral Eq. , the reduction of the involved 1s GTO into a unique center can be done by means of the previous discussion on the GGPT [17], as it has been commented. By a simple algebraic manipulation one arrives to a symmetrical formulation involving the integrand in the triple 1s GTO integral Eq. as:

$$\Theta_{IJK} = N_{IJK} \exp\left(-\left(\frac{\alpha_I \alpha_J R_{IJ}^2 + \alpha_I \alpha_K R_{IK}^2 + \alpha_K \alpha_J R_{JK}^2}{\alpha_I + \alpha_J + \alpha_K}\right)\right)$$

and the integral Eq. is easily evaluated next by means of the transformed GTO centers and the auxiliary integrals Eq. :

$$\langle \gamma_I \gamma_J \gamma_K \rangle = \Theta_{IJK} \iiint_{-\infty}^{+\infty} e^{-(\alpha_I + \alpha_J + \alpha_K)|r-Q|^2} dr = \Theta_{IJK} \left(I_0\left(\alpha_I + \alpha_J + \alpha_K\right)\right)^3,$$

an expression which permits to arrive to the final compact expression of the integral Eq. as:

$$\langle \gamma_I \gamma_J \gamma_K \rangle = \Theta_{IJK} \left(\frac{\pi}{\alpha_I + \alpha_J + \alpha_K}\right)^{\frac{3}{2}}.$$

Taking now into account that one can write explicitly the product of the three normalization factors, as shown in the first steps of the above discussion, then, finally the triple 1s GTO's integral general expression can be explicitly written as follows:

$$S_3 = \langle \gamma_I \gamma_J \gamma_K \rangle =$$
$$\left(\frac{8\alpha_I\alpha_J\alpha_K}{\pi^3}\right)^{\frac{3}{4}}\left(\frac{\pi}{\alpha_I+\alpha_J+\alpha_K}\right)^{\frac{3}{2}}\exp\left(-\left(\frac{\alpha_I\alpha_J R_{IJ}^2+\alpha_I\alpha_K R_{IK}^2+\alpha_J\alpha_K R_{JK}^2}{\alpha_I+\alpha_J+\alpha_K}\right)\right) \quad (4)$$

12.3 TRIPLE DENSITY QUANTUM SIMILARITY INTEGRALS

Having studied the expression of the triple integrals involving three 1s GTO's and its generalization, it is not difficult to imagine a TDQS integral (TDQSI), which can be defined as the integral, typically a measure, of the product of three DF, which can be symbolically written as:

$$\langle \rho_P \rho_Q \rho_R \rangle = \int_D \rho_P(r)\rho_Q(r)\rho_R(r)\,dr$$

Taking into account the usual way to describe DF within LCAO MO theory for a given molecular structure *P*, one can write:

$$\rho_P(\mathrm{r}) = \sum_\alpha \sum_\beta D_{\alpha\beta}^P |\alpha\rangle\langle\beta|$$

where $D_P = \{D_{P;\alpha\beta}\}$ is the density coordinates matrix[*4] and $\{|\alpha\rangle = \chi_\alpha(r)\}$ is the chosen AO basis set functions. Therefore the TDQSI can be rewritten as:

$$\langle \rho_P \rho_Q \rho_R \rangle = \sum_\alpha \sum_\beta D_{P;\alpha\beta} \sum_\gamma \sum_\kappa D_{Q;\gamma\kappa} \sum_\lambda \sum_\mu D_{R;\lambda\mu} \langle \alpha\beta\gamma\kappa\lambda\mu \rangle$$

[*4] There has been a systematic confusion in the literature about the name given to this matrix. Here it is proposed the logical name *density coordinates matrix* instead the common name of density matrix, which can be confused with the matrix constructed by means of the DF itself, also called density matrix.

where the integral over six AO basis functions can be defined in turn by means of the integral symbol:

$$\langle \alpha\beta\gamma\kappa\lambda\mu \rangle = \int_D \chi_\alpha(r)\chi_\beta(r)\chi_\gamma(r)\chi_\kappa(r)\chi_\lambda(r)\chi_\mu(r)dr$$

The analytical expression of such an integral, involving six general GTO's, can be constructed by simple integration from the results obtained within the GGPT [17 treatment.

12.3.1 ASA DF APPROXIMATION

In order to have a not so demanding *ab initio* computational structure at the moment to evaluate TDQSI, as commented beforehand, the so-called molecular ASA approximation [30–34] can be employed. The molecular DF in this approach is expressed as a linear combination of ASA functions associated to some atom *A*, defined in turn by means of the linear combination:

$$\sigma_A(r) = \sum_{I \in A} \omega_{A,I} \left| s_{A,I}\left(r \middle| \alpha_{A,I}\right)\right|^2 .$$

Where $\{\omega_{A,I}\}$ is a set of convex coefficients[*5] and $\left\{s_{A,I}\left(r \middle| \alpha_{A,I}\right)\right\}$ a normalized set of 1s GTO functions respectively. Both sets are fitted to some corresponding HF *ab initio* atomic DF calculation [32, 33]. Some of such families of fitted ASA functions can be downloaded from a public website [45]. A choice of three ASA basis sets can be also obtained from the first program of the molecular quantum similarity program suite (MQSPS), described in CQS3 [14].

[*5] That is fulfilling: $\{\omega_{A,I}\} \subset R^+ \wedge \sum_I \omega_{A,I} = 1$

In fact, due to a construction choice, for the whole set of ASA DF, the Minkowski norms can become unity: $\langle\sigma_A(r)\rangle = 1$. In practice, for the purpose of the ASA construction one can use the normalized 1s GTO's functions Eq. , as defined at the beginning of this study, but squared, that is:

$$\left|s_{A,I}\left(r - R_A \middle| \alpha_{A,I}\right)\right|^2 = \left(\frac{2\alpha_{A,I}}{\pi}\right)^{\frac{3}{2}} e^{-2\alpha_{A,I}|r-R_A|^2} \tag{5}$$

Where it is supposed that the atom A is centered at the position vector: R_A. With the fitted ASA functions handy, the so-called promolecular approximation for a molecular DF can be easily put forward; it can be written in a general way as:

$$\rho_P^{ASA}(r) = \sum_{A\in P} Q_{P,A}\sigma_A(r), \tag{6}$$

where the set of constants $\{Q_{P,A}\}$ can be chosen in different ways, but in any case possessing the constraint property: $N_e^P = \sum_{A=1}^{N_a^P} Q_{P,A}$; with the integers $\{N_e^P; N_a^P\}$ representing the number of electrons and atoms of molecule *P*, respectively. The constant coefficients $\{Q_{P,A}\}$ appearing into the promolecular approximation Eq. can be selected, for instance, as the atomic numbers or as any kind of overlap atomic populations, like Mulliken's [46] or NBO [47], associated to each atom *A* of the molecule *P*. See for more application details the papers on the CQS series [12–14] and the recently published papers on DF41 and EMP [37, 38, 43].

12.3.2 TDQSI IN THE ASA FRAMEWORK

In this manner, within the ASA description of a DF, the TDQSI can be easily written using the following formulation in terms of a triple sum:

$$\langle \rho_P \rho_Q \rho_R \rangle \approx \langle \rho_P^{ASA} \rho_Q^{ASA} \rho_R^{ASA} \rangle = \sum_{A \in P} \sum_{B \in Q} \sum_{C \in R} Q_{P,A} Q_{Q,B} Q_{R,C} \langle \sigma_A \sigma_B \sigma_C \rangle$$

The triple integral $\langle \sigma_A \sigma_B \sigma_C \rangle$, involving three ASA atomic functions can be rewritten now in terms of the ASA 1s GTO basis set as:

$$\langle \sigma_A \sigma_B \sigma_C \rangle = \sum_{I \in A} \sum_{J \in B} \sum_{K \in C} \omega_{A,I} \omega_{B,J} \omega_{C,K} \left\langle \left| s_{A,I} s_{B,J} s_{C,K} \right|^2 \right\rangle.$$

The new triple GTO integral $\left\langle \left| s_{A,I} s_{B,J} s_{C,K} \right|^2 \right\rangle$ is just the overlap integral involving a triple product of 1s GTO, like the one discussed at the beginning of this paper; it can be computed as in Eq. , although it has to be somehow modified, because of Eq. , which slightly redefines the 1s GTO basis functions needed in the ASA framework.

That is, for instance, when the basic integrals needed to evaluate QS measures in triple 1s GTO overlap integrals have to be evaluated, one arrives to the alternative expression:

$$\left\langle \left| s_{A,I} s_{B,J} s_{C,K} \right|^2 \right\rangle = \left(\frac{32 \alpha_{A,I} \alpha_{B,J} \alpha_{C,K}}{\pi^2 \left(\alpha_{A,I} + \alpha_{B,J} + \alpha_{C,K} \right)} \right)^{\frac{3}{2}} \times \exp\left(-2 \left(\frac{\alpha_{A,I} \alpha_{B,J} R_{AB}^2 + \alpha_{A,I} \alpha_{C,K} R_{AC}^2 + \alpha_{B,J} \alpha_{C,K} R_{BC}^2}{\alpha_{A,I} + \alpha_{B,J} + \alpha_{C,K}} \right) \right) \quad (7)$$

12.3.3 SOME NOTATION REMARKS

In the same manner as the usual DDQS measure can be shortly written as: $\langle \rho_P \rho_Q \rangle \equiv \langle PQ \rangle \equiv z_{PQ}$, involving two DF, the TDQS integrals of the type: $\langle \rho_P \rho_Q \rho_R \rangle$ or the equivalent ASA approach: $\langle \rho_P^{ASA} \rho_Q^{ASA} \rho_R^{ASA} \rangle$, can be also shortly written with the simplified symbol: $\langle PQR \rangle \equiv z_{PQR}$, using the context to know if they correspond to an exact TDQS expression or to an ASA one.

12.4 QS OF THREE MOLECULES AND OPTIMAL TDQSI

Within any approximation level, TDQS measures will depend on the relative positions in space of the three molecular structures involved. An integral maximum will be reached when the three molecules are optimally superposed, in a similar way as previously discussed within the CQS3 of the CQS series involving two molecular structures.

The algorithms, which can be applied in the present case, in order to optimize this kind of integrals with respect of their relative molecular positions in three-dimensional spaces, have been already discussed schematically in the CQS3 paper. It is time, then, to give more details and provide with appropriate codes the calculation of TDQS integrals, at least within the ASA approach. More sophisticated approaches will only differ in the way the involved integrals $\langle PQR \rangle$ are calculated.

The TDQS integrals of TD self similarity (TDSS) type, like $\langle PPP \rangle$ do not have any molecular superposition (MS) difficulty as the integral will possess the maximal value, when the involved unique molecular structure possesses the same atomic coordinates in the three involved densities.

The TDQS integrals of the kind: $\langle PPQ \rangle = \langle PQP \rangle = \langle QPP \rangle$ in principle will behave in the same way as the usual QS measures involving two molecules: $\langle PQ \rangle = \langle QP \rangle$ when optimized by means of the MS described for this situation in CQS3. Thus, the discussed algorithms in CQS3, will apply without modification in this particular case, except that a TDQS integral has to be calculated, instead of the DDQS integral. It is not difficult to use the previous bimolecular superposition (BMS) coordinates for molecules *P* and *Q* in order to obtain this TDQS integral kind.

However, the TDQS integrals of the type: $\langle PQR \rangle$ with the three involved molecules different, have to follow an extended optimization algorithm, which takes necessarily into account the relative position coordinates of the three structures simultaneously.

The ideas of the CQS3 can be easily applied in the TDQS integral case, as it has been already commented. However, instead of the unique squared distances matrix (SqDM), when optimizing TDQS integrals there will be needed three possible ones, whose complete sums shall be kept minimal. That is, the complete sums of the three SqDM: $\left\{ D_{PQ}^{(2)}; D_{QR}^{(2)}; D_{RP}^{(2)} \right\}$, bearing the squared distances between the atoms of one molecule and the atoms of the other, must be sought to be simultaneously as minimal as possible.

This could be obtained with a process, according to the CQS3 findings, which first prepares the three molecular structures with a common origin, excluding the hydrogen atoms of every one of the three molecules involved. When in these positions, the sum of the respective atomic coordinates for every molecule is computed. Namely, if for molecule *P* the actual translated atomic coordinates are written: $\{R_A^P | \forall A \in P\}$, then a vector sum is defined as: $R_P = \sum_{A \in P} R_A^P$; similar definitions are adopted for the other two molecular structures, yielding the vector sums: R_Q and R_R.

Once known the three translated coordinate vector sums, the three complete SqDM sums: $\{\langle D_{PQ}^{(2)} \rangle ; \langle D_{QR}^{(2)} \rangle ; \langle D_{RP}^{(2)} \rangle\}$, will become minimal, when the three crossed scalar vector products: $\{\langle R_P | R_Q \rangle ; \langle R_Q | R_R \rangle ; \langle R_R | R_P \rangle\}$ become maximal, according to the exposition of CQS3. However, this sum vector reorientation can be easily obtained when the three, once translated molecular frames, are rotated afterwards (translated-rotated: TR) in such a way that the sums of atomic coordinate vectors: $\{R_P ; R_Q ; R_R\}$ are lying in the same direction, the x-axis for instance, yielding the x-axis vectors: $\{L_P ; L_Q ; L_R\}$, say.

Then, as it was already commented, the application of the CQS3 **Algorithm 1**, extended in order to superpose three molecular structures is straightforward to construct. It is a matter of adding a third pair of loops to the BMS algorithm. It can be written as a new Algorithm:

Algorithm TDQS: *Maximal value of a TDQS*$\langle PQR \rangle$ *integral between three molecular structures:*

!

Let: $Z_{PQR} = 0$

Loop over atoms of molecule *P*: $A=1,N_P$

Translate the atomic coordinates of *P*: $\forall K \in P : {}^{(\tau)}R_K^P \rightarrow R_K^P - R_A^P$

Compute the vector sum: ${}^{(\tau)}R_P = \sum_{K=1}^{N_P} {}^{(\tau)}R_K^P$

Search for cosine and sine to rotate: $L_P \leftarrow {}^{(\tau)}R_P$

Rotate all coordinates of molecule *P*: $\forall K \in A : L_K^A \leftarrow {}^{(\tau)}R_K^A$

!

Loop over atoms of molecule *Q*: $B=1,N_Q$

Translate the atomic coordinates of Q: $\forall L \in Q: {}^{(\tau)}R_L^Q \rightarrow R_L^Q - R_B^Q$

Compute the vector sum: ${}^{(\tau)}R_Q = \sum_{L=1}^{N_Q} {}^{(\tau)}R_L^Q$

Search for cosine and sine to rotate: $L_Q \leftarrow {}^{(\tau)}R_Q$

Rotate all coordinates of molecule Q: $\forall L \in Q: L_L^Q \leftarrow {}^{(\tau)}R_L^Q$

!

Loop over atoms of molecule R: C=*1*,N_R

Translate the atomic coordinates of R: $\forall M \in R: {}^{(\tau)}R_M^R \rightarrow R_M^R - R_C^R$

Compute the vector sum: ${}^{(\tau)}R_R = \sum_{M=1}^{N_R} {}^{(\tau)}R_M^R$

Search for cosine and sine to rotate: $L_R \leftarrow {}^{(\tau)}R_R$

Rotate all coordinates of molecule R: $\forall M \in R: {}^{(\tau)}R_M^R \rightarrow R_M^R - R_C^R$

!

Compute the TDQS integral: $\langle \rho_P \rho_Q \rho_R \rangle$ using the TR atomic coordinates: $\{L_K^P\}\{L_L^Q\}\{L_M^R\}$.

If $\langle \rho_P \rho_Q \rho_R \rangle > Z_{PQR}$ then: $Z_{PQR} \leftarrow \langle \rho_P \rho_Q \rho_R \rangle$
!
End ***Loop*** *R*
End ***Loop*** *Q*
End ***Loop*** *P*
!
End of Algorithm TDQS

When comparing the **Algorithm TDQS** with the **Algorithm 1** of the CQS3 paper, it can be seen that nothing has changed between both of them, but a pair of loops, involving the atomic coordinates of a third molecular structure, have been added. Of course, the DDQS integral present in the CQS3 has been changed here to the corresponding TDQS integral measure.

12.4.1 SOME PRACTICAL CONSIDERATIONS ON TDQS ALGORITHM

Nevertheless, the above-described **Algorithm TDQS** computational structure contains a large amount of operations to be performed. In fact, for a molecular set, whose average number of non-hydrogen atoms is: N_a, the number of operations to be done within the algorithm is about: N_a^3, precluding a high computational cost, even within the simplified ASA framework.

For molecules with a large number of atoms, like the Cramer steroids [48] where

$N_a = 20$, it is not advisable to run **Algorithm TDQS**, as the primary aim of the CQS series is to permit QS calculations on portable and desktop computers. In cases of studying MQOS with elements possessing a large number of atoms, the best is to use a highly parallel computing tool with teraflop facilities if available. In molecular sets formed of molecules with $N_a \leq 5$ there should not be a problem though.

The extension to higher DF QS measures not only becomes the straightforward task of adding two more do loops for one QS measure order, but because of this modular structure of the **Algorithm TDQS**, the nested do loop programming technique [49–51] can be applied for the purpose of generating such a code, which could be associated to any kind of QS computation level and becomes completely adapted to parallel machines.

12.4.2 SIMPLIFIED TRIPLE MS AND APPROXIMATE OPTIMAL TDQS MEASURES

Nowadays, in order to make feasible the computation of TDQS measures involving large cardinality MQOS, possessing molecular elements with large N_a values, the compulsory need to use a feasible approximate computational structure has lead towards the finding and description of a simplified TDQS procedure, which has been implemented as follows.

The TDQS measures of type $\langle PQR\rangle$ can be easily associated to the set of the three bimolecular measures: $\{\langle PQ\rangle,\langle PR\rangle,\langle QR\rangle\}$, for which the superposition of the implicit bimolecular triad, can be supposedly computed beforehand using, for example CQS3 instructions and programs. As these BMS coordinates are also interesting for the integrals of type $\langle PQQ\rangle$ the best strategy is to have already computed with the CSQ3 techniques the BMS coordinates associated to the optimal $\langle PQ\rangle$ MQS measures. This will provide the program with a set of six possible relative TR optimal coordinates for every molecule present in the involved molecular triad, which, in turn, can be named as:

$$\left\{R_{P(Q)};R_{P(R)};R_{Q(P)};R_{Q(R)};R_{R(P)};R_{R(Q)}\right\}$$

where the notation: $R_{A(B)}$ means the atomic coordinates of molecule A already TR in front of molecule B. Then, one can choose 2^3 combinations of the already TR coordinates to construct eight TDQS measures. For instance, one can write the eight TD integral terms needed, like in the following list:

$$\left\{\begin{matrix}\langle P(Q)Q(P)R(P)\rangle;\langle P(Q)Q(P)R(Q)\rangle;\langle P(Q)Q(R)R(P)\rangle;\langle P(Q)Q(R)R(Q)\rangle\\ \langle P(R)Q(P)R(P)\rangle;\langle P(R)Q(P)R(Q)\rangle;\langle P(R)Q(R)R(P)\rangle;\langle P(R)Q(R)R(Q)\rangle\end{matrix}\right\},$$

where the symbols: $\langle A(B)B(A)C(A)\rangle$ mean the TDQSI has been computed for the three different molecules A, B and C, using the respective geometries BMS optimized in front of molecules: B, A and A, respectively. After the computation of the eight TDQS measures in this way, the larger value can be adopted for the partially optimized integral: $\langle PQR\rangle$. Computing time can be lowered dramatically in this fashion and make reasonable involved and cheaper the computation of TDQS measures for large molecular sets with a large number of non-hydrogen atoms. The limitation now will be the one imposed by the computation of the BMS geometries.

The mathematical structure of the collection of TDQS integrals, when computed over a molecular quantum object set (MQOS)

MQOS

The **Algorithm TDQS** or the previously described approximate procedure based on BMS can be both applied in practice over the elements of any set of molecules $\mathrm{M}=\{m_1;...m_P;...m_M\}$ of cardinality *M*, the *object set*, whenever it is attached in parallel to a known set of DF, the *tag set*, in a one-to-one correspondence, see for example Ref. [22] to obtain extended and detailed information on these definitions. That is, for every molecule in M a well-defined DF has to be known, belonging in turn to another functional set, which can be termed the DF tag set: $\mathrm{P}=\{\rho_1;...\rho_P;...\rho_M\}$. Thus, both object and DF tag sets are related in a one-to-one correspondence:

$$\forall m_P \in \mathrm{M} \rightarrow \exists \rho_P \in \mathrm{P} \leftrightarrow \forall \rho_P \in \mathrm{P} \rightarrow \exists m_P \in \mathrm{M} \cdot \tag{8}$$

The Cartesian product of both sets: $\mathrm{O}=\mathrm{M}\times\mathrm{P}$ is called a *molecular quantum object set* (MQOS) [22, 52–54]. This MQOS definition constitutes a way to put together and resume into a unique structure the Eq. .

12.4.3 DISCRETE REPRESENTATION OF MQOS ELEMENTS IN TDQS FRAMEWORK

Due to the fact that, every TDQS integral: $\langle PQR \rangle$ can be considered a MQS measure, weighted by one of the three DF entering the integrand considered as an operator; then, this point of view can be formally written in any of the following ways:

$$\langle PQR \rangle = \langle Q|P|R \rangle = \langle P:QR \rangle = Z_{P;QR} \tag{9}$$

Meaning that for every density function $\rho_P(r)$ contained into the MQOS one can define a $(M \times M)$ symmetric QS matrix like:

$$\forall P: Z_P = \{Z_{P;QR}\} = \{\langle PQR \rangle\} \rightarrow (Z_P)^T = Z_P, \tag{10}$$

because:

$$\langle PQR \rangle = \langle PRQ \rangle \leftrightarrow Z_{P;QR} = Z_{P;RQ}\,.$$

Then, all these matrices, attached to every DF belonging to the tag set $\mathbb{P}$, can be collected into a hyper vector array or third order tensor, which can be expressed formally like:

$$|Z\rangle = |Z_1;...Z_P;...Z_M\rangle \in \mathbb{P} \otimes \mathbb{P} \otimes \mathbb{P} \tag{11}$$

The interesting situation consists now in that every TDQS hyper vector (TDQSH) matrix element has a MQS matrix-like structure. Such a matrix can be admitted to be a discrete representation of every one of the elements of the DF tag set $\mathbb{P}$, represented by means of the space generated by the tensorial basis set: $\mathbb{P} \otimes \mathbb{P}$. See for more details Refs. [18] and [23].

This situation in TDQS theory contrasts with the usual DDQS matrix representations, where the columns of the associated QS matrix, $Z = \{Z_{PQ} = \langle PQ \rangle\} = \{|\mathbf{z}_Q\rangle\}$, which is a representation of an element of $\mathbb{P} \otimes \mathbb{P}$, can be considered in turn discrete vector representations only of an element of $\mathbb{P}$, with respect the space generated by the DF tag set $\mathbb{P}$ itself.

Consequently, it seems that the TDQS representations can bear more information than the DDQS ones. This constitutes a subject which has not been yet studied and which is obviously worth to analyze here.

12.4.4 TDQS MATRICES AND COMPUTER MEMORY COMPACT STORAGE

In general, as the TDQSH submatrices are symmetric, the elements made by integrals of the type:

$$\langle PPQ \rangle = \langle PQP \rangle = \langle QPP \rangle,$$

will be present three times in the corresponding hyper vector, while the elements with three different indices:

$$\langle PQR \rangle = \langle PRQ \rangle = \langle RPQ \rangle = \langle QPR \rangle = \langle QRP \rangle = \langle RQP \rangle,$$

correspond to the six permutations of such three indices. Quite a large amount of memory space can be obviously spared, by means of storing just the final TDQSH compactly as, for instance using the implicit following do loop scheme:

$$\{\langle PQR\rangle(P=1,M;Q=1,P;R=1,Q)\}.$$

Considering the super symmetric nature of the resultant TDQSH integral elements, upon transposition of the molecular indices, then the amount of TDQS intrinsically different integrals needed to be connected to the molecular set M it is easily calculated as:

$$6^{-1}\left(M\left(M+1\right)\left(M+2\right)\right).$$

12.4.5 A SIMPLE ALBEIT ILLUSTRATIVE EXAMPLE OF THE TDQSH STRUCTURE

Anot so obvious example can be furnished by three p-like GTO functions. This example, as it has been previously commented, will also provide a clue on how QS techniques can discriminate between degenerate states.

Starting from the three p-type GTO functions and in order to condense their representation, they can be expressed into a unique vector written as follows:

$$\begin{aligned}|p\rangle = N|r\rangle\exp\left(-ar^2\right) \leftarrow |r\rangle = \left(x \quad y \quad z\right)^T \\ \wedge r^2 = |r|^2 = \langle r|r\rangle = x^2+y^2+z^2\end{aligned} \tag{12}$$

while N is a common normalization factor, such that every p-type GTO function has unit Euclidean norm.

When using these three functions and considering them as a set of degenerate wave functions in QS manipulations, there is needed to obtain the square of everyone, which will then act as a set of three associated DF, the basic elements in QS calculations.

That is, Eq. produces a set of DF, which can be also written compactly as a vector:

$$|\rho\rangle = N^2\left|r^{[2]}\right\rangle\exp\left(-2ar^2\right) \leftarrow \left|r^{[2]}\right\rangle = \left(|r\rangle * |r\rangle\right) = \left(x^2 \quad y^2 \quad z^2\right)^T \tag{13}$$

Whose elements are coincident with the diagonal terms of the d-type GTO's. In Eq. there has been used an *inward matrix product*, described

in CQS3. The three elements obtained DF in Eq. can be renamed as: $\{X,Y,Z\}$.
With the aid of the generalization of integral [45]:

$$I_n(a)=\int_{-\infty}^{+\infty}x^{2n}\exp\left(-ax^2\right)dx=\frac{(2n-1)!!}{(2a)^n}\sqrt{\frac{\pi}{a}} \tag{14}$$

then the nature of the three simplified wave functions Eq. and their associated DF Eq. preclude that there will be only three possible kinds of distinct TDQS integrals, namely:

$$\begin{aligned}(XXX)&=(YYY)=\ldots=\alpha\\(XXY)&=(XYY)=\ldots=\beta\\(XYZ)&=(YZX)=\ldots=\gamma\end{aligned} \tag{15}$$

There is no need to give explicitly the expressions of the parameter triplet: $\{\alpha,\beta,\gamma\}$; because with the aid of the integral Eq. , it is a matter of trivial mathematical manipulation to obtain them. It is easy to write, after calling $v=N^6$:

$$\alpha=vI_3(6a)\left[I_0(6a)\right]^2;\beta=vI_2(6a)I_1(6a)I_0(6a);\gamma=v\left[I_1(6a)\right]^3$$

For the purposes of the following use of these three parameters, it is not necessary to know explicitly them with more detail than in the above expression.

Once the relationships Eq. are constructed, a hyper vector can be written representing the three p-type functions:

$$\left\langle Z^{(3)}\right|=\{Z_X;Z_Y;Z_Z\}=\left\{\begin{pmatrix}\alpha&\beta&\beta\\\beta&\beta&\gamma\\\beta&\gamma&\beta\end{pmatrix};\begin{pmatrix}\beta&\beta&\gamma\\\beta&\alpha&\beta\\\gamma&\beta&\beta\end{pmatrix};\begin{pmatrix}\beta&\gamma&\beta\\\gamma&\beta&\beta\\\beta&\beta&\alpha\end{pmatrix}\right\} \tag{16}$$

Because of the fact, which can be easily checked, that the three submatrices in the hyper vector Eq. are equivalent upon permutations of rows and columns, it can be also effortlessly seen that the three determinants of the hyper vector matrix components are the same:

$$\forall I = X,Y,Z : Det\left|Z_I\right| = (\alpha + 2\gamma)\beta^2 - (2\beta^3 + \alpha\gamma^2),$$

but considered as mathematical objects, the three hyper vector components are different though. They are linearly independent vectors, therefore can be considered as a basis set belonging to a (3×3) dimensional matrix vectorial space.

12.5 TRANSFORMATION OF THE TDQS HYPER VECTORS: GENERALIZED CARBY SIMILARITY INDICES

12.5.1 PRELIMINARY CONSIDERATIONS

Once known the TDQSH: $|\mathrm{Z}\rangle = \{\langle PQR\rangle\}$ as defined in Eq. and in the previous paragraph discussion example, one can envisage to transform it in an equivalent way as DDQS matrices can be manipulated; for example, defining a transformation leading to a TD Carbó similarity index (CSI) matrix (CSIM).

This kind of transformation in the framework of QS integrals involving two densities has been proposed in the early times of QS development, see for example the initial work1on QS and has been submitted to varied discussions [5, 55, 56] along the QS theoretical progress. CQS3 provides up to date theoretical and computational details on CSIM. Note also a recent description in reference [57], which shows a direct relationship between CSIM's and DDQS matrices.

According to the CSI transformation philosophy, TDQSH's can be also brought into another hyper vector, whose elements are defined in the range $[0,1]$ in the same way as bimolecular CSIM elements are defined in this range too. A recent discussion related to the definition of generalized scalar products and QS measures can be invoked for the mathematical background [58] of this CSI extended definition, which can be employed in TDQSH and higher order QS hyper matrices as well [21]. In order to perform with ease such a transformation, a vector containing TDSS has to be defined first. That is, using the notation:

$$\left|\mathrm{s}^{(3)}\right\rangle = \{\langle PPP\rangle | P = 1, M\} \tag{17}$$

say, where the supraindex in the vector symbol stands for the number of DF involved or the order of the QS measures taken into account and M corresponds to the cardinality of the MQOS which is being represented via the TDQS measures. The elements of the TDQSH Eq. can be considered generalized scalar products involving three vectors of a functional vector semispace*6 [21, 22, 27, 59]. Hence, the generic TDQS measure $\langle PQR \rangle$ can be associated to such a generalized scalar product; in the same way as the elements of the following self similarity vector, defined within the DDQS domain:

$$\left| \mathrm{s}^{(2)} \right\rangle = \left\{ \langle PP \rangle \middle| P = 1, M \right\}$$

can be considered the equivalent of some Euclidian norms evaluated over the elements of the DF tag set $\mathbb{P}$. Due to the positive non-negative nature of the basis DF tag set, the DDQS, TDQS and higher order DF QS measures, considered as generalized scalar products, will in any case yield essentially positive definite real numbers as a result.

12.5.2 DEFINITIONS OF THE TD CARBY SIMILARITY INDEX

For every element of the TDQSH and using the TDSS vector Eq. elements, there can be also defined a TD Carbó similarity index (TDCSI) as follows:

$$r_{PQR} = \frac{\langle PQR \rangle}{\left(\langle PPP \rangle \langle QQQ \rangle \langle RRR \rangle \right)^{\frac{1}{3}}} \tag{18}$$

It is easy to see that the maximal value of the TDCSI Eq. will reach the unity when: $P = Q = R$; otherwise the TDCSI values will be less than one, in other words one will have, as in the bimolecular case:

$$r_{PQR} \in [0,1].$$

---

*6A vector semispace is considered here as a vector space defined over the positive real numbers. In this way the additive group of the semispace is a semigroup, a group without reciprocal elements. DF sets like the quantum object tags already discussed in MQOS definition are subsets of some Hilbert semispace.

Such transformed elements of any TDQSH will yield a new TDCSI hyper vector (TDCSIH), which will possess unit elements in the corresponding self similarity measure hyper vector sites and values less than the unit elsewhere.

12.5.3 THE P-LIKE GTO FUNCTIONS EXAMPLE

In order to fill the corresponding hyper vector elements with some numerical structure, the former simple example involving the three p-like GTO can be used again to obtain a concrete TDCSIH. From the hyper vector Eq. one can easily construct the associated TDCSI submatrices. Taking into account the relationships Eq. and the structure of this particular TDSS vector: $\left|s^{(3)}\right\rangle = (\alpha \quad \alpha \quad \alpha)^T$, then all the TDQSH elements in equation might be divided by α; so, using the new parameters:

$$b = \alpha^{-1}\beta \wedge c = \alpha^{-1}\gamma$$

one can write the corresponding TDCIH as:

$$\left\langle \mathrm{R}^{(3)} \right| = \left\{ \begin{pmatrix} 1 & b & b \\ b & b & c \\ b & c & b \end{pmatrix}; \begin{pmatrix} b & b & c \\ b & 1 & b \\ c & b & b \end{pmatrix}; \begin{pmatrix} b & c & b \\ c & b & b \\ b & b & 1 \end{pmatrix} \right\} \tag{19}$$

12.5.4 GEOMETRIC INTERPRETATION OF THE GENERALIZED CSI

It is now interesting to interpret the resulting CSI elements within any TDCIH in general, or particularly, these present in the hyper vector expression Eq. . As the bimolecular CSI formalism indicates, one can interpret the number:

$$r_{PQ} = \frac{\langle \rho_P \rho_Q \rangle}{\left(\langle \rho_P \rho_P \rangle \langle \rho_Q \rho_Q \rangle \right)^{\frac{1}{2}}} \tag{20}$$

as a cosine of the angle subtended by two vectors, in this case by two DF. The elements of TDCIH too can be associated to a cosine. This is so due that for every submatrix element of the generic form: $r_{P;QR}$, one can heuristically consider the resulting value , as the cosine of the solid angle subtended by the three DF's, when one considers the cone with vertex at the origin and the base of such a cone as a circle circumscribing the three vertices of the triangle, defined in turn by the three DF's. Without doubt, it is better to visualize the bimolecular cosine or this cone generated by three DF, using shape functions (SHF)[*7] instead of the original DF. Thus, using the fact that SHF can be considered as origin centered vectors, ending into the surface of a sphere of unit radius, independently of the number of SHF considered. The value of the cosines computed in a generalized CSI way is invariant [27] whether DF or SHF being employed:

$$r_{PQ} = \frac{\langle \sigma_P \sigma_Q \rangle}{\left(\langle \sigma_P \sigma_P \rangle \langle \sigma_Q \sigma_Q \rangle \right)^{\frac{1}{2}}}$$

Obviously enough, for higher order QS integrals, this kind of reduced CSI is difficult to visualize. However, some hint permitting the same interpretation as in the lower order CSI can be found, generalizing the two and three DF cases. In any p-tuple DF case, the considered set of p DF, if associated to p different molecules, they must be linearly independent. Thus, this DF set generates a p-dimensional vector semispace.

Thinking in terms of SHF, two of these generate a line over a circumference, three a triangle over the surface of a 3-dimensional sphere, four SHF generate a four vertex complex[*8] over the surface of a four dimensional sphere,... a p-tuple of SHF generate a complex with p vertices over the surface of a p-dimensional sphere.

[*7] SHF are nothing else that DF scaled by the number of electrons, like the ASA basis set as commented before. If ρ is a DF with Minkowski norm equal to the number of electrons: $\langle \rho \rangle = N_e$, then the associated SHF is defined as: $\sigma = N_e^{-1}\rho$ and then: $\langle \sigma \rangle = 1$. See for more details about SHF and CSI properties Ref. [27].

[*8] Consider a polyhedral figure with p-vertices embedded into a (p-1)-dimensional space. If the distances between all the vertices are the same the polyhedron is called a *simplex*. A tetrahedron is a simplex with four vertices defined in 3-dimensional spaces. The three p-type GTO example of the present CQS4 paper generates a simplex in a 2-dimensional space. A *complex* corresponds to a similar geometrical object with different vertex distances.

12.5.5 SOME GENERAL REMARKS ABOUT TDCSI AND HIGHER DF ORDER CSI

As occurs in the bimolecular case, it must be also considered that the use of shape functions instead of the DF themselves, they will produce invariant TDCSI values, which it is a well-known fact in the bimolecular CSI case [27]. This invariant property of generalized CSI can be now easily included into TDQS framework too.

12.5.6 ALTERNATIVE LOCAL CSI DEFINITION

There are still other alternative possible transformations of the TDQSH resulting into CSI-like elements. For instance, one can start considering every $(M \times M)$ submatrix of the TDQSH separately and suppose one is able to locally transform any generic submatrix:

$$\forall P : \mathrm{Z}_P = \{\langle P;QR\rangle\} \tag{21}$$

A local CSI definition for every TDQSH submatrix, like the one described in Eq. , can be set up in the same fashion as it is usually done in the bimolecular case:

$$\forall P : R_P^{(L)} = \left\{ r_{P;QR}^{(L)} = \frac{\langle P;QR\rangle}{\left(\langle P;QQ\rangle\langle P;RR\rangle\right)^{\frac{1}{2}}} \right\}$$

With such a local CSI definition, the diagonal of each local TDCSIH submatrix will be the unit matrix.

12.5.7 THE LOCAL TDCSI TRANSFORMATION FOR THE P-LIKE GTO FUNCTIONS EXAMPLE

Using the p-like GTO example again, with the local definition of the TDCSI, and using now the symbols:

$$b=\sqrt{\frac{\beta}{\alpha}}\wedge c=\frac{\gamma}{\beta}$$

one will straightforwardly arrive to the following result for such kind of local CSI:

$$\left\{\begin{bmatrix}\begin{pmatrix}1 & b & b\\ b & 1 & c\\ b & c & 1\end{pmatrix};\begin{pmatrix}1 & b & c\\ b & 1 & b\\ c & b & 1\end{pmatrix};\begin{pmatrix}1 & c & b\\ c & 1 & b\\ b & b & 1\end{pmatrix}\end{bmatrix}\right\}$$

The use of shape functions instead of DF will leave these local CSI values invariant too, as in the previous global TDCSI case.

12.6 MQOS TENSORIAL REPRESENTATIONS AND THEIR POSSIBLE COMPARISON RELATIONSHIPS

When considering TDQSH as third order tensors, as it has been previously commented, the hyper vector elements can be ordered into as many submatrices as every molecule belonging to the working background MQOS $\mathbb{M}$, and every TDQS measure will be included within such submatrices, according to the Eqs. and .

In turn, each submatrix can be considered a member of a novel discrete tag set, easily defined and ordered by means of the submatrix set: $\mathbb{Z}=\{Z_P|P=1,M\}$, which can be also associated to a $(M\times M)$ symmetric matrix collection. To every member of the new tag set one can attach the meaning of being a discrete molecular description, associated to every element of the MQOS object set $\mathbb{M}$. These matrices, as it has been defined several years ago [52–54], can be interpreted as tag elements, substituting in a one-to-one correspondence fashion the DF tag set $\mathbb{P}$ in the original MQOS. Yielding in this way a new MQOS, possessing discrete tags instead of continuous DF ones, that is: constructing a discrete MQOS (DMQOS) [18, 22, 23]. Considered as tags, the elements of the set $\mathbb{Z}$ can be also seen as a vector subset of some vector semispace constructed in turn by $(M \times M)$ symmetric matrices with positive elements as vectors. Whenever the elements of the MQOS are different, their DF tags are linearly independent and the associated TDMQSH submatrices necessarily have to be linearly independent too, providing a discrete representation of the MQOS tags, distinct for every QO.

Because of this evident interpretation of the matrix set $\mathcal{Z}$, one can try to find out a possible way to compare the molecular QO's, described now by the discrete tags of the new DMQOS. In fact, a scalar product involving two discrete tags belonging to $\mathcal{Z}$ can be easily defined by means of a generalized scalar product again. This prospect can be ready made, involving the inward product of both elements followed by a complete sum of the resulting matrix[*9]:

$$\forall P,Q : \langle \mathbf{Z}_P | \mathbf{Z}_Q \rangle = \langle \mathbf{Z}_P * \mathbf{Z}_Q \rangle = \sum_R \sum_S Z_{P;RS} Z_{Q;RS} \qquad (22)$$

such scalar products will in any case yield positive definite scalars, being all the involved TDQSI positive definite.

12.6.1 THE P-LIKE GTO FUNCTIONS EXAMPLE

Considering once more the former p-like GTO example, one can construct the symmetric (3×3) matrix, resulting from the scalar products of the three TDQSH submatrices as defined in Eq. .

Because of the characteristic peculiarities of the elements of the TDQSH Eq. submatrices, there can only be two kinds of results, as their scalar products can be expressed by means of the two kinds of diagonal and off-diagonal equalities only, yielding:

$$\langle \mathbf{Z}_X | \mathbf{Z}_X \rangle = \langle \mathbf{Z}_Y | \mathbf{Z}_Y \rangle = \langle \mathbf{Z}_Z | \mathbf{Z}_Z \rangle = \alpha^2 + 6\beta^2 + 2\gamma^2 = A$$

[*9] Inward matrix products and complete sums are well documented in several previous places [21,61–65] and in CQS3.

and

$$\langle Z_X|Z_Y\rangle=\langle Z_Y|Z_X\rangle=\langle Z_X|Z_Z\rangle=\langle Z_Z|Z_X\rangle=$$
$$=\langle Z_Y|Z_Z\rangle=\langle Z_Z|Z_Y\rangle=2\alpha\beta+3\beta^2+4\gamma\beta=B$$

From here one can easily deduce a scalar product matrix of the TDQSH submatrices, which can be expressed like:

$$Z=\begin{pmatrix} A & B & B \\ B & A & B \\ B & B & A \end{pmatrix} \rightarrow Det|Z| = A^3+2B^3-3AB^2 \qquad (23)$$

The determinant of the matrix Z will be null if and only if: $A=B$, and that cannot be so, unless: $\alpha=\beta=\gamma$. This occurrence will never happen, because the parameters $\{\alpha;\beta;\gamma\}$ are the only possible three kinds of TDQS integrals with distinct values only within this p-like GTO calculation example. Therefore, the three TDQSH submatrices associated to the degenerate p-like GTO functions are linearly independent vectors; thus, TDQS measures of three degenerate wave functions can perfectly distinguish their densities in this way. The matrix Z has to be always a positive definite metric matrix.

Moreover, the eigenvalues of the above matrix Z are related to the ones of the topological matrix, which in turn can be associated to an equilateral triangular graph[*10]:

$$T=\begin{pmatrix} 0 & 1 & 1 \\ 1 & 0 & 1 \\ 1 & 1 & 0 \end{pmatrix} \rightarrow Det|T-\lambda I| = \lambda^3-3\lambda-2=0$$

furnishing a spectrum: $\{2,-1,-1\}$. In terms of the parameters A and B associated to the metric matrix in Eq. , the eigenvalues of Z become:

[*10] This topological matrix form bears the QS symmetry topological image, which can be associated to the three studied degenerate p-like GTO DF. It is coincident with the Hückel matrix for a triangular C_3 structure. Such geometrical image, which can be associated to an equilateral triangle, becomes coherent with this matrix form, and in accord with the hologram-like nature of QS representations [23].

$\{A+2B; A-B; A-B\}$; thus, if the matrix Z must be positive definite, this property necessarily results from $A > B$.

12.6.2 DDQS OVER THE SIMPLIFIED P-LIKE GTO EXAMPLE

It will be interesting to obtain the DDQS matrix for the case of three p-type GTO's, in order to compare the result with the previous findings, associated in turn to the similarity matrix obtained in Eq. . Observing the double density QS construction only two kinds of integrals are different, namely:

$$A^{(2)} = \langle XX \rangle = \langle YY \rangle = \langle ZZ \rangle = \mu I_2(4\alpha)\left[I_0(4\alpha)\right]^2$$
$$B^{(2)} = \langle XY \rangle = \langle XZ \rangle = \langle YZ \rangle = \langle YX \rangle = \langle ZX \rangle = \langle ZY \rangle = \mu\left[I_1(4\alpha)\right]^2 I_0(4\alpha)$$

adding the ancillary definition: $\mu = N^4$. This DDQS integral pattern provides the corresponding QS matrix, which bears formally the same structure, equivalent to the one previously obtained in Eq. , as a result of the TDQSH elements manipulation. This result indicates without doubt the coherent topological and algebraic structure, reflecting in fact the hologram contents [23] of the discrete descriptions provided by QS at various complexity levels, even when highly degenerate wave functions and the corresponding DF are compared.

12.6.3 SOME COMMENTS ABOUT THE LINEAR INDEPENDENCE OF TDQSH SUBMATRICES

According to this, QS measures involving DF can be employed to distinguish QO, which could be indistinguishable from the sole point of view of their wave functions and energies. Although this result is here attached to a quite simple case, there is obvious that can be the same in other, as complex as possible, occurrences one can deal with. In fact, early in the QS theoretical evolution it has been shown that MO's could be ordered just employing their associated DF set [67].

In the TDQS framework this is simply so, each TDQSH submatrix is a discrete representation of a molecule, clearly distinct from other elements of the generic MQOS, as every DF is linearly independent of the rest, if the molecular QO are different. Then the representation of every molecule relies on the DF representation of such molecule in the basis set of all the elements of the tag set of the MQOS. As all MQOS elements are distinct, so the tag set elements are; accordingly from the algebraic point of view, every representation in form of TDQSH submatrix is distinct of the rest. Thus, the set of TDQSH submatrices has to be linearly independent and their metric matrices are accordingly positive definite when properly constructed. This problem, which can be named the molecular QS *Aufbau principle* and it is well documented for the bimolecular case in reference 60and also discussed in CQS3.

12.6.4 CARBY SIMILARITY INDICES OVER CONDENSED TDQSH

Likewise, the scalar products between submatrices of the TDQSH, as defined in Eq. , entering the metric matrices similar to the Eq. , can be easily employed to construct a new $(M \times M)$ matrix constituted by bimolecular CSI-like elements. This new CSI matrix construction can be achieved by using the already presented scalar product definition, involving the submatrices as defined earlier in Eq. , that is:

$$\forall P,Q : R_{PQ}^{(2)} = \frac{\langle Z_P | Z_Q \rangle}{\left(\langle Z_P | Z_P \rangle \langle Z_Q | Z_Q \rangle \right)^{\frac{1}{2}}} \tag{24}$$

Due to the nature of the $(M \times M)$ matrix elements defined in Eq. , it is not difficult to see that: $R_{PQ}^{(2)} \in [0,1]$. Thus, these matrix elements can be interpreted as cosines of the angle subtended by the pairs of submatrices $\{Z_P; Z_Q\}$ within the semispace where they belong; moreover, as a consequence of the definition of Eq. , the diagonal elements of this CSI matrix will become unity: $\forall P : R_{PP}^{(2)} = 1$.

Such a discrete CSI construction procedure permits to envisage that one can progress with the present CSI-like matrix, as defined in Eq. : $R^{(2)} = \left\{ R_{PQ}^{(2)} \right\}$ in the same way, as it was explained when dealing with CSI matrices in the previous paper CQS3. The generated matrices presently computed in this manner, can be introduced in the appropriate MQSPS programs described in CQS3 to obtain Kruskal Tree orderings, for instance.

12.6.5 THE P-LIKE SIMPLE APPLICATION EXAMPLE

In the former three p-like GTO functions example, which it has been used thru this discussion, such manipulation of the metric matrix in Eq. will result into a new matrix, which will look like:

$$\mathrm{R} = \begin{pmatrix} 1 & R & R \\ R & 1 & R \\ R & R & 1 \end{pmatrix} \leftarrow R = \frac{B}{A}$$

And as it has earlier been found that $A > B$, in order to ensure the positive definiteness of the metric matrix in Eq. ; this fact will certainly yield the unique new parameter value into the range: $0 < R < 1$. Moreover, this property also indicates anew the equilateral triangle topological structure of the three involved submatrices. A behavior tuned with the similar symmetric structure of the associated DF set.

12.6.6 DISTANCES USING TDQSH SUBMATRICES

The possibility to obtain scalar products between the TDQSH submatrices obtained with the algorithm shown in Eq. , permits to construct Euclidian distances between the quantum objects discretely described. The square of the Euclidian distance between two TDQSH submatrices can be easily written as:

$$\forall P,Q: D_{PQ}^{2} = \left\langle \left| Z_P - Z_Q \right|^{*2} \right\rangle = \left\langle \left(Z_P - Z_Q \right) * \left(Z_P - Z_Q \right) \right\rangle =$$
$$\left\langle Z_P \middle| Z_P \right\rangle + \left\langle Z_Q \middle| Z_Q \right\rangle - 2\left\langle Z_P \middle| Z_Q \right\rangle$$

Such a classical expression can be also easily generalized with the definition:

$$\forall P,Q: D_{PQ}^{(\nu)} = \left\langle \left| Z_P - Z_Q \right|^{*\nu} \right\rangle = \left\langle \left| \left(Z_P - Z_Q \right) * \left(Z_P - Z_Q \right) * \ldots \left(Z_P - Z_Q \right) \right|^{*} \right\rangle \quad (25)$$

The adequate distance is obtained with the ν −th root of the previous expression and the symbol $|A|^{*} = \left\{ \left| A_{PQ} \right| \right\}$ corresponds to the inward absolute value of a matrix or second rank tensor. A Minkowski distance maybe of easy definition as corresponds to the first order expression of Eq. :

$$\forall P,Q: M_{PQ} = D_{PQ}^{(1)} = \left\langle \left| Z_P - Z_Q \right|^{*} \right\rangle.$$

One must note here that distances in general and Euclidian distances in particular are essentially two object similarity indices, contrarily to the Carbó index indices which can be easily generalized to an arbitrary number of objects comparison, as it will be commented below. Recent work performed to overcome this apparent limitation can be somehow overcome, as already published [61].

12.6.7 EUCLIDIAN DISTANCES BETWEEN THE P-TYPE GTO TENSOR REPRESENTATIONS

As an application example it can be calculated the matrix of the Euclidian distances between the p-type GTO orbitals discrete representation used until now to provide simple examples of the similarity operations one can obtain with the TDQS framework.

Starting from the hyper vector described in Eq. , one can compute the Euclidian norms first:

$$\langle Z_X|Z_X\rangle=\langle Z_X*Z_X\rangle=\left\langle\begin{pmatrix}\alpha&\beta&\beta\\\beta&\beta&\gamma\\\beta&\gamma&\beta\end{pmatrix}*\begin{pmatrix}\alpha&\beta&\beta\\\beta&\beta&\gamma\\\beta&\gamma&\beta\end{pmatrix}\right\rangle=\left\langle\begin{pmatrix}\alpha^2&\beta^2&\beta^2\\\beta^2&\beta^2&\gamma^2\\\beta^2&\gamma^2&\beta^2\end{pmatrix}\right\rangle$$
$$=\alpha^2+6\beta^2+2\gamma^2$$

but it is easy to see that the Euclidian norms of the other two p-orbitals will bear the same value, thus it can be written:

$$\langle Z_X|Z_X\rangle=\langle Z_Y|Z_Y\rangle=\langle Z_Z|Z_Z\rangle=\alpha^2+6\beta^2+2\gamma^2.$$

The cross scalar products will be, for instance:

$$\langle Z_X|Z_Y\rangle=\left\langle\begin{pmatrix}\alpha&\beta&\beta\\\beta&\beta&\gamma\\\beta&\gamma&\beta\end{pmatrix}*\begin{pmatrix}\beta&\beta&\gamma\\\beta&\alpha&\beta\\\gamma&\beta&\beta\end{pmatrix}\right\rangle=\left\langle\begin{pmatrix}\alpha\beta&\beta^2&\beta\gamma\\\beta^2&\alpha\beta&\beta\gamma\\\beta\gamma&\beta\gamma&\beta^2\end{pmatrix}\right\rangle$$
$$=(2\alpha+4\gamma+3\beta)\beta$$

But it is easy to see that all cross products have the same value, thus one can write:

$$\langle Z_X|Z_Y\rangle=\langle Z_X|Z_Z\rangle=\langle Z_Y|Z_Z\rangle=(2\alpha+4\gamma+3\beta)\beta$$

Therefore the three non-zero non-redundant squared Euclidian distances will possess the same value:

$$D_{XY}^2=D_{XZ}^2=D_{ZY}^2=2\left(\alpha^2+3\beta^2+2\gamma^2\right)-2(2\alpha+4\gamma)\beta$$

A coherent result with the fact that the p-type GTO's representation is topologically equivalent to an equilateral triangle, as it has been previously commented.

12.6.8 HIGHER ORDER CARBY SIMILARITY INDICES

In the same way as it has been proceeded to use the TDQSH submatrices as DMQOS tags and then continued to perform a bimolecular comparison as described in the previous section, one can think to use them to obtain CSI involving discrete trimolecular comparisons. This can be easily done just using the generalization power of the scalar product definition in Eq. , for a general mathematical discussion see Ref. [58]. Then, for three discrete submatrix tags instead of two, one can easily define the triple matrix scalar product as follows:

$$\forall P,Q,R:\left\langle Z_P Z_Q Z_R \right\rangle = \left\langle Z_P * Z_Q * Z_R \right\rangle = \sum_S \sum_T Z_{P;ST} Z_{Q;ST} Z_{R;ST}$$

This is sufficient to define a new discrete trimolecular CSI-like index, which can be seen as the discrete counterpart of the TDCSI :

$$\forall P,Q,R: R_{PQR}^{(3)} = \frac{\left\langle Z_P Z_Q Z_R \right\rangle}{\left(\left\langle Z_P Z_P Z_P \right\rangle \left\langle Z_Q Z_Q Z_Q \right\rangle \left\langle Z_R Z_R Z_R \right\rangle\right)^{\frac{1}{3}}} \quad (26)$$

When the generalized discrete CSI elements in Eq. are conveniently ordered, they will construct a hyper vector: $\mathrm{R}^{(3)} = \left\{ R_{PQR}^{(3)} \right\}$, whose hyperdiagonal elements will be the unit, just in the same way as in the continuous case in Eq. .

In no way there will appear any problem with negative signs in the cubic root of Eq. , because of the already commented vector semispace nature of all the involved tags, which bear everywhere positive definite elements.

12.7 QUANTUM QSPR (QQSPR) AND THE ORIGIN OF TDQS MEASURES

Considerations relying on QS have been employed since the very beginning of its appearance as a potential source of ordering quantum objects, like MO's [67], conformers [68] or to find out properties of stereoisomers [69, 75]. However, one of the applications most used as time passed by corresponds to the possibility to directly develop quantum QSPR (QQSPR)[69,71].

The origin of the QQSPR equations is simply the quantum mechanical statistical calculation of observable expectation values; see for instance Ref. [11] for more information. Indeed, as quantum mechanics admits, knowing the DF tag for a given molecular quantum object P: $\rho_P(\mathrm{r})$, the observables, associated to some Hermitian operator $\Omega(\mathrm{r})$ of such submicroscopic object can be formally obtained as the expectation values:

$$\pi_P = \int_D \Omega(\mathrm{r})\rho_P(\mathrm{r})d\mathrm{r} = \langle \Omega\rho_P \rangle \tag{27}$$

If the property leading to the expectation value π_P is quite a complicated one, that is: which cannot be easily attachable to a simple Hermitian operator, then the basic QQSPR theory postulates that, knowing a DF tag set from a given MQOS: $\mathrm{P} = \{\rho_I(\mathrm{r}) | I = 1, M\}$, additionally one can consider the following approximate construction of the QQSPR operator:

$$\Omega \approx \omega_0 + \sum_I \omega_I \rho_I + \sum_I \sum_J \omega_I \omega_J \rho_I \rho_J + O(3)$$

The coefficient set $\{\omega_I\}$ has to be calculated in some way, however once determined by some feasible procedure, certainly the property values sought can be obtained as the expectation values in Eq. , producing:

$$\pi_P = \langle \Omega\rho_P \rangle = w_0 N_P + \sum_I w_I \langle \rho_I \rho_P \rangle + \sum_I \sum_J w_I w_J \langle \rho_I \rho_J \rho_P \rangle + O(3) \tag{28}$$

Thus, in the above expectation value expression [Eq.], the second order terms in the unknown QQSPR operator correspond to a bilinear form, involving a TDQSI array. As previously commented, the implied integrals between two DF can be collected into the columns of the so-called DDQS matrix of dimension $(M \times M)$:

$$Z^{(2)} = \{ z_{IP} = \langle \rho_I \rho_P \rangle | I, P = 1, M \} = (|z_1\rangle; |z_2\rangle; \dots |z_P\rangle; \dots |z_M\rangle)$$

with the P-th column defined as: $|z_P\rangle = \{ z_{IP} | I = 1, M \}$.

In the same manner, the TDQS measures can be collected into a hyper vector, which can be reordered with M elements made by symmetric $(M \times M)$ submatrices $\{Z_P\}:\langle Z^{(3)}| = (Z_1;...Z_P;...Z_M)$, as has been also commented before. Consequently, one can rewrite the expectation value in Eq. as:

$$\forall P : \pi_P = w_0 N_P + \langle w | z_P \rangle + \langle w | Z_P | w \rangle + O(3) \tag{29}$$

whenever the vector containing the QQSPR operator coefficients is defined as: $|w\rangle = \{w_I\}$.

This theoretical formalism constitutes an important result, where the parallelism between tensorial representation of molecular structures and quantum mechanical complex[*11] QQSPR operator description, produces a step forward to understand and rationalize the framework of QSPR.

Classically, within the usual parameterization or using the first order QQSPR operator description, a unique structure of the operator is found for all the elements of the MQOS. Instead, at the second order level as it has been discussed before, QQSPR provides an operator structure, which is varying with every element of the MQOS, as the usual changing form taken by the quantum mechanical operators associated to expectation values.

12.8 BINARY STATES TD REPRESENTATIONS AND ORIGIN SHIFTS

At the light of recent studies about the origin shift which can be performed on the QOS tag sets [24], several interesting results have been obtained. At the same time, some work has been done, using this origin shift point of view, on stereoisomers [75]. These previous studies are relevant at this stage of the present study and thus some analysis will be performed on the question of binary QOS and the connection with the TD tensorial representation of such systems when origin shifts are done on the constituent tag set DF.

[*11] In the sense of complicated, entangled.

In binary conformer or two state sets of quantum objects one has to deal with just two linearly independent DF, acting as tag set elements; say: $\{\rho_A;\rho_B\}$.

Such DF set provides the following triple density matrices for the two QO elements. The tensor representation appears in form of two (2×2) matrices and their computed elements can be written as follows and simplified, as shown below:

$$Z_A=\begin{pmatrix} Z_{AAA} & Z_{AAB} \\ Z_{AAB} & Z_{ABB} \end{pmatrix} \wedge Z_B=\begin{pmatrix} Z_{AAB} & Z_{ABB} \\ Z_{ABB} & Z_{BBB} \end{pmatrix} \Rightarrow Z_A=\begin{pmatrix} \alpha & \beta \\ \beta & \gamma \end{pmatrix} \wedge Z_B=\begin{pmatrix} \beta & \gamma \\ \gamma & \eta \end{pmatrix}$$

Apparently, from this point of view, the tensorial description of both states or molecular structures yields a linearly independent framework, represented by the matrix pair as described above, which can be further employed to deal with the binary difference associated to the original DF.

However, when origin shifting the original densities by the centroid DF, it is obtained in fact a unique function, because the resulting shifted DF: $\{\theta_A;\theta_B\}$ become collinear[*12]:

$$\theta_A=\rho_A-\frac{1}{2}(\rho_A+\rho_B)=\frac{1}{2}(\rho_A-\rho_B)\wedge\theta_B=-\theta_A.$$

Any other origin shift will result in a similar situation, where a function degree of freedom will be lost, see for example Ref. [76].

Therefore, the triple shifted DF tensor elements can be obtained in the shifted framework like:

$$\Theta_{AAA}=\langle\theta_A\theta_A\theta_A\rangle=\frac{1}{8}\langle|\rho_A-\rho_B|^3\rangle=\frac{1}{8}(Z_{AAA}-3Z_{AAB}+3Z_{ABB}-Z_{BBB})$$

$$\Theta_{AAB}=\langle\theta_A\theta_A\theta_B\rangle=-\frac{1}{8}\langle|\rho_A-\rho_B|^3\rangle=-\Theta_{AAA}$$

$$\Theta_{ABB}=\langle\theta_A\theta_B\theta_B\rangle=\frac{1}{8}\langle|\rho_A-\rho_B|^3\rangle=\Theta_{AAA}$$

$$\Theta_{BBB}=\langle\theta_B\theta_B\theta_B\rangle=-\frac{1}{8}\langle|\rho_A-\rho_B|^3\rangle=-\Theta_{AAA}$$

[*12] *The shifted set $\{\theta_A;\theta_B\}$ cannot be considered any longer as a DF set, as the shifted functions are not non-negative definite as the original DF set elements are.

Thus, one now can construct the two attached matrices representing each molecular structure, which can be written as:

$$\Theta_A = \begin{pmatrix} \Theta_{AAA} & \Theta_{AAB} \\ \Theta_{AAB} & \Theta_{ABB} \end{pmatrix} \wedge \Theta_B = \begin{pmatrix} \Theta_{AAB} & \Theta_{ABB} \\ \Theta_{ABB} & \Theta_{BBB} \end{pmatrix}$$

$$\Rightarrow \Theta_A = \Theta_{AAA} \begin{pmatrix} 1 & -1 \\ -1 & 1 \end{pmatrix} \wedge \Theta_B = \Theta_{AAA} \begin{pmatrix} -1 & 1 \\ 1 & -1 \end{pmatrix} = -\Theta_A$$

The squared distance and the Euclidian distance between both representation matrices can be written in terms of the original triple density matrix elements as:

$$\begin{aligned} D_{AB}^2 &= 16\Theta_{AAA}^2 = \left(Z_{AAA} - 3Z_{AAB} + 3Z_{ABB} - Z_{BBB}\right)^2 \\ \to D_{AB} &= 4\Theta_{AAA} = Z_{AAA} - 3Z_{AAB} + 3Z_{ABB} - Z_{BBB} \\ &= \alpha - 3\beta + 3\gamma - \eta \end{aligned}$$

When considering *R, S* optical isomers, using: $A = R, B = S$, for instance, then one will have:

$$Z_{RRR} = Z_{SSS} \wedge Z_{RRS} = Z_{RSS} \to D_{RS} = 0 .$$

The best is to consider Θ_{AAA} as a distance of rank 3, but it will amount the same for stereoisomers, as it will become null. In general binary systems it can be of some interest though, becoming a unique numerical representation of the whole system behavior.

So, the triple density representation will not provide extra information on binary systems of stereoisomer kind. Such a situation is due to the fact that at any similarity level, the comparison of two densities becomes irrelevant and can be associated to a unique representation, which from a TDQSM point of view is represented by the triple selfsimilarity integral: Θ_{AAA}. This situation coherently follows from the fact that, upon origin shift, the two DF can be brought into a unique function only.

12.9 DESCRIPTION OF A DEVOTED PROGRAM SUITE

In CQS3 [14], a set of programs was presented: the MQSPS, such that both source codes and appropriate examples are available for free downloading in the web site of Ref. [77]. Following this trend, within the present CQS4 study a TDQS program suite (TDQSPS) has been also constructed in order to illustrate specially the computation of optimal TDQSI. The CQS4codes within the TDQSPS are written in F95 language and can be found accompanied with input and output assorted examples in the free downloading web site of Ref. [78].

The present TDQSPS benefits of the structure of the MQSPS, formerly described in CQS3, in such a way that the present TDQSPS can be considered an extension of it. Thus, input of molecular structures and the prepared molecular information output of the *MQSS 1* code in MQSPS can be used to construct from a given molecular set the associated TDQS hyper vector (TDQSHV).

12.10 CONCLUSIONS

In the present CQS4 discussion the TDQS theoretical and computational aspects have been studied as exhaustively as possible. In this way, the possibility of a discrete tensorial description of quantum objects has been put forward; constituting a step in front of the classical vectorial context, usually employed to describe molecules by means of empirical parameters or in the built in characteristics of DDQS and, of course, in the usual classical descriptor-structured framework applied to empirical QSPR procedures.

Among the themes treated in the present CQS4 paper one can quote the TDQS basic mathematical development within the ASA framework, including the triple 1s GTO's overlap integrals, a basic tool for the computational implementation of approximate TDQS measures. After this the triple MS problem has been studied, both from the point of view extending the algorithm put forward in double MS described in CQS3 and in the approximate computationally feasible approximate way, employing previously computed MS performed among molecular pairs.

The TDQS tensor structure over MQOS has also been studied in deep, providing simple three element MQOS examples to observe in this schematic, but sufficiently illustrative case, all the possible characteristics this kind of discrete molecular representations can have. Also a handy numerical example of this cardinality kind has been chosen to fill up this three-element MQOS scheme: the set of three degenerate p-type GTO functions. Such an example not only fulfilled a proper TDQS scheme, associated to a MQOS with cardinality three, but permitted to visualize how QS could handle degenerate QO states.

Manipulation of the TDQSH by means of the construction of CSI has been also performed throughout devoted sections of CQS4. Generalized scalar products based on inward matrix products have been profusely employed for this purpose. The results show a wide variety of possible CSI definitions, which can be employed into MQOS description and ordering.

Among the possible TDQSH elements an interesting result appears. The possible construction of a positive definite similarity matrix, involving the dimensions associated to the attached MQOS cardinality, which can be manipulated in the same way as the ones issued from the DDQS framework, but has as origin the DMQOS associated to the submatrices of the TDQSH. The definite positive nature of such QS matrices, constructed by means of scalar products of the discrete molecular DF matrix representations, indicates the adequate and coherent discretization of the DF tags within the TDQS framework. Moreover, this characteristic permits to back up the definition of discrete MQOS based on the TDQSH submatrix elements.

The origin of the tensorial description of molecular sets has been proved appearing as a consequence of quantum mechanical expectation value computation, via approximate QSPR operators. TDQS measures constitute a second order building block set for such QSPR operators. TDQS measures provide, at this second order approximation level, QSPR operator unconstrained optimal structures, which comply with the usual quantum mechanical form of the Hermitian operators, associated to the observables for every molecular structure.

A set of program sources in F95 with some application examples has been made available in a devoted web site. Programming extension to higher n-tuple QS measures and highly parallel computers becomes well defined.

The issue and feasibility of both theoretical and computational discrete tensorial descriptions of quantum objects in general and molecular structures in particular have been proven.

KEYWORDS

- **Aufbau principle**
- **Carby index**
- **Euclidian norms**
- **Kruskal Tree**
- **résumé**

REFERENCES

1. Carbó-Dorca, R.; Leyda, L.; Arnau, M. *Int. J. Quantum. Chem.* **1980**, *17*, 1185.
2. Carbó-Dorca, R.; Robert, D.; Amat, L. ; Gironés, X.; Besalú E. In *Molecular Quantum Similarity in QSAR and Drug Design.* Lecture Notes in Chemistry, Vol. 73; Springer Verlag: Berlin, 2000.
3. Carbó, R.; Calabuig, B. *Int. J. Quant. Chem.* **1991**, *42*, 1681.
4. Carbó, R.; Calabuig, B. *J. Chem. Infra. Comp. Sci.* **1992**, *32*, 600.
5. Carbó, R.; Besalú, E.; Calabuig, B.; Vera, L. *Adv. Quant. Chem.* **1994**, *25*, 255.
6. Besalú, E.; Carbó, R.; Mestres, J.; Solà, M. In *Topics in Current Chemistry: Molecular Similarity I*, Springer-Verlag: Berlin, **1995**, *173*, 31.
7. Carbó-Dorca, R.; Besalú, E. *J. Mol. Struct. Theochem.* **1998**, *451*, 11.
8. Carbó-Dorca,R.; Besalú, E. *Int. J. Quant. Chem.* **2002**, *88,* 167.
9. Carbó-Dorca, R. *J. Math. Chem.* **2004**, *36*, 241.
10. Bultinck, P.; Gironès, X.; Carbó-Dorca, R. In *Rev. Comput. Chem.* Lipkowitz, K. B.; Larter, R.; Cundari, T., Eds.; John Wiley & Sons, Inc.: Hoboken (USA), **2005**, *21*, 127.
11. Carbó-Dorca, R.; Gallegos, A. In Encyclopedia of Complexity and Systems Science, Vol. 8; Meyers, R., Ed.; Springer: New York, 2009; pp. 7422–7480.
12. Carbó-Dorca, R.; Mercado, L. D. *J. Comp. Chem.* **2010**, *31*, 2195.
13. Carbó-Dorca, R.; Besalú E. *J. Comp. Chem.* **2010**, *31*, 2452.
14. Carbó-Dorca, R.; Besalú, E.; Mercado, L. D. *J. Comp. Chem.* **2011**, *32*, 582.
15. Carbó-Dorca, R.; Gallegos, A.; Sánchez, A. J. *J. Comp. Chem.* **2008**, *30*, 1146.
16. Carbó-Dorca, R.; Gallegos, A. *J. Comp. Chem.* **2009**, *30*, 2099.
17. Besalú, E.; Carbó-Dorca, R. *J. Math. Chem.* **2011**, *49*, 1769.

18. Mercado, L. D.; Carbó-Dorca, R.; *J. Math. Chem.* **2011**, *49*, 1558.
19. Carbó-Dorca, R. *J. Math. Chem.* **2012**, *50*, 741.
20. Carbó-Dorca, R. *J. Math. Chem.* ***2012**, 50, 734.*
21. Carbó-Dorca, R. *J. Math. Chem.* **2011**, *49*, 2109.
22. Carbó-Dorca, R.; Besalú, E. *J. Math. Chem.* **2012**, *50*, 210.
23. Carbó-Dorca, R. *J. Math. Chem.* **2012**, DOI: 10.1007/s10910-012-0034-6.
24. Carbó-Dorca, R.; Besalú, E. *J. Math. Chem.* **2012**, *50*, 1161.
25. Carbó, R.; Calabuig, B.; Besalú, E.; Martínez, A. *Mol. Eng* .**1992**, *2*, 43.
26. Robert, D.; Carbó-Dorca, R. *J. Chem. Inf. Comp. Sci.* **1998**, *38*, 620.
27. Bultinck, P.; Carbó-Dorca, R. *J. Math. Chem.* **2004**, *36*, 241.
28. Carbó-Dorca, R.; Besalú, E. *J. Math. Chem.* **2010**, *48*, 914.
29. Besalú, E.; Carbó-Dorca, R. *J. Math. Chem.* **2011**, *49*, 836.
30. Constans, P.; Carbó, R. *J. Chem. Inf. Comp. Sci.* **1995**, *35*, 1046.
31. Constans, P.; Fradera, X.; Amat, Ll.; Carbó, R. In *Quantum Molecular Similarity Measures (QMSM) and the Atomic Shell Approximation (ASA)* in Proceedings of the 2nd. Girona Seminar on Molecular Similarity. July 1995. Advances in Molecular Similarity. JAI PRESS INC. Greenwich (Conn.) **1996**, *1*, 187.
32. Amat, Ll.; Carbó-Dorca, R. *J. Comput. Chem.* **1997**, *18*, 2023.
33. Amat, Ll.; Carbó-Dorca, R. *J. Comput. Chem.* **1999**, *20*, 911.
34. Amat, Ll.; Carbó-Dorca, R. *J. Chem. Inf. Comput. Chem. Sci.* **2000**, *40*, 1188.
35. Gironés, X.; Carbó-Dorca, R.; Mezey, P. G. *J. Mol. Graph Mod.* **2001**, *19*, 343.
36. Gironés, X.; Amat, L.; Carbó-Dorca, R. *J. Chem. Inf. Comput. Sci.* **2002**, *42*, 847.
37. Carbó-Dorca, R.; Besalú, E.; *J. Math Chem.* **2012**, *50,* 981.
38. Besalú, E.; Carbó-Dorca, R.; In *Softened Electrostatic Molecular Potentials* IQC Technical Report TR-2012-3 (2012).
39. Bonnacorsi, R.; Scrocco, E.; Tomasi, J. *J. Chem. Phys.* **1970**, *52*, 5270.
40. Mezey, P. G. *Mol. Phys.* **1999**, *96*, 169.
41. Besalú, E.; Carbó-Dorca, R. *J. Chem. Theor. Comp.* **2012**, 8, 854.
42. See for example: a) Thurston, W. P. "Three-Dimensional Geometry and Topology", Vol. 1, Princeton University Press, Princeton, NJ, 1997. p. 57. b) Hazewinkel, M. (Ed.); "Encyclopedia of Mathematics", Vol. 8, Kluwer Academic Publishers, Dordrecht, 1992. pp. 530–531. c) Boas, M. L.; "Mathematical Methods in the Physical Sciences", John Wiley & Sons Inc., Hoboken, NJ, 2006.
43. Besalú, E.; Carbó-Dorca, R. "Softened Electrostatic Molecular Potentials" IQC Technical Report TR-2012-3; *J. Mol. Graph. Mod.* (2012) DOI:10.1016/j.bbr.2011.03.031.
44. Gröbner, W.; Hofreiter, N.; "Integraltaffeln" Zweiter Teil, Springer Verlag: Wien, 1966.
45. http://iqc.udg.es/cat/similarity/ASA/.
46. Mulliken, R. S. *J. Chem. Phys.* **1955**, *23*, 1833.
47. Reed, A. E.; Weinstock, R. B.; Weinhold, F. *J. Chem. Phys.* **1985**, *83*, 735.
48. Cramer III, R. D.; Patterson, D. E.; Bunce J. D. *J. Am. Chem. Soc.* **1988**, *110*, 5959.
49. Carbó, R.; Besalú, E. *Comput. Chem.* **1994**, *18*, 117.
50. Besalú, E.; Carbó, R. *J. Math. Chem.* **1995**, *18*.
51. Besalú, E.; Carbó, R. "Applications of Nested Summation Symbols to Quantum Chemistry:Formalism and Programming Techniques" In *Strategies and Applications in Quantum Chemistry: from Astrophysics to Molecular Engineering* An Hommage

to Prof. G. Berthier. M. Defranceschi, Y. Ellinger (Editors) Kluwer Academic Pub. (Amsterdam) **1996**, 229.
52. Carbó-Dorca , R. *J. Math. Chem.* **1998**, *23*, 353.
53. Carbó-Dorca , R. *J. Math. Chem.* **1998**, *23*, 365.
54. Carbó-Dorca, R. *Adv. Molec. Simil.* **1998**, 2, 43.
55. Carbó, R.; Besalú, E.; Amat, Ll.; Fradera, X. *J. Math. Chem.* **1996**, *19*, 47.
56. Robert, D.; Carbó-Dorca, R. *J. Chem. Inf. Comp. Sci.* **1998**, *38*, 469.
57. Ayers, P.W.; Carbó-Dorca, R. *J. Math. Chem.* DOI 10.1007/s10910-010-9737-8.
58. Carbó-Dorca, R. *J. Math. Chem.* **2010**, *47*, 331.
59. Carbó-Dorca, R. *J. Math. Chem.* **2008**, *44*, 628.
60. Carbó-Dorca, R. *J. Math. Chem.* **2008**, *44*, 228.
61. Carbó-Dorca, R. In *Collective Euclidian Distances and Quantum Similarity.* IQC Technical Report TR-2012-6; *J. Math. Chem.* (2012) DOI: 10.1007/s10910-012-0086-7.
62. Sen, K.; Carbó-Dorca, R. *J. Mol. Struct. Theochem.* **2000**, *501*, 173.
63. Carbó-Dorca, R. *J. Math. Chem.* **2010**, *47*, 331.
64. Carbó-Dorca, R. *Int. J. Quant. Chem.* **2003**, *91*, 607.
65. Carbó-Dorca, R. *J. Math. Chem.* **2001**, *30*, 227.
66. Carbó-Dorca, R. *J. Mol. Struct. Teochem.* **2001**, *537*, 41.
67. Carbó, R.; Domingo, Ll. *Intl. J. Quant. Chem.* **1987**, *32*, 517.
68. Carbó, R.; Calabuig, B. *Intl. J. Quant. Chem.* **1992**, *42*, 1695.
69. Mezey, P.G.; Ponec, R.; Amat, Ll.; Carbó-Dorca, R. *Enantiomers* **1999**, 4, 371.
70. Carbó, R.; Besalú, E.; Amat, Ll.; Fradera, X. *J. Math. Chem.* **1995**, *18*, 237.
71. Carbó-Dorca, R.; Besalú, E. *Contributions to Science* **2000**, 1, 399.
72. Carbó-Dorca, R. *SAR & QSAR Env. Res.* **2007**, *18*, 265.
73. Carbó-Dorca, R.; Van Damme, S. *Theor. Chem. Acc.* **2007**, *118*, 673.
74. Carbó-Dorca, R.; Van Damme, S. *Intl. J. Quant. Chem.* **2007**, *108*, 1721.
75. Carbó-Dorca, R. "Quantum Similarity" In *Concepts and Methods in Modern Theoretical Chemistry, Vol. 1,* Ghosh, S. K.; Chattaraj, P. K., eds. Taylor & Francis.
76. Carbó-Dorca, R. "Mathematical Aspects of the LCAO MO First Order Density Function (5): centroid shifting of MO ShF basis set, properties and applications" IQC Technical Report TR-2012-7; *J. Math. Chem.* (2012) DOI 10.1007/s10910-012-0083-x.
77. http://iqc.udg.edu/~quantum/software/MQSPS/.
78. http://iqc.udg.edu/~quantum/software/TDQSPS/.

INDEX

D

F

G

H

P

S

T

V

W

X

Y

Z